# CHINA–ASEAN RELATIONS

Economic and Legal Dimensions

# CHINA–ASEAN RELATIONS

## Economic and Legal Dimensions

*editors*

### John Wong
*East Asian Institute, National University of Singapore*

### Zou Keyuan
*East Asian Institute, National University of Singapore*

### Zeng Huaqun
*International Economic Law Institute, Xiamen University*

NEW JERSEY · LONDON · SINGAPORE · BEIJING · SHANGHAI · HONG KONG · TAIPEI · CHENNAI

*Published by*

World Scientific Publishing Co. Pte. Ltd.
5 Toh Tuck Link, Singapore 596224
*USA office:* 27 Warren Street, Suite 401-402, Hackensack, NJ 07601
*UK office:* 57 Shelton Street, Covent Garden, London WC2H 9HE

**Library of Congress Cataloging-in-Publication Data**
China-ASEAN relations : economic and legal dimensions / John Wong, Zou Keyuan and
  Zeng Huaqun, editors.
    p. cm.
  Papers of a conference organized by the East Asian Institute and Xiamen University and
held in Singapore in December 2004.
  Includes bibliographical references and index.
  ISBN-13 978-981-256-657-7
  ISBN-10 981-256-657-0
    1. China--Foreign economic relations--Southeast Asia--Congresses. 2. Southeast
Asia--Foreign economic relations--China--Congresses. I. Wong, John, 1939–   II. Zou,
Keyuan. III. Zeng, Huaqun.

HF1604.Z4S652 2006
337.51059--dc22

2005056973

**British Library Cataloguing-in-Publication Data**
A catalogue record for this book is available from the British Library.

Copyright © 2006 by World Scientific Publishing Co. Pte. Ltd.

*All rights reserved. This book, or parts thereof, may not be reproduced in any form or by any means, electronic or mechanical, including photocopying, recording or any information storage and retrieval system now known or to be invented, without written permission from the Publisher.*

For photocopying of material in this volume, please pay a copying fee through the Copyright Clearance Center, Inc., 222 Rosewood Drive, Danvers, MA 01923, USA. In this case permission to photocopy is not required from the publisher.

Typeset by Stallion Press
Email: enquiries@stallionpress.com

Printed in Singapore

# Contents

List of Editors and Contributors     vii

## Part I: Introduction

1. New Dimensions in China-ASEAN Relations     3
   *John WONG, ZOU Keyuan and ZENG Huaqun*

## Part II: Changing China-ASEAN Relations in Perspective

2. China-ASEAN Relations: An Economic Perspective     17
   *John WONG*

3. ASEAN-China Relations: An International Law Perspective     33
   *ZENG Lingliang*

## Part III: China-ASEAN on Non-traditional Security Issues

4. Regional Cooperation in Epidemic Prevention:
   China and ASEAN     59
   *LAI Hongyi*

5. Anti-Piracy Cooperation Dilemma: ASEAN and China     75
   *XU Ke*

## Part IV: Towards a Free Trade Area

6. WTO Rules and China-ASEAN FTA Agreement     93
   *ZENG Huaqun*

7. The International Legal Personality of ASEAN and the Legal
   Nature of the China-ASEAN FTA     111
   *Jiangyu WANG*

8.  China-ASEAN FTA: An Investment Perspective  133
    *CHEN Huiping*

## Part V: Issues in China-ASEAN Regional Cooperation

9.  China's Border Trade with Newer ASEAN Members:
    Problems and Prospects  161
    *LIAO Shaolian*

10. Transboundary Environmental Issues of the Mekong River:
    Cooperation or Conflicts among the Riparian Countries  179
    *LU Xixi*

11. Are China and ASEAN Competing for FDI?  199
    *CHEN Wen*

## Part VI: The South China Sea and Maritime Issues

12. Regional Cooperation for Conservation and Management of
    Fishery Resources in the South China Sea  219
    *Kuen-chen FU*

13. Prospects for Joint Development in the South China Sea  245
    *ZOU Keyuan*

14. Maintaining Maritime Safety in Southeast Asia:
    Regional Cooperation  263
    *CHEN Haibo*

## Part VII: China-ASEAN Relations in Regional Perspectives

15. China-ASEAN FTA and Korean FTA Policies  293
    *Moon-Soo CHUNG*

16. China's Ties with Southeast Asia in the Post-Cold War Era:
    Japan's Response  309
    *LAM Peng Er*

Index  327

# List of Editors and Contributors

## Editors

John WONG, Professor and Research Director, East Asian Institute, Singapore. He obtained PhD in economics from the University of London in 1966 and was Director of the Institute of East Asian Political Economy (1990–1997). He had taught economics at the University of Hong Kong (1966–1970) and the National University of Singapore (1970–1990). He was also Visiting Professor at Florida State University, Harvard's Fairbank Center, Yale's Economic Growth Center, Oxford's St. Antony College, University of Toronto and Stanford University. He had served as a consultant to many international organizations. He has written and edited over 18 books, and published numerous articles and papers.

ZENG Huaqun, Professor and Director at the International Economic Law Institute, Xiamen University, China; Visiting Scholar at East Asian Institute (1998–1999), SOAS, University of London (1996, 1998), Swiss Institute of Comparative Law (1995–1996) and College of Law, Willamette University, USA (1987–1988); Vice Chairman, Chinese Society of International Economic Law; Member of Standing Council, Chinese Society of International Law; a legal consultant to the People's Government of Fujian Province; Deputy Director of the Commission of Home Affairs and Justice, People's Congress of Xiamen Municipality.

ZOU Keyuan, Senior Research Fellow, East Asian Institute, Singapore. He has published numerous articles in international journals. His recent publications include *Law of the Sea in East Asia* (London/New York: Routledge, 2005), *China's Marine Legal System and the Law of the Sea* (Leiden/Boston: Martinus Nijhoff, 2005). Dr. Zou is Editorial Member of the *International Journal of Marine and Coastal Law*, *Ocean Development and International Law*, and *Journal of International Wildlife Law and Policy*, as well as Deputy Editor-in-Chief of the *Chinese Journal of International Law*.

## Contributors

CHEN Haibo, Lecturer, Law School, Xiamen University, China. She received LLM from Dalian Maritime University in 1997 and is currently Ph.D. candidate in international law in Xiamen University. She practices law part-time at Fujian Xiamen Lianhe Xinshi Law Firm, and a member of the China Maritime Law Association, and the Law Association of Fujian Province. She has contributed to several books and published a number of journal articles.

CHEN Huiping, Associate Professor, Law School, Xiamen University, China. She received LLB, LLM and PhD in international law from Xiamen University. She has been working in the Law School of Xiamen University since 1993. Her research interest lies in international economic law and has published articles and books on multilateral investment law, regional investment agreements and other related legal subjects. She is also Secretary-General of the Administrative Council of the Xiamen Academy of International Law.

CHEN Wen, Lecturer, Institute of Economics, Xiamen University, China. She has PhD in economics. Her research interests are international economics and development economics. She has conducted several research projects including the research project funded by the National Social Science Foundation of China.

Moon-Soo CHUNG, Advisor to the President for Economic Policy, Republic of Korea. Prior to joining the presidential office in 2005, he was Professor of International Trade at Inha University, having served as Dean of Graduate School of International Trade and Regional Studies during 2004–2005. From 1984 to 1994, he worked as a Counsel/Senior Counsel in the Office of the General Counsel in the Asian Development Bank and served as Chairman of the Korean Trade Commission, Government of Korea between 1998 and 2001. Dr. Chung earned his LLM and SJD from University of Michigan and was admitted into the State of New York Bar and American Bar Association in 1984.

Kuen-chen FU, Professor, Xiamen University Law School, China. He is the Chief Editor of the *China Oceans Law Review*, and has been Executive Deputy Director of the Xiamen University Center for Oceans Policy and Law since 2002. Dr. Fu obtained his LLM and SJD from the University of Virginia Law School. His main interests are international law, the law of the sea, and Anglo-American law of contracts. He has authored and co-authored 21 books in either Chinese or English.

LAI Hongyi, Research Fellow, East Asian Institute, Singapore. He received PhD in political science from UCLA and researches on China's political economy. He has over a dozen of published or forthcoming book chapters published by RoutledgeCurzon and World Scientific, and peer-refereed articles appearing in *Modern China, Third World Quarterly, Issues & Studies, Provincial China, American Asian Review*, and *Asian Journal of Political Science*.

LAM Peng Er, Senior Research Fellow, East Asian Institute, Singapore. His articles have appeared in journals such as the *Pacific Affairs, Asian Survey* and the *Japan Forum*. Lam's forthcoming edited book is *Japan's Relations with China: Facing a Rising Power* (London: RoutledgeCurzon). Dr. Lam is researching on the impetus and impediments to political change in Japan, and Tokyo's political activism in East Asia.

LIAO Shaolian, Deputy Director and Professor at Center for Southeast Asian Studies, Xiamen University, China. He is also Vice President of China Association of Southeast Asian Studies. He was educated at Zhongshan University in the 1960s, and later studied world economy at Fudan University, and development economics at Cornell University in the early 1980s. He has published four books and more than 70 articles on economic development and cooperation in Southeast Asia and the Asia Pacific region.

LU Xixi, Assistant Professor, Department of Geography, National University of Singapore, Singapore. He obtained his PhD from the University of Durham, and was an NSERC post-doctoral fellow in the University of Western Ontario, Canada. As a trained geomorphologist, his research interests lie in the investigation of soil erosion, sediment delivery and deposition in the large Asian rivers including the Yangtze River, the Pearl River, the Mekong River and the Irrawandy River where he had done extensive field work. His research has been funded by various organizations such as NUS, START SARCS, and China "973" Project.

WANG Jiangyu, Assistant Professor, Faculty of Law, National University of Singapore, Singapore. He obtained SJD and LLM from University of Pennsylvania, M.Juris from University of Oxford, LLM from Peking University and LLB from China University of Political Science and Law. He specializes in Chinese law and international economic law. He practiced law in the Legal Department of the Bank of China and Chinese and American law firms. He is a member of the Chinese Bar Association and the New York State Bar Association.

XU Ke, Ph.D. candidate, Program of Southeast Asian Studies, National University of Singapore, Singapore. He is currently completing a doctoral thesis on maritime piracy in Southeast Asia. He gained his Master degree in economics from Xiamen University, China, and was an assistant fellow in the Research School of Southeast Asian Studies in Xiamen University before he was enrolled in NUS.

ZENG Lingliang, Dean and Professor of Law School, Jean Monnet Chair of European Union Law, Wuhan University, China. He received LLM and SJD from Wuhan University, and LLM from University of Michigan. He is Panelist of the Indicative List of DSB, WTO. He was a researcher at the European University Institute (1989–1990), High Fulbright Scholar at the University of Denver (1996–1997), Guest Professor at the University of Birmingham (1998), Distinguished Visiting Professor at Kent College of Law of Illinois Institute of Technology (2004). His representative monographs are *European Communities and Modern International Law* (1992), *Law of the WTO* (1996) and *International Law and China in the Early 21st Century* (2005).

# Part I
## Introduction

# Chapter 1

# New Dimensions in China-ASEAN Relations

*John WONG, ZOU Keyuan and ZENG Huaqun*

On account of China's dynamic economic growth, its relations with the Association of Southeast Asian Nations (ASEAN) states to its south have expanded rapidly in recent years, culminating in the conclusion of the landmark China-ASEAN Comprehensive Economic Cooperation Agreement in 2002. Beyond trade and economic activities, China-ASEAN cooperation has been broadened to cover environment, science and technology, non-traditional security areas and related legal issues. Thus, China's relations with ASEAN have reached a new era where the two sides have established an economic, political and legal framework for their comprehensive cooperation.

Closer cooperation between China and ASEAN is mutually beneficial economically, providing them with an additional source of economic growth and catalyzing the process of economic integration in the whole East Asian region. In the longer run, the process will also bring about significant geo-political and geo-economic transformation of the region. Cognizant of such new developments, the East Asian Institute, hub of China studies in Singapore, and Xiamen University, hub of Southeast Asian studies and international economic law, jointly organized a symposium, which was held in December 2004 in Singapore, to encourage and promote studies on new developments in the China-ASEAN relations. This book is based on the papers generated from this joint symposium.

The papers selected for this book cover the five most important areas in China-ASEAN relations for examination and discussion. They include the general framework of this bilateral cooperation; non-traditional security issues; free trade area arrangements; regional economic development; the South China Sea and maritime security; and regional perspectives on ASEAN-China relations.

## General Framework of China-ASEAN Relations

The China-ASEAN relations are characterized as economic cooperation at first hand. Such relationship has been developing from economic cooperation to other areas of cooperation. As manifested by the recent developments of the bilateral relations between China and ASEAN, a comprehensive cooperative framework has gradually evolved and been formulated, thus extending such relations from economic to political, security and legal fields and touching on many areas of cooperation between the two sides.

Of course, economic cooperation is most important in China-ASEAN relations. As shown in John WONG's paper, the economic cooperation and integration between China and ASEAN members is most remarkable in terms of trade volume as well as depth of cooperation. Since the economic reform and open-door policy in the late 1970s, China's economy has become more closely integrated with its neighboring economies, including ASEAN. As China continues its export-oriented economic development, there is competition between China and ASEAN countries in attracting foreign direct investment (FDI) and in exporting manufactured products to the same third-country markets. In order to avoid any disruption in China-ASEAN cooperation, China initiated the arrangement for a free trade area (FTA) agreement with ASEAN that was designed to turn competition into complementation for the benefits of both sides. The details of the FTA arrangement will be further discussed in the other chapters. While addressing the general framework of China-ASEAN cooperation from an economic perspective, the author also touches on geo-political implications of the economic rise of China. In his conclusion, China's fast-growing economy, averaging at 9.3% for the past 25 years, has become a new engine of economic growth not only for ASEAN but for the whole East Asian region. It has also catalyzed the process of integration among East Asian economies.

The other paper in this section contributed by ZENG Lingliang deals with the general framework of the China-ASEAN relations and cooperation from a legal perspective. The author lists several important documents between China and ASEAN as a legal basis for their bilateral general framework, including the 1994 Exchange of Letters; the 1997 Joint Statement; the 2002 Framework Agreement on Comprehensive Economic Cooperation; the 2002 Joint Declaration on Cooperation in the Field of Non-Traditional Security Issues; the 2002 Declaration on the Conduct of the Parties in the South China Sea; and the 2003 Joint Declaration of Strategic Partnership. In addition, China acceded to the Treaty of Amity and Cooperation in Southeast Asia. Being guaranteed by these documents, China and ASEAN are able to develop their cooperative relations smoothly and rapidly. On the other hand, according to Zeng, there is still room for improvement in the China-ASEAN relationship. For example, there are a few hard-law sources regulating the

China-ASEAN relations. The documents listed above can only be categorized as soft-law in the legal sense. Therefore, it is desirable that these soft-law documents be converted into hard-law so as to consolidate the legal framework of China-ASEAN relations. It is also necessary to lay down the implementation measures in order to realize the goals and objectives embodied in these existing documents.

## Non-Traditional Security Issues

Non-traditional security issues come to the forefront of the world community after the September 11 terrorist attack in USA. Naturally, they also become subject-matters to be dealt with within the China-ASEAN cooperative framework. In November 2002, the Joint Declaration of ASEAN and China on Cooperation in the Field of Non-Traditional Security Issues was adopted, which initiated full cooperation between ASEAN and China in the field of non-traditional security issues and listed the priority and form of cooperation. The priorities at the current stage of cooperation include "combating trafficking in illegal drugs, people-smuggling including trafficking in women and children, sea piracy, terrorism, arms-smuggling, money-laundering, international economic crime and cyber crime." As to multilateral and bilateral cooperation, it aims to "strengthen information exchange, strengthen personnel exchange and training and enhance capacity building, strengthen practical cooperation on non-traditional security issues, strengthen joint research on non-traditional security issues, and explore other areas and modalities of cooperation".[1] While there are a number of non-traditional security issues existing in this region, this section includes two papers dealing with pandemic disease and piracy at sea respectively.

The most salient pandemic in recent years is, of course, the severe acute respiratory syndrome (SARS), as discussed by LAI Hongyi. The SARS outbreak occurred in early 2003 in southern China and then spread to other regions of China as well as other countries. While it posed a serious health crisis for human beings, it also marked a new avenue of cooperation between China and ASEAN in fighting epidemics. In order to make up for its early mismanagement of SARS and repair its damaged ties with ASEAN, China from April 2003 onwards initiated vigorous measures to fight the epidemics jointly with ASEAN, which was also receptive to China's olive branch for epidemic cooperation. Forums, coordination measures and funds were launched; and cooperative mechanisms were established and further developed. The joint efforts made by ASEAN and China in their common fight

---

[1] See Joint Declaration of ASEAN and China on Cooperation in the Field of Non-Traditional Security Issues, Sixth ASEAN-China Summit, Phnom Penh, Cambodia, 4 November 2002, http://www.aseansec.org/13185.htm (accessed 2 May 2005).

against the SARS set a good example for future cooperation against the spread of epidemics, which are not limited by national boundaries. As Lai has pointed out, when an outbreak of avian flu hit Southeast Asia and some Chinese provinces in 2004, China re-opened the established cooperation avenues with ASEAN. Hence, the two sides have made unprecedented progress in pandemic disease control. The experiences in this respect no doubt can be instructive for handling other non-traditional security issues between China and ASEAN nations.

XU Ke's paper turns to the maritime security by addressing the prominent issue of piracy in Southeast Asia. It is generally accepted that piracy, though having a long history, has become one of the most serious non-traditional security issues in the post-Cold War era. According to Xu, since piracy in Southeast Asian waters is a trans-national crime, any unilateral or bilateral anti-piracy action is not sufficient to suppress piracy, and suppression of piracy requires consistent and comprehensive multilateral and international cooperation. He devoted his discussion mainly to anti-piracy cooperation between ASEAN and China. While they have made important progress in their joint anti-piracy efforts, there are still a number of difficulties existing in this anti-piracy cooperation.

With the recent conclusion (still subject to ratification) of the Regional Anti-Piracy Agreement among 16 Asian countries, including Bangladesh, Brunei, Cambodia, China, India, Indonesia, Japan, Laos, Malaysia, Myanmar, the Philippines, Sri Lanka, Singapore, South Korea, Thailand and Vietnam, regional cooperation in anti-piracy has indeed entered a new phase. The agreement obliges contracting states: to prevent and suppress piracy and armed robbery against ships; to arrest pirates or persons who have committed armed robbery against ships; to seize ships or aircraft used for committing piracy or armed robbery against ships; and to rescue victim ships and victims of piracy or armed robbery against ships.[2] For cooperation purposes, the contracting parties should endeavor to render mutual legal assistance as well as extradition for piracy suppression and punishment. In addition, the Agreement establishes an Information Sharing Center to be located in Singapore.

## Free Trade Area Arrangements

The free trade area (FTA) arrangements mark one of the most important breakthroughs in China-ASEAN economic cooperation. There is abundant literature about it; but papers from a legal perspective are rare. Accordingly, this section of

---

[2] Article 3 of the Regional Cooperation Agreement on Combating Piracy and Armed Robbery against Ships in Asia.

the book serves to fill important gaps in the existing literature dealing with the legal aspects of the FTA activities.

ZENG Huaqun's paper addresses this legal subject matter in the context of the relevant rules of the World Trade Organization (WTO) and attempts to compare the 2002 Framework Agreement on Comprehensive Economic Cooperation between ASEAN and China with the WTO rules, as many provisions of the Framework Agreement are governed by the latter. Within ASEAN, there are WTO and non-WTO members whose legal rights and obligations are different. However, after all the non-WTO members in ASEAN acceded to the WTO, the rules laid down by this international organization can be more widely implemented in this region. As the author concludes, the Framework Agreement is a milestone in the development of China-ASEAN comprehensive economic cooperation and also groundwork for the establishment of the China-ASEAN FTA. At the same time, WTO rules provide the legal basis for and have a close link to the development of the Framework Agreement and the China-ASEAN FTA process.

The paper contributed by Jiangyu WANG deals with the same subject matter but from a different angle. He discusses the international legal personality of ASEAN and its impact on the negotiation and implementation of the China-ASEAN Free Trade Agreement first, and then turns to address the legal nature of the China-ASEAN free trade agreement. As he perceives it, the proliferation of regional (free) trade agreements — often known as "trade regionalism" in the past decade — has posed serious challenges to the multilateral trading system and raised significant questions in international law. In East Asia, one of the most salient developments in trade regionalism is the negotiation on a free trade agreement between China and ASEAN, known as the China-ASEAN Free Trade Agreement (CAFTA). This chapter examines the CAFTA in the context of international law and provides a brief introduction to the development of the CAFTA proposal and the current negotiating process as well as the results achieved. The author regards the CAFTA as a multilateral agreement signed by 11 nation-states. In its nature, it mainly contains bilateral obligations between China and individual ASEAN countries. This legal reality will have a profound impact on the enforcement of treaty obligations as well as on the resort to the dispute settlement mechanism in the implementation of the CAFTA.

The last paper in this section deals with the FTA issue by focusing mainly on the investment field. The CAFTA negotiations between China and ASEAN began with the investment issue, which becomes an integral part of the CAFTA negotiations. As CHEN Huiping rightly points out, international economy has two pillars: transnational trade and transnational investment, which are mutually complementary as well as growth-promoting. The concept of comprehensive economic cooperation naturally incorporates both free trade and free movement of FDI (foreign direct investment). And that is well reflected in the China-ASEAN FTA. The real purpose of the investment regime as contained in the CAFTA is to facilitate intra-Area FDI flows and also foreign investment into the Area, so as to promote economic development of both China and ASEAN. Thus, the chapter ends with the observation that the CAFTA is expected to increase intra-regional trade and intra-regional FDI flows, and to improve the economic efficiency of the region through greater scale economies as a result of market and resources. Ultimately, this will lead to greater opportunities for business for all.

## Regional Economic Development

Apart from the comprehensive economic cooperation centered on the FTA arrangements between China and ASEAN members, bilateral or sub-regional economic cooperation has been also developing rapidly and smoothly in China-ASEAN relations. This section contains three papers, focusing specifically on the less well-known but no less important area of border trade between several of China's southwestern provinces and the newer ASEAN members, cooperation and conflict between China and riparian ASEAN countries of the Mekong River, and FDI in China and ASEAN countries.

China shares land borders with several ASEAN countries, including Vietnam, Laos and Myanmar, and the border trade between China and these ASEAN countries is deeply rooted in history. LIAO Shaolian's paper deals specifically with this subject matter by looking at the favorable factors, development process, characteristics and positive aspects of the border trade between China and newer ASEAN members. While border trade has been developing very rapidly in recent years, there remain some problems, such as the weak mechanism of administration, different policies of neighboring countries having different implementation results, irregularity of accounting settlement, and the burdens of taxes and fees. However, as stressed by the author, existing problems may not prevent the development of the border trade, as countries in the region are likely to step up their trade liberalization and facilitation, including the unification of trade policies implemented in different regions. The author concludes by saying that although border trade has benefited both sides, it is undergoing changes including the expansion from exchange market trade and small-volume trade to various other forms of economic relations, including entrepôt trade, transnational economic and technological cooperation. Border trade will continue to contribute to the comprehensive development and advancement of the border regions in China and new ASEAN members.

LU Xixi's paper addresses a different issue from a different perspective but within the same border area between China and ASEAN member countries. The Mekong River runs 4,800 kilometers from its headwaters on the Tibetan Plateau through Yunnan Province of China, Myanmar, Thailand, Cambodia, Lao PDR and Vietnam. Being the 12th longest river in the world,[3] it is regarded as the heart and soul of mainland Southeast Asia. The upper stream of the river within the Chinese border is called Lancang River, which is an international river. Lu's paper is focused on some environmental issues resulting from the construction of the dams along the river. Environmental impacts can be found in water discharge and water level fluctuations; sediment concentration and sediment flux; channel and river bank stability; and aquatic ecology. These environmental issues are trans-boundary in nature, involving riparian countries sharing a single resource. For example, China's hydropower projects on the Lancang River are likely to create extensive environmental effects on lower river basin of the Mekong River. There are also political and social consequences resulting from such environmental impacts. Therefore, as the author concludes, cooperation among all the countries in the Mekong River basin is imperative in order to avoid any potential conflict as a result of the trans-boundary environmental issues.

CHEN Wen's paper tries to answer a recent hotly-debated issue on whether China is competing with ASEAN for FDI after China's entry into WTO. By analyzing the sources and the extent of China's FDI inflow, China's "round-tripping" phenomenon of investment, factors for the decrease of ASEAN inward FDI, and Chinese outward direct investment in ASEAN, the author finds that there is no obvious evidence that the rise of China has squeezed out the FDI inflow to ASEAN. However, the author admits that the competition for FDI between China and ASEAN does exist and suggests that in order to be more attractive to FDI, ASEAN economies should improve their overall investment environment and expedite the regional economic integration process within ASEAN. Economic integration between China and ASEAN, on the other hand, can also bring about more opportunities to ASEAN in terms of attracting FDI from China as well as from other countries.

## The South China Sea and Maritime Security

The South China Sea, a semi-enclosed sea surrounded by a number of Southeast Asian countries as well as China, is one of the most important areas in

---

[3] In terms of water capacity and volume, it is the tenth largest river in the world and the third largest in Asia, see Ed Lee, "The Mekong River Basin Agreement", *Asia Pacific Journal of Environmental Law*, Vol. 1, pp. 134–139 (1996).

China-ASEAN relations in terms of political and security cooperation as well as the maintenance of regional peace and security. It is also a major disputed area in China-ASEAN relations. There are maritime and territorial disputes in the South China Sea, in particular around and regarding the Spratly Islands, which have been claimed by five countries and six parties (Brunei, China, Malaysia, Philippines, Taiwan and Vietnam). Such multiple claims have caused tensions between/among relevant claimants on the one hand and make cooperation difficult on the other.

Under international law, States in dispute should resolve their problems through peaceful means. This is embodied in the Charter of the United Nations and many other international political and legal documents, including the 1955 Bandung Declaration (which included the principles of Pancha Shila for promoting international peace and security). While resolution cannot be reached at present, countries concerned should seek cooperative measures to defuse tensions and to conduct joint programmes. The 1982 United Nations Convention on the Law of the Sea, which has been ratified by all South China Sea countries, also requires coastal States adjacent to a semi-enclosed sea to carry out regional cooperation. In compliance with international norms, ASEAN countries and China signed the milestone Declaration on the Conduct of the Parties in the South China Sea in 2002,[4] promising to resolve their territorial and jurisdictional disputes by peaceful means, without resorting to the threat or use of force, and pledging to explore or undertake cooperative activities in the South China Sea. For that purpose, resource use and management could be one of the most feasible cooperative areas between China and ASEAN countries.

Kuen-chen FU's paper is focused on such cooperation in fisheries resources. As he puts it, conservation and management of the South China Sea fishery resources is a complicated issue, which is not possible for a single State among the South China Sea countries to resolve alone. A joint effort is thus essential, particularly in consideration that the state of the South China Sea fisheries gets worsened but the demand for fisheries has escalated. He suggests that there is an urgent need of a more effective regional cooperation scheme for fishery resources conservation and management. Based on the 1982 United Nations Convention on the Law of the Sea and the 1995 Straddling Fish Stocks Agreement, and the theory of the "Three-Level South China Sea", Fu discusses the possible ways of creating such a scheme and the basic structures of such a cooperation scheme. His discussion is also expanded to include a possible model for cooperation between China and Taiwan. According to him, if the proposed scheme could be realized, it would constitute a sound

---

[4] Text is available at http://www.aseansec.org/13163.htm (accessed 28 April 2005).

basis for a more comprehensive cooperation scheme among all the South China Sea countries.

On the other hand, ZOU Keyuan's paper is focused on non-living resources, i.e., oil and gas and a possible cooperation mechanism between China and ASEAN countries. He uses the legal concept of joint development as a basis to argue that a joint development arrangement should be made for the South China Sea mineral resources exploration and exploitation. As illustrated by his paper, there are quite a few precedents of joint development arrangements around the world and some of them are in fact in existence in the East Asian region. They are good examples that the South China Sea countries can learn from, and their useful experiences can be transferred to facilitating the China-ASEAN cooperation in this regard. It is strongly suggested that a preferable joint development scheme, once established, should be multilateral in nature, rather than bilateral, due to the fact that multilateral claims exist in the South China Sea, involving all the interested ASEAN members.

Related to the South China Sea is the critical issue of maritime security. Though the issue of piracy has been discussed in the section of non-traditional security issues above, CHEN Haibo's paper is concerned more with the safety of sea-lane navigation. She argues, quite rightly, that maritime safety could only be assured by cooperation among all the interested countries in line with international law, including the 1982 United Nations Convention on the Law of the Sea and numerous International Maritime Organization conventions, guidelines and directives. However, by examining the status of ASEAN states and China (including Hong Kong) in this global cooperation framework, Chen points out four disadvantages, e.g., non-enforcement of some international treaties in Southeast Asia, which can block maritime safety cooperation in Southeast Asia. She also provides three forms of regional cooperation on maritime safety, such as regional bilateral cooperation, as the foundation for ASEAN-China regional maritime safety cooperation. She also discusses the process of such cooperation and suggests in the end that certain maritime safety cooperation measures and projects should be given priorities.

## Regional Perspectives on ASEAN-China Relations

How China and ASEAN's neighbors and countries in other regions perceive the China-ASEAN relations is important. This section accommodates two papers respectively from Korean and Japanese perspectives. It is natural for South Korea and Japan to take a close look at every development in the China-ASEAN relations, bearing in mind particularly that South Korea and Japan are members of the ASEAN+3 process.

Moon-Soo CHUNG's paper examines the China-ASEAN FTA and Korean FTA policies. As he perceives it, the China-ASEAN FTA can have indirect but significant impact on the Korean economy. In the long run, the FTA will accelerate the pace of industrial restructuring and add further economies of scale to China's economy. On the other hand, the free access to the vast Chinese market will induce more FDI to ASEAN and also facilitate its industrial restructuring. The interaction between China and ASEAN will also affect the globalization process in East Asia. Chung concludes that FTAs are only second best options to multilateralism based on the principle of most favored nations. In that sense, an FTA in East Asia can be viewed as a defense against rampant regionalism in the rest of the world and hence the author urges the promotion of multilateralism, which has proven beneficial to all.

The paper by LAM Peng Er is about the Japanese perspective on China-ASEAN relations. This paper first examines why China's relations with Southeast Asia was not perceived to be a serious challenge to Japan's position in the region during the Cold War era. Then, it analyzes the changes in the post-Cold War environment that led to Beijing's offer of an FTA to ASEAN and this, in turn, has triggered off Tokyo's bid for a closer partnership with the Southeast Asian region. In addition, the chapter examines Japan's multi-faceted responses to China's growing ties with Southeast Asia. These include not only the counter-offer of an economic partnership but also new Japanese initiatives to play a key political role in peace-building in areas suffering from ethnic conflicts, such as Aceh and Mindanao, and in combating piracy in the Southeast Asia seas, especially the Straits of Malacca. Lam concludes with an overall assessment of Japan's responses to China's diplomatic overtures to individual ASEAN countries and their implications for Tokyo's quest for a political leadership role in the region. According to Lam, the Japanese should feel reassured by the continuing presence of its ally, the United States, the dominant military power in the region and the balance diplomacy played by the ASEAN countries in their relations with great powers. He argues that Japan is likely to remain much more affluent and technologically more advanced than China and thus still an important actor in Southeast Asia in many years to come.

## The Way Forward

It is sufficiently clear that China-ASEAN relations will continue to develop in greater breadth and depth. Economically, China has set forth the target of US$200 billion annual trade with ASEAN countries by 2010 — their two-way trade in 2004 had already breached the US$100 billion level — and of steady progress towards the goal of the China-ASEAN FTA. This goal was reiterated by Hu Jintao when he made a speech before the joint session of the Philippine Congress on

27 April 2005.[5] The latest data from the Chinese Customs shows that for the first time, ASEAN has become China's fourth trade partner in 2005 in the first two months of 2005, with a total turnover of US$17 billion.[6] Related to this is the establishment of the China-ASEAN FTA. According to the agreement signed by the two sides, China and ASEAN would launch tariff cuts process as of 1 July 2005 and a zero tariff will be offered for bilateral trade by 2010.

In the political and security arena, the 11th China-ASEAN High Rank Official Consultation held in Shanghai on 29 April 2005 deliberated the implementation of the Action Plan to Implement the Joint Declaration of China-ASEAN Strategic Partnership, and identified five new areas, i.e., transport, energy, culture, tourism and public health, as important cooperative priorities. In addition, the meeting also agreed to hold the first working group meeting to implement the Declaration on the Conduct of the Parties in the South China Sea.[7] However, in comparison with the level of cooperation in the economic sphere, cooperation in the political and security areas have been lagging behind. As shown by relevant chapters in this book, most of the documents signed between China and ASEAN still remain to be effectively implemented.

Finally, in the context of the China-ASEAN relations, the question of whether the "peaceful rise" of China could benefit ASEAN countries should be addressed. The term "peaceful rise" was first introduced by Zheng Bijian (former executive vice president of China's Central Party School) in his speech at the Boao Forum in November 2003, and was later endorsed by the Chinese top leaders Hu Jintao and Wen Jiabao, as a national development strategy, particularly relating to China's link to the international arena.[8] According to this, China is to pursue its soft power in the world by mainly using its economic leverage and seeking a peaceful environment for its economic development.

Clearly, the doctrine of "peaceful rise" is used to counter-balance the prevailing perception of "China threat", a notion that has been actively promoted by many Western journalists and some Western scholars. To ease the concerns of its neighbors, China has on many occasions reiterated that it will never become hegemonic. China's leadership has often taken great pains to explain to its neighbors that China has historically never harboured territorial ambitions beyond its present borders.

---

[5] See Qin Jize, "14 Agreements Signed in Manila", *China Daily*, 28 April 2005, p. 1.
[6] "ASEAN Becomes China's 4th Trade Partner", 12 April 2005, http://www.chinadaily.com.cn/english/doc/2005-04/12/content_435584.htm (accessed 5 May 2005).
[7] See "The 11th China-ASEAN High Rank Official Consultation Held", *People's Daily* (in Chinese), 30 April 2005, p. 3.
[8] For more details, see Bo Zhiyue, "Elite Politics and the 'Peaceful Rise' of China", *EAI Bulletin*, Vol. 6(2), pp. 1 and 10 (2004).

However, while taking China's gesture as embodying a good intention, China's rise, whether peaceful or not, still worries its neighbors as its military capability has been fast expanding along with the pace of its economic growth.

This can be seen from Japan's ambivalence to China's rise. On the one hand, Japan has benefited from China's economic growth through trade and investment; but on the other, Japan is truly concerned about China's rise, particularly under the circumstances when the two countries have not yet reached reconciliation towards the issue of history. For China, peaceful rise is really a good term, favorable for its economic development; but for its neighbors, particularly the smaller ASEAN countries to its south, China has already become their "peaceful" rival in competing for markets and FDI. In this sense, a more comprehensive cooperation and a more progressive integration between China and ASEAN, such as embodied in the CAFTA scheme, is crucial for a "win-win" future for both, especially in terms of peace and economic prosperity.

# Part II

## Changing China-ASEAN Relations in Perspective

# Chapter 2

# China-ASEAN Relations: An Economic Perspective

*John WONG*

## Introduction

The structure and pattern of China's economic relations with countries in Southeast Asia that now constitute the Association of Southeast Asian Nations (ASEAN) have been shaped by many complicated factors and have undergone significant changes.

Since the early 1980s, China's economy has experienced spectacular growth as a result of its successful economic reform and open-door policy. In the process, China's economy has also become more closely integrated with its neighboring economies, including ASEAN. China's dynamic economic growth and its integration process carry both positive and negative spillover effects on the region.

As China continues to press ahead with its export-oriented development strategies, it starts to cast a large shadow on the ASEAN economies, which have to compete with China in attracting FDI (foreign direct investment) and in exporting manufactured products to the same third-country markets. The rise of China is therefore considered a disruptive force to ASEAN's economic growth, at least in the short run.

To allay ASEAN's growing apprehension of China, Beijing took a bold step in arranging a FTA (free trade agreement) with ASEAN to increase the region's trade and investment to the benefit of both sides. Signed in November 2002, this landmark FTA deal has exerted tremendous pressure on Japan and Korea to follow suit and to intensify their economic relations with ASEAN under the general regional cooperation umbrella of the "ASEAN + Three" (ASEAN with China, Japan and Korea) scheme. With China as a rising regional political and economic power, its FTA initiative with ASEAN has thus sparked off a process of what may be called the New-Age economic integration in East Asia.

In the longer run, the economic rise of China can be seen not only as a new engine for the region's economic growth, but also as a powerful force for the economic integration of East Asia. The rise of China may rekindle new hopes of making the 21st century the Pacific Century.

## The Economic Rise of China

The Chinese economy has displayed spectacular performance since it embarked on economic reform and open-door policy some 25 years ago. The average rate of growth for the period 1978–2003 was 9.4%. In 2003, despite disruptions caused by SARS and global economic recession, China's economy still chalked up a hefty 9.3% growth.

China's exports have also been growing very rapidly, averaging at 16% for over two decades, rising from US$9.8 billion in 1978 to US$438 billion in 2003. China is now the world's fourth largest exporter after USA, EU and Japan. For FDI, China has since the early 1990s become the world's most favored destination in comparison with all other developing countries. By the end of 2003, China had attracted a total of US$677 billion in FDI. In fact, China in recent years has consistently captured more than half of all FDI in Asia. Not surprisingly, over 80% of the world's 500 largest companies have set up businesses in China.[1] Above all, on account of its strong external balance, China's foreign exchange reserves recently soared to US$400 billion to become the world's second largest. This has led to mounting international pressure on China to revalue its *Renminbi*.[2]

In 2003, China's total GDP reached 11.7 trillion yuan (US$1.24 trillion)[3] — or more than twice the combined GDP of Indonesia, Malaysia, the Philippines, Singapore and Thailand. China's per-capita GDP in 2003, at US$1,000, is about the same as that in the Philippines but higher than that of Indonesia. In nominal terms, China is the world's fifth largest economy. In terms of purchasing power parity (PPP), the Chinese economy today is already the world's second largest after the USA — one needs, of course, to be aware of the problem of overstating China's real GDP by the PPP measure.[4] (Table 2.1)

Indeed, as a result of its rapid industrialization progress, China is fast becoming the world's foremost manufacturing base. In 2003, China produced 222 million tons of steel, 65 million sets of color TV, 50 million air-conditioners, 22 million

---

[1] "China Trade Surplus Adds Up to US$13.3 Billion", *The Standard* (Hong Kong), 12 July 2002.
[2] See Lu Ding, "Why and How China Maintains RMB's Stability", *EAI Background Brief No. 178* (16 January 2004).
[3] "GDP Growth Last Year at 9.1%", *Renmin Ribao* (People's Daily), 21 January 2004.
[4] World Bank, *World Development Report 2000/2001* (New York: Oxford University Press, 2000). Also, *Mingpao* (Hong Kong), 20 September 2000.

refrigerators, and 32 million PCs.[5] In 2003, China became the world's third largest automobile manufacturer, with a total output of 4.4 million units, after USA and Japan.[6] By 2003, China surpassed the USA as the world's largest telephone market (263 million fixed lines plus 269 million mobile phones).[7] Also by mid-2003, China's registered Internet users reached 68 million to form the world's second largest "Web population".[8] When a huge country like China is industrializing so rapidly, every item of its economic activities inevitably turns out to be a jumbo number due to the combined effect of scale and speed.

Accordingly, the meteoric rise of China's economy has become a "hot" topic in international and regional media.[9] Many Asian economies are concerned about the potential displacement effect of China becoming the world's factory. Others even point fingers at China for their own economic woes, including the accusation of China exporting deflation to them. Even the Japanese were worried by China's recent dynamic industrial expansion. The noted Japanese economist Kenichi Ohmae even used a sensational title "Asia's next crisis: 'Made in China'" to talk alarmingly about the rise of China.[10] Not surprisingly, many Asian economies, especially the smaller ASEAN countries, are watching the rise of China with apprehension.

## East Asian Growth and Interdependence

China's economic performance in the past two decades has indeed been breathtaking. Viewed in the overall East Asian context, however, China's high economic growth is actually not so exceptional. Nor is it unprecedented, as many other Asian economies have experienced such rapid growth before the emergence of China.

East Asia (EA) is commonly defined as comprising Japan, China, the four Newly Industrialized Economies (NIEs) of South Korea, Taiwan, Hong Kong

---

[5] National Bureau of Statistics, "Statistical Communique of the People's Republic of China on National Economic and Social development in 2003", 26 February 2004. *China Daily* (27 February 2004).
[6] "Slowdown in Vehicle Output Pace Expected", *China Daily* (16 January 2004).
[7] National Bureau of Statistics, op cit.
[8] "China's Netizens See Rapid Growth", *China Daily* (22 July 2003).
[9] Recently at the World Economic Forum in Davos, "everything is China, China, China", according to one observer. "The Talk of the Town at Davos: China", *International Herald Tribune* (26 January 2004). China's emergence as the world's manufacturing powerhouse after two decades of dynamic growth has invited prominent worldwide attention. The international media have recently portrayed China's economic resurgence as an economic threat. David Roche, a famous Wall Street economist, commented on China being a source of current global recession with its mass production of a wide range of low-priced manufactured products for the world market. In early 2003, Japan's *Nikkei Weekly* reported about China setting pace in markets for commodities around the world. The Chinese media and academia have since come out to defend China's position.
[10] "Asia's Next Crisis: 'Made in China'", *The Straits Times* (Singapore), 2 August 2001.

and Singapore, and the four Association of Southeast Asian Nations (ASEAN) of Indonesia, Malaysia, the Philippines, and Thailand — the original ASEAN members.[11] Situated on the western rim of the Pacific, many of these East Asian economies (EAEs) have displayed dynamic growth for a sustained period until 1997 when they were hit, in varying degrees, by the regional financial crisis. The World Bank in its well-known study referred to this high-growth phenomenon as the "East Asian Miracle".[12]

Why have the EAEs been able to sustain high growth for so long? The EAEs have generally committed a high proportion of their GDP (at around 30%) to domestic investment during their critical periods of industrial take-off. Furthermore, their high levels of gross domestic investments were largely matched by their equally high levels of domestic savings.[13] In short, high investment, as the single most important neo-classical explanation of high growth, has worked on these EAEs much like a "virtuous circle": high rates of investment induce high export growth, and corresponding high GDP growth, high savings and finally high investments again. Other explanations include intensive human resource development for certain EAEs which are resource-poor. In terms of policy, the EAEs share the salient common feature of operating effective export-oriented development strategies, as reflected in their generally high export-GDP ratios and their relative high shares in the world export markets. Their export orientation has propelled them to high growth through reaping the gains from international trade and specialization.[14]

Historically speaking, the EA growth process is marked by three waves. Japan was the first non-Western country to become industrialized. Its high growth dated back to the 1950s after it had achieved its rapid post-war recovery, and carried the growth momentum over to the 1960s and much of the 1970s. Japan's economic growth engine was initially based on the export of labor-intensive manufactured products; but it was soon forced by rising wages and increasing costs to shed its comparative advantage for labor-intensive manufacturing in favor of the four NIEs, which started their industrial take-off in the 1960s. These four NIEs, once dubbed

---

[11] Singapore is historically and geographically an integral part of Southeast Asia, and politically a member of ASEAN. However, economically and socially, Singapore is more akin to the other East Asian NIEs, and hence commonly labelled as one of the four East Asian NIEs.

[12] *The East Asian Miracle* (New York: Oxford University Press, 1994).

[13] See World Bank, *World Development Report* (various years) and Asia Development Bank, *Asian Development Outlook* (various years), which provide data on investment and savings rates of the EAEs for various years.

[14] For further discussion of this topic, see John Wong, "The East Asian Phenomenon and the Implications for Economic Development", in Basant K. Kapur *et al.* (eds), *Development, Trade and the Asia-Pacific, Essays in honour of Professor Lim Chong Yah* (Singapore: Prentice Hall, 1996).

"Asia's Four Little Dragons", were arguably the most dynamic economies in Asia, as they had sustained near double-digit rates of growth for three decades, from the early 1960s to the 1980s. The rise of the NIEs constituted the second wave of the region's growth and integration.

By the early 1980s, high costs and high wages had also caught up with these four NIEs, which had to restructure their economies towards more capital-intensive and higher value-added activities by passing their comparative advantage in labor-intensive products to the late-comers of China and the four ASEAN economies, and thereby spreading economic growth to the latter. In this way, China and some ASEAN economies were able to register high growth through the 1980s and the 1990s. Many Japanese scholars like to depict this pattern of development in Asia as the "Flying Geese" model.[15] (See Table 2.1)

During the past two decades, the Chinese economy on account of its successful market reform has consistently chalked up near double-digit rates of growth while many ASEAN economies, due to a number of institutional and structural constraints, were losing growth dynamism. The rise of China promises to usher in the third wave of growth and integration for the region, with even greater geo-political and geo-economic implications than the previous two because of China's vast size and diversity.

In other words, the rise of China will ensure that the EA region as a whole will not lose its economic dynamism. As the EAEs keep on growing, they will also increase their economic interaction with each other. Thus, an important feature of these EAEs is their growing economic interdependence. The EAEs, despite their inherent political, social and economic divergences, can actually integrate quite well economically as an informal and loosely-constituted regional grouping. This is essentially the underlying meaning of the "flying geese" principle. To start with, Japan is the natural economic leader of the group and has in fact been the prime source of capital and technology for other EAEs, first the NIEs followed by China and ASEAN. The resource-based ASEAN-4 complement well the manufacturing-based NIEs while both are also complementary with the more developed Japanese economy. Then the huge potential of China, with its vast resource base and diverse needs, offers additional opportunities for all.

Accordingly, the EA region has already developed a fairly high level of intra-regional trade. As shown in Table 2.2, the EA region in 2001, despite its economic slowdown, still absorbed 41% of China's total exports; 43% of Korea's, 49% of

---

[15] The "flying geese" concept of development was coined by a Japanese economist, Kaname Akamatzu. ("A Historical Pattern of Economic Growth in Developing Countries", *Developing Economies*, Vol. 1, March/August, 1962).

Table 2.1  East Asian Economies: Performance Indicators

| | Population (Mn) 2002 | GNP per capita (US$) 2002 | PPP estimates of GNP per capita (US$) 2001 | Total GDP (US$ bn), 2002 | Growth of GDP (%) | | | | | | | | | Gross Domestic Investment as % of GDP 2001 | Annual Export Growth (%) 1999–2000 | Mfg Exports as % of total exports 2001 | Exports as % of GDP 2001 |
|---|---|---|---|---|---|---|---|---|---|---|---|---|---|---|---|---|---|
| | | | | | 1960–70 | 1970–80 | 1980–90 | 1990–2001 | 1998 | 1999 | 2001 | 2002 | 2003 | | | | |
| China | 1,285 | 952 | 3,950 | 1,266 | 5.2 | 5.5 | 10.3 | 10.0 | 7.8 | 7.1 | 7.3 | 8.0 | 9.3 | 39 | 14.5 | 89 | 26 |
| Japan | 127 | 33,550 | 25,550 | 3,973 | 10.9 | 4.3 | 4.1 | 1.3 | −1.1 | 0.8 | 0.4 | −0.3 | 2.1 | 26 | 4.1 | 93 | 10 |
| **NIEs** | | | | | | | | | | | | | | | | | |
| South Korea | 48 | 10,014 | 15,060 | 476 | 8.6 | 10.1 | 8.9 | 5.7 | −6.7 | 10.9 | 3.0 | 6.3 | 2.7 | 27 | 10.1 | 91 | 43 |
| Taiwan | 23 | 12,900 | n.a. | 282 | 9.2 | 9.7 | 7.9 | 5.7 | 4.6 | 5.4 | −2.2 | 3.6 | 3.2 | 17 | 7.9 | n.a. | 51 |
| Hong Kong | 7 | 24,532 | 25,560 | 162 | 10.0 | 9.3 | 6.9 | 3.8 | −5.3 | 3.0 | −0.2 | 2.3 | 3.0 | 27 | 8.3 | 95 | 144 |
| **ASEAN-5** | | | | | | | | | | | | | | | | | |
| Indonesia | 211 | 780 | 2,830 | 173 | 3.9 | 7.2 | 6.1 | 3.8 | −14.2 | 0.8 | 3.3 | 3.7 | 4.1 | 22 | 8.1 | 56 | 41 |
| Malaysia | 25 | 3,609 | 7,910 | 95 | 6.5 | 7.9 | 5.3 | 6.5 | −7.4 | 6.1 | 0.4 | 4.2 | 4.7 | 24 | 12.2 | 80 | 116 |
| Philippines | 82 | 1,034 | 4,070 | 78 | 5.1 | 6.0 | 1.0 | 3.3 | −0.6 | 3.4 | 3.2 | 4.6 | 4.3 | 18 | 18.8 | 91 | 49 |
| Singapore | 4 | 20,613 | 22,850 | 87 | 8.8 | 8.3 | 6.7 | 7.4 | −0.1 | 6.9 | −2.0 | 2.2 | 0.8 | 24 | 9.9 | 85 | 174 |
| Thailand | 63 | 1,960 | 6,230 | 126 | 8.4 | 7.1 | 7.6 | 3.8 | −10.5 | 4.4 | 1.8 | 5.0 | 6.2 | 24 | 10.5 | 74 | 66 |

Notes: (1) 2003 GDP growth figures for China and Singapore are official figures; the rest are real GDP growth estimates from EIU.
(2) 1998–2001 GDP growth rates extracted from *Regional Outlook: Southeast Asia*, 2003–2004.
(3) Per capita GNP figure for Taiwan extracted from *Taiwan Statistical Data Book*, 2001.
(4) 1996–2002 GDP growth rates for Japan represent real GDP growth rates.
(5) Figures for Gross Domestic Investment (% of GDP) derived from *Asian Development Outlook* 2003.

Sources: *World Development Report* 1995; *World Development Report* 2000/2001; *World Development Report* 2002; *World Development Indicators* 2003; World Bank website; ISEAS, *Regional Outlook: Southeast Asia* 2003–04; Taiwan Ministry of Economic Affairs, http://www.moea.gov.tw; EIU DataServices; *Taiwan Statistical Data Book*, 2002.

China-ASEAN Relations: An Economic Perspective   23

Table 2.2a  Intra-regional Trade in East Asia

| East Asian Economy | Year | Total Exports (US$ Million) | Share of Total Exports Destined For (%) |||||||||
|---|---|---|---|---|---|---|---|---|---|---|---|
| | | | USA | EU | Japan | China | Korea | Taiwan | Hong Kong | Singapore | ASEAN-4 | East Asia |
| Japan | 1980 | 130,441 | 24.5 | 14.6 | | 3.9 | 4.1 | — | 3.7 | 3.0 | 7.0 | 21.7 |
| | 1988 | 264,856 | 34.1 | 19.7 | | 3.6 | 5.8 | 5.4 | 4.4 | 3.1 | 4.9 | 27.2 |
| | 1992 | 339,885 | 28.5 | 19.7 | | 3.5 | 5.2 | 6.2 | 6.1 | 3.8 | 8.1 | 32.9 |
| | 1996 | 410,901 | 27.5 | 14.3 | | 5.3 | 7.1 | 6.3 | 6.2 | 5.1 | 12.4 | 42.4 |
| | 2000 | 479,249 | 30.0 | 16.4 | | 6.3 | 6.5 | 7.5 | 5.7 | 4.3 | 9.5 | 39.8 |
| | 2001 | 403,496 | 30.4 | 16.0 | | 7.7 | 6.3 | 6.0 | 5.8 | 3.6 | 9.3 | 38.7 |
| China | 1980 | 18,099 | 5.4 | 13.7 | 22.3 | | — | — | 24.1 | 2.3 | 4.3 | 53.0 |
| | 1988 | 47,540 | 7.1 | 10.4 | 16.9 | | — | — | 38.4 | 3.1 | 2.8 | 61.2 |
| | 1992 | 80,517 | 10.7 | 10.8 | 13.8 | | 2.9 | 0.8 | 44.2 | 2.5 | 2.8 | 67.0 |
| | 1996 | 151,197 | 17.7 | 13.1 | 20.4 | | 5.0 | 1.9 | 21.8 | 2.5 | 3.4 | 55.0 |
| | 2000 | 249,297 | 20.9 | 15.3 | 16.7 | | 4.5 | 2.0 | 17.9 | 2.3 | 3.7 | 47.1 |
| | 2001 | 266,620 | 20.4 | 15.3 | 11.0 | | 4.7 | 1.9 | 17.5 | 2.2 | 3.8 | 41.1 |
| Korea | 1980 | 17,505 | 26.4 | 16.3 | 17.4 | — | | — | — | 1.5 | 4.6 | 23.5 |
| | 1988 | 60,696 | 35.4 | 14.7 | 19.8 | — | | 1.6 | 5.9 | 2.2 | 2.8 | 32.3 |
| | 1992 | 76,632 | 23.7 | 12.8 | 15.1 | 3.5 | | 3.0 | 7.7 | 4.2 | 7.0 | 40.5 |
| | 1996 | 129,715 | 16.9 | 11.4 | 12.2 | 8.8 | | 3.1 | 8.6 | 5.0 | 9.3 | 47.0 |
| | 2000 | 172,268 | 21.9 | 13.6 | 11.9 | 10.7 | | 2.0 | 6.2 | 3.3 | 7.2 | 41.3 |
| | 2001 | 150,439 | 20.8 | 13.1 | 11.0 | 12.1 | | 3.9 | 6.3 | 2.7 | 6.8 | 42.8 |
| Taiwan | 1980 | — | — | — | — | — | — | | — | — | — | — |
| | 1988 | 60,667 | 28.9 | 17.7 | 10.9 | 3.7 | 1.4 | | 18.9 | 3.1 | 6.9 | 41.2 |
| | 1992 | 81,470 | 23.2 | 13.6 | 11.8 | 12.9 | 2.3 | | 23.1 | 4.0 | 8.3 | 50.1 |
| | 1996 | 115,942 | 23.5 | 14.8 | 11.2 | 17.9 | 2.6 | | 21.1 | 3.7 | 7.4 | 48.8 |
| | 2000 | 148,321 | 22.5 | 14.8 | 10.4 | 16.9 | 2.7 | | 21.9 | 3.3 | 7.2 | 49.4 |
| | 2001 | 122,866 | | | | 17.9 | | | | | | |

Table 2.2b  Intra-regional Trade in East Asia

| East Asian Economy | Year | Total Exports (US$ Million) | Share of Total Exports Destined For (%) | | | | | | | | |
|---|---|---|---|---|---|---|---|---|---|---|---|
| | | | USA | EU | Japan | China | Korea | Taiwan | Hong Kong | Singapore | ASEAN-4 | East Asia |
| Hong Kong | 1980 | 19,730 | 26.1 | 24.5 | 4.6 | 6.3 | 1.2 | — | | 4.4 | 6.8 | 23.3 |
| | 1988 | 63,163 | 24.8 | 16.9 | 5.9 | 27.0 | 2.6 | 3.6 | | 2.8 | 3.2 | 41.9 |
| | 1992 | 119,512 | 23.1 | 17.1 | 5.2 | 29.6 | 1.6 | 3.5 | | 2.6 | 3.1 | 45.6 |
| | 1996 | 180,750 | 21.2 | 12.7 | 6.5 | 34.3 | 1.6 | 2.4 | | 2.7 | 3.7 | 51.2 |
| | 2000 | 201,860 | 23.3 | 15.3 | 5.5 | 34.6 | 1.9 | 2.5 | | 2.3 | 3.3 | 50.1 |
| | 2001 | 189,894 | 22.3 | 14.5 | 5.9 | 36.9 | 1.8 | 2.4 | | 2.0 | 3.3 | 52.3 |
| Singapore | 1980 | 19,375 | 12.5 | 12.5 | 8.1 | 1.6 | 1.5 | — | 7.7 | | 20.8 | 39.7 |
| | 1988 | 39,306 | 23.6 | 13.5 | 8.6 | 3.0 | 2.0 | 2.8 | 6.2 | | 20.3 | 42.9 |
| | 1992 | 63,483 | 16.6 | 11.9 | 4.4 | 1.8 | 2.6 | 2.4 | 7.2 | | 14.3 | 32.7 |
| | 1996 | 125,014 | 18.4 | 12.7 | 8.2 | 2.7 | 3.8 | 3.9 | 8.2 | | 25.5 | 52.3 |
| | 2000 | 137,804 | 17.3 | 13.2 | 7.5 | 3.9 | 3.6 | 6.0 | 7.9 | | 24.9 | 53.8 |
| | 2001 | 121,751 | 15.4 | 13.4 | 7.7 | 4.4 | 3.9 | 5.1 | 8.9 | | 24.2 | 54.2 |
| ASEAN-4 | 1980 | 47,100 | 18.8 | 13.8 | 34.5 | 1.1 | 1.7 | — | 1.9 | 11.8 | | 51.0 |
| | 1988 | 80,080 | 16.4 | 12.6 | 19.5 | 2.2 | 2.8 | 2.0 | 2.9 | 9.0 | | 38.4 |
| | 1992 | 112,788 | 21.0 | 17.6 | 21.9 | 2.6 | 2.9 | 3.1 | 3.9 | 13.6 | | 48.0 |
| | 1996 | 204,270 | 18.6 | 13.7 | 17.8 | 3.3 | 3.5 | 3.4 | 5.1 | 14.0 | | 47.1 |
| | 2000 | 269,099 | 20.4 | 14.8 | 16.0 | 3.4 | 3.8 | 4.2 | 4.2 | 12.5 | | 44.0 |
| | 2001 | 250,656 | 20.0 | 14.7 | 16.1 | 4.4 | 3.7 | 3.8 | 4.1 | 11.8 | | 43.9 |

Note: ASEAN-4 refers to Malaysia, Indonesia, Thailand and the Philippines.
Taiwan's indirect trade with China is calculated from data available at http://www.seftb.org.

Source: IMF, *Direction of Trade Statistics Yearbook* (1993, 1987 and 2002 issues).

Taiwan's, 52% of Hong Kong's, 54% of Singapore's, and 44% of the average of the ASEAN-4, though only 39% of Japan's — still unusually high for Japan as a global trading power.

Table 2.2 also describes the process of EA's growing export dependence over the past two decades. It shows Japan's highly remarkable shift in export orientation over the years towards greater regional focus, with its export share to the EA region increasing from 22% in 1980 to 39% in 2001. The four NIEs have similarly made significant shifts in the same period by re-orienting their exports towards the region, mainly as a result of the opening up of China: Korea from 24% to 43%, Taiwan from 41% (1992) to 49%, Hong Kong from 23% to 52%, and Singapore from 40% to 54%. China, on the other hand, has moved in the opposite direction, slightly reducing its export dependence on the region to one which is, in recent years, geared more to the US and EU markets. Likewise, the ASEAN-4 has also shown a slight reduction of export dependence on the EA region, from 51% in 1980 to 44% in 2001.

Apart from intra-regional trade, intra-regional FDI flows have also operated as a powerful integrating force for the EA region, especially since a great deal of regional FDI is trade-related in nature. The EAEs, as essentially open and outward-looking economies, are highly dependent on foreign trade and foreign investment for their economic growth. Both China and ASEAN have devised various incentive schemes to vie for FDI, which is generally treated not just as an additional source of capital supply but, more importantly, as a means of technology transfer and export market development.

In particular, China in recent years has become the most favored destination of all developing economies for FDI. As can be seen from Table 2.3, the EA region, especially Hong Kong, Taiwan, Japan, Singapore and South Korea, accounted for an overwhelming share of FDI inflow to China. Such regional predominance has been declining in recent years, as China has made efforts to attract more technology-intensive FDI from North America and the EU. By 2001, the East Asian share of FDI in China declined to 57%, down from 88% in 1992. Suffice it to say that the rise of China has completely altered the FDI landscape of East Asia.

## A Source for Growth and a Catalyst for Integration

It is thus clear that China's economic growth fits in quite well with the overall EA growth patterns. Since the EA region absorbs about 50% of China's exports and supplies three quarters of China's FDI, it is not hard to see that China's rapidly growing economy since 1978 has impacted significantly on many EAEs to each other's advantage. On the one hand, China has been able to harness the region's vast trade and investment opportunities to facilitate its own economic growth. At

Table 2.3  Foreign Direct Investment in China (US$ Million)

| | 1992 | | 1993 | | 1997 | | 1998 | | 2001 | | 2002 | | 2003 | |
|---|---|---|---|---|---|---|---|---|---|---|---|---|---|---|
| | Actual sum of capital | % | Actual sum of capital | % | Actual sum of capital | % | Actual sum of capital | % | Actual sum of capital | % | Actual sum of capital | % | Actual sum of capital | % |
| Total | 11292 | 100 | 27771 | 100 | 45257 | 100 | 45463 | 100 | 45984 | 100 | 51585 | 100 | 51493 | 100 |
| Asia Pacific | 9900 | 88 | 23333 | 84 | 30389 | 67.2 | 26626 | 58.6 | 26197 | 57.1 | 28744 | 55.7 | 30620 | 59.5 |
| Hong Kong | 7706 | 68 | 17445 | 63 | 20632 | 45.6 | 18508 | 40.7 | 16717 | 36.4 | 17861 | 34.6 | 17700 | 34.4 |
| Taiwan | 1053 | 9 | 3139 | 11 | 3289 | 7.3 | 2915 | 6.4 | 2980 | 6.5 | 3971 | 7.7 | 3377 | 6.6 |
| Japan | 748 | 7 | 1361 | 4.9 | 4326 | 9.6 | 3400 | 7.5 | 4348 | 9.5 | 4191 | 8.1 | 5054 | 9.8 |
| South Korea | 120 | 1 | 382 | 1.4 | 2142 | 4.7 | 1803 | 4.0 | 2152 | 4.7 | 2721 | 5.3 | 4489 | 8.7 |
| ASEAN | 271.6 | 2.4 | 1006 | 3.6 | 3418 | 7.6 | 4197 | 9.2 | 2970 | 6.5 | 3201 | 6.2 | 2853 | 5.5 |
| Indonesia | 20.2 | 0.2 | 65.8 | 0.2 | 80 | 0.2 | 69 | 0.2 | 160 | 0.3 | 122 | 0.2 | 150 | 0.3 |
| Malaysia | 24.7 | 0.2 | 91.4 | 0.3 | 382 | 0.8 | 340 | 0.7 | 263 | 0.6 | 368 | 0.7 | 251 | 0.5 |
| Philippines | 16.6 | 0.2 | 122.5 | 0.4 | 156 | 0.3 | 179 | 0.3 | 209 | 0.5 | 186 | 0.4 | 220 | 0.4 |
| Singapore | 125.9 | 1.1 | 491.8 | 1.8 | 2606 | 5.8 | 3404 | 7.5 | 2144 | 4.7 | 2337 | 4.5 | 2058 | 4.0 |
| Thailand | 84.3 | 0.8 | 234.4 | 0.8 | 194 | 0.4 | 205 | 0.5 | 194 | 0.4 | 188 | 0.4 | 174 | 0.3 |
| USA | 519 | 5 | 2068 | 7.4 | 3239 | 7.2 | 3898 | 8.6 | 4433 | 9.6 | 5424 | 11 | 4199 | 8.2 |
| Others | 873 | 8 | 2370 | 8.5 | 8192 | 18 | 10729 | 23.6 | 12062 | 26.2 | 19642 | 38 | 13821 | 26.8 |

Note: Percentages may not be exact due to rounding up or down.
Sources: *Statistical Yearbook of China* (various issues); *China Monthly Statistics*.

the same time, China's economic growth and increasing integration with the region also provide new opportunities to enhance the region's overall growth potential and new impetus for regionalism.

However, the actual impact of the fast-growing Chinese economy on the EAEs is quite uneven. China's dynamic economic growth has produced both positive and negative effects for the individual EAEs in the region. By and large, Japan and the four NIEs have been able to benefit from China's open-door policy by exporting more high-tech products, and by investing in China, as shown in Table 2.3.

Indeed, for the past two decades, as shown in the trade matrix of Table 2.2, Japan's share of exports to China almost doubled from 3.9% in 1980 to 7.7% in 2001; Korea's share from no direct trade in the 1980s to 12% in 2001; Taiwan's share also soared from zero direct trade in the early 1980s to 18% in 2001; Hong Kong's share from 6% in 1980 to a stunning 37% in 2001; and Singapore's share from 1.6% to 4.4%. In time to come, these five economies, which are inherently complementary with China, are set to be even more closely integrated with China. Hong Kong today, and to some extent Taiwan, has already become highly dependent on China for its economic growth.

Table 2.4, using "trade intensity index" to measure the actual strength of trade relationship between two countries relative to all their respective trade partners,

Table 2.4  Intra-trade Intensity Index for the East Asian Region, 1998–2000

| | Japan | China | Korea | Taiwan | Hong Kong | Singapore | Indonesia | Malaysia | Philippines | Thailand | ASEAN | USA |
|---|---|---|---|---|---|---|---|---|---|---|---|---|
| Japan | — | 1.8 | 2.5 | 3.5 | 1.6 | 2.0 | 2.3 | 2.2 | 3.6 | 3.0 | 2.3 | 1.6 |
| China | 2.8 | — | 1.7 | 0.8 | 6.2 | 1.1 | 1.5 | 0.7 | 1.1 | 0.7 | 1.0 | 1.3 |
| Korea | 1.9 | 3.2 | — | 2.3 | 1.9 | 1.5 | 3.2 | 2.1 | 3.9 | 1.3 | 2.3 | 1.1 |
| Taiwan | 1.8 | — | 1.0 | — | 6.6 | 1.7 | 2.0 | 1.9 | 3.6 | 2.0 | 2.1 | 1.4 |
| Hong Kong | 1.0 | 11.1 | 0.7 | 1.0 | — | 1.2 | 0.8 | 0.7 | 1.8 | 1.0 | 1.1 | 1.2 |
| Singapore | 1.3 | 1.2 | 1.5 | 2.1 | 1.8 | — | 4.3 | 14.0 | 4.4 | 4.8 | 8.1 | 1.0 |
| Indonesia | 3.8 | 1.6 | 3.0 | 2.0 | 0.9 | 5.7 | — | 2.5 | 2.4 | 2.0 | 3.7 | 0.8 |
| Malaysia | 2.1 | 0.9 | 1.4 | 2.3 | 1.4 | 8.9 | 2.9 | — | 2.9 | 3.7 | 5.9 | 1.2 |
| Philippines | 2.6 | 0.5 | 1.3 | 4.0 | 1.5 | 3.6 | 0.7 | 3.5 | — | 2.8 | 2.9 | 1.8 |
| Thailand | 2.7 | 1.2 | 0.8 | 1.9 | 1.6 | 4.6 | 3.7 | 3.2 | 2.9 | — | 4.1 | 1.2 |

Note: The trade intensity index is defined as:

$$T_{ij} = [x_{ij}/X_{it}]/[x_{wj}/X_{wt}]$$

where $x_{ij}$ and $X_{wj}$ are the values of $i$'s exports and world exports to $j$, $X_{it}$ is $i$'s total exports and $X_{wt}$ are total world exports. As such, the index reflects the ratio of the share of country $i$'s exports going to country $j$, relative to the share of world trade destined for country $j$.

Source: Computed from data in IMF, *Direction of Trade Statistics* (various issues).

brings out a clearer picture of East Asian intra-regional trade than that indicated by simple trade share analysis. When the index exceeds "one", it means that these countries are trading with each other above the "normal" level. Thus, for the period of 1998–2000, Japan's trade with ASEAN was more intensive than with China while Japan was trading more intensively with China than with USA. Korea and Hong Kong (which embody Taiwan's indirect trade with China) have a very intense trade relationship with China, more than with ASEAN. The ASEAN countries, on the other hand, are trading far more intensively with each other than with China, even though Singapore, Indonesia and Thailand are still trading with China above their "normal" levels. In short, China has become a strong trade partner with Japan and the three Asian NIEs of Korea, Taiwan and Hong Kong, which are economically complementary with China.

## Pressure on ASEAN-China Relations

China's fast-growing economy has inevitably produced significant impact on its neighboring economies in the Asia-Pacific region, which absorb 50% of China's exports and supply about three quarters of China's FDI. Broadly speaking, the spillovers of China's economic growth will produce both positive and negative effects on the region. For Japan and other East Asian NIEs of Korea, Taiwan and Hong Kong, they may lose their comparative advantage for many of their manufactured exports. But they can also capture the benefits of the growing Chinese economy by exporting high-tech products and by investing in China. The East Asian economies have since become more closely integrated with China.

However, the economies of China and ASEAN (minus Singapore) at their present stages of development tend to be more competitive than complementary with each other. In many ways, China's dynamic economic growth has exerted strong competitive pressure on the ASEAN economies, which are vying for FDI with China as well as competing head-on with China's manufactured exports in the developed-country markets.[16]

Initially, China's success in economic reform and development had produced very little impact on the ASEAN countries to its south. Sino-ASEAN trade was very small — in fact, only a small fraction of each other's total trade and with a large part being centred in Singapore (see Chart 2.1). Even by the early 1990s, when massive FDI began to flow into China, there was no evidence that China had "sucked" in a lot of capital from the ethnic Chinese in Southeast Asia.[17]

However, it was a different scenario towards the end of the 1990s. While many ASEAN countries were plagued by persistent economic crises and domestic political instability, China has been intent on its single-minded pursuit of economic

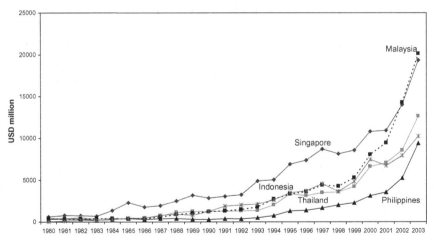

Sources: United Nations, *Statistical Yearbook for Asia and the Pacific*, 1991 and 2000; *China Statistical Yearbook* (various issues).

Chart 2.1  China's Trade with ASEAN-5, 1980–2003

---

[16] For further discussion of this topic, see Prakash Loungani, "Comrades or Competitors? Trade Links between China and Other East Asian Economies", *Finance & Development*, June (2000).
[17] See John Wong, "Southeast Asian Ethnic Chinese Investing In China", *EAI Working Paper No. 15* (23 October 1998).

modernization. This has resulted in the further narrowing of development gaps between ASEAN and China. In fact, the ASEAN region risks being left behind by China's relentless economic growth. Not surprisingly, many ASEAN economies are watching the recent economic rise of China with apprehension.

## China's Bold FTA Initiative

Mindful of ASEAN's worries over the possible disruptive effects of its rapid economic growth, China in recent years has been under mounting pressures to dispel the "China threat" fears by improving its overall relations with its ASEAN neighbors. During the 1997 Asian financial crisis, Beijing's steadfast refusal to devalue its *Renminbi* was much appreciated by ASEAN, as such a move would have aggravated the region's economic crisis. But the single most important step ever undertaken by China in recent years to upgrade its long-term political and economic relations with the ASEAN region is China's bold FTA scheme.

At the ASEAN-China Summit in November 2001, former Chinese Premier Zhu Rongji proposed the creation of a free trade area between China and ASEAN within ten years. On November 4, 2002, China and the ASEAN countries formally signed a landmark framework agreement in Cambodia to establish a FTA by 2010.[18] The formation of the China-ASEAN FTA signifies the creation of an economic region of 1.7 billion consumers with a combined GDP of US$2 trillion. It offers an effective means for smaller ASEAN states to overcome their disadvantage of smallness by pooling resources and combining markets. This will in time lead to greater economic integration between China and ASEAN, clearly a win-win situation for both sides.[19]

In the short run, however, ASEAN has to deal with the initial risks of a potential trade diversion effect and related structural adjustment.[20] In general, the FTA scheme will give rise to an uneven distribution of costs and benefit different industries, different sectors, and even different ASEAN countries. After the initial process of adjustment, individual ASEAN economies will then develop their own niches in their economic relations with China.

---

[18] The framework agreement signed by the 11 nation states sets out a road map for trade liberalization in goods and services for most countries by 2010 and for the less developed ASEAN nations (namely Cambodia, Laos, Myanmar and Vietnam) by 2015.

[19] For further discussion of this topic, see John Wong and Sarah Chan, "China-ASEAN Free Trade Agreement: Shaping Future Economic Relations", *Asian Survey*, Vol. XLIII, No. 3, May/June (2003).

[20] Trade diversion occurs when members of a free trade grouping trade more among themselves than with other non-member countries, due to a lowering of tariffs or non-tariff barriers within the FTA. Structural adjustments occur because when intra-regional barriers are dismantled, industries will expand in some countries and contract in others as industries relocate in response to differences in factor endowments. The costs of adjustment resulting from such relocation of economic activity can be asymmetrical since some economies will incur higher costs in the short run than others.

With China continuing its dynamic economic growth, opportunities will arise for the ASEAN countries to exploit China's growing market. Apart from its primary commodities, ASEAN's resource-based products will be in great demand in China. China is such a vast and disparate market that East China, South China and Southwest China can individually offer different opportunities to different ASEAN producers. Beyond merchandise trade, FTA also promotes trade in services, including tourism. China may generally have strong comparative advantage in manufacturing because it enjoys the economies of scale, which, however, may not apply to many service activities. In fact, a lot of China's service activities, on account of their socialist legacies, are known to be more backward than those in ASEAN.[21]

Many Asia-Pacific economies have started to experience the positive spillovers of China's economic growth. Apart from the surge in Chinese tourists to other Asian countries, in recent years China's imports from Japan, Korea, Taiwan, Singapore, Indonesia, Malaysia, the Philippines, Thailand, India and Australia had exceeded its exports to these economies, thereby incurring trade deficits with them, which were offset by China's trade surplus with USA and EU (see Chart 2.2). This means that

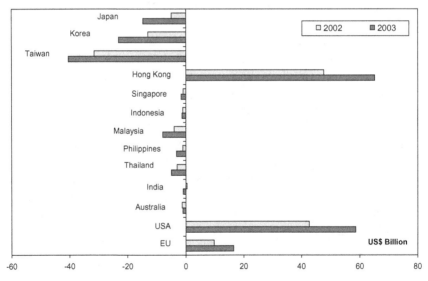

Source: *China Monthly Customs Statistics* (December 2002); Ministry of Commerce website, http://www.mofcom.gov.cn.

Chart 2.2  China's Trade Balance with Selected Countries

---

[21] See John Wong and Ruobing Liang, "China's Service Industry (II): Gearing Up for WTO Challenges", *EAI Background Brief No. 163* (28 July 2003).

these Asian economies are tapping China's vast domestic market and its growing demand for consumer goods, capital equipment and raw materials.

Over the years as the FTA scheme is gradually phased in, multinationals in the region will gradually restructure their supply chains and rationalize their production networks by taking China and ASEAN together as a single market. This will eventually lead to a reshuffle of regional production networks and hence a redistribution of FDI inflow. The new regional production patterns will be based on a bigger and more diverse market. In short, both trade and FDI in the region should continue to grow under the impact of the ASEAN-China FTA. This will certainly be a boon to both ASEAN and China.

## New Impetus for ASEAN+3

Besides creating a new source of economic growth for the region, China is also seen as a new force for revitalizing the region's economic integration process. China's FTA with ASEAN had exerted tremendous pressure on Japan and Korea to follow suit, prompting similar responses from them. Indeed, in the wake of China-ASEAN FTA, Japan signed a joint declaration with ASEAN to draw up a general framework for a FTA in ten years that would comprise Japan's bilateral FTA arrangements with individual ASEAN member countries.

In June 2003, China signed the Closer Economic Partnership Arrangement (CEPA) with Hong Kong (and subsequently with Macau). CEPA is obviously aimed at the eventual integration of these Greater China economies after the inclusion of Taiwan in future.[22] Prior to this, China had agreed to initiate a joint study with Japan and Korea on possible Northeast Asian economic cooperation. In October 2003, Premier Wen Jiabao attended the Ninth ASEAN Summit in Bali, where he signed with the heads of government from Japan and Korea the Joint Declaration on the Promotion of Tripartite Cooperation among these three Northeast Asian countries. This tripartite cooperation is not just for the promotion of economic cooperation and peace dialogue in Northeast Asia, but also aimed at strengthening the process of ASEAN economic integration with other EAEs, i.e., a more concrete way of accelerating the realization of the ASEAN+3 plan. At the Summit, Premier Wen also signed the Treaty of Amity and Cooperation (TAC) with ASEAN. In concluding this historic treaty, China has signaled to the ASEAN countries its acceptance of ASEAN's norms and values. But this move once again pressurises Japan into considering its accession to a similar TAC.

---

[22] See John Wong and Sarah Chan, "China's Closer Economic Partnership Arrangement (CEPA) with Hong Kong: A Gift from Beijing?" *EAI Background Brief No. 177* (12 December 2003).

Viewed in a larger context, China's FTA initiative with ASEAN not only marks the most important first step in the "ASEAN+3" scheme, but in fact also plays a crucial catalytic role in galvanizing what may be called the New-Age economic integration process for the East Asian region as a whole. As long as China's economy could sustain its dynamic growth, its regional integration initiatives would remain effective. In short, the spate of new cooperation initiatives in recent years has shown that such an economic integration scheme for East Asia is no longer an abstract notion, but something that is achievable once major players like China and Japan are serious about it.

## Broader Geo-Political Implications

As a rising regional political and economic power, China is destined to play an important role in the growth and development of the ASEAN region. However, there is still a great deal of uncertainty as to: (1) how will China play out its geo-political role in the region? (2) what kind of new security architecture will the region develop? and (3) will China push for a greater leadership role in the region in order to counter Western (American) influences?

It is commonly assumed that as China grows stronger, it will also become more assertive in its dealings with neighboring countries. On the other hand, if China were able to manage its rise as a gradual process of its "peaceful ascendancy" (*heping jueqi*), the total spillover effect on the region would be much less disruptive.[23] On balance, ASEAN should have no problem adjusting to the rise of China, particularly when China's economy is operating as another engine of economic growth. China, on its part, is likely to continue its warm relations with ASEAN so long as the latter subscribes to the One-China principle over the Taiwan issue.

In the meanwhile, both China and ASEAN still need to step up the process of consensus building and continuing dialogues. For this, Singapore should be able to play an important role. First, Singapore plays a pivotal role in the growing China-ASEAN economic relations as Singapore accounts for a significant proportion of China-ASEAN trade and their two-way FDI flows. Secondly, Singapore's political relations with China are maturing and ready to move forward, thereby providing potentially useful guideposts for some ASEAN countries over their burgeoning relations with China. Thirdly, Singapore can provide China with useful Southeast Asian perspectives, helping China to forge consensus with ASEAN.

---

[23]Yoichi Funabashi, "China is Preparing a 'Peaceful Ascendancy'", *International Herald Tribune* (30 December 2003). See also Bruce Klingner, "'Peace Rising' Seeks to Allay 'China Threat'", http:\\www.atimes.com.

# Chapter 3

# ASEAN-China Relations: An International Law Perspective

*ZENG Lingliang*

## Introduction to ASEAN-China Relations: A Neglected Topic in the Study of International Law

Since the 1990s, the relationship between ASEAN and China has been developing in both horizontal and vertical ways. As a result, the related studies on politics, military, economy, culture and technology are in full swing; however, the research of ASEAN-China relations in the perspective of law, especially in international law, is out of favor, at least that is the case in mainland China.

Therefore, the international symposium "China's Relations with ASEAN: New Dimensions", co-sponsored by the East Asian Institute of the National University of Singapore and the Research Institute of International Economic Law of Xiamen University, is of vital significance. Many experts in the related disciplines such as politics, economics and law were invited to attend it. The multi- and interdisciplinary studies will greatly benefit the respective research. Researchers can take the advantage of each other's methodology and achievements. Moreover, in this way, the study on ASEAN-China relations will become more comprehensive, systematic and transparent.

Undoubtedly, international law plays an indispensable part in ASEAN-China relations. In a logical way, the functions of law can be summarized as follows. Firstly, it establishes a certain legal relationship between the legal subjects. Secondly, it serves to consolidate the established relations between the legal subjects, and finally it also promotes the development of relations between them. In the same logic, as a category of law, the basic functions of international law can be simply generalized as "establishment, consolidation and promotion" in ASEAN-China relations, especially in the spheres of politics, economy, culture, science and technology. For ASEAN and China, the periodic political consultation as well as the cooperation in the non-traditional security field must be established through some instruments in international law. Both the establishment of ASEAN-China FTA and the development of further cooperation in economy, technology and social affairs must be centered on the international agreements, which have been concluded or are being negotiated and to be concluded. To sum up, international law is the platform and means for the establishment and development of various friendly co-operations in ASEAN-China relations. Therefore, the same importance must be attached to the study of international law in relation to studies of politics, military, economics, culture, science and technology embodied in ASEAN-China relations.

With that in mind, this paper first makes a survey over the present framework and system of ASEAN-China relations from the perspective of international law. Next comes the analysis of the declarative or implied principles of international law through the studies of various legal instruments in ASEAN-China relations. Finally, based on the present instruments and realities, some remaining legal defects or imperfections are figured out and followed with respective suggestions for improvement.

## Overview of the Present Legal Framework of ASEAN-China Relations

### *Exchange of Letters 1994: Founding of Formal Cooperative Relationship*

Generally speaking, the relationship between ASEAN and China started with China's normalization of the diplomatic relationship with Indonesia and the founding of diplomatic relationship with Myanmar and Singapore.[1] The development of ASEAN-China relations is based on the establishment and resumption of diplomatic relationship with the member states of ASEAN. It is widely accepted that the formal diplomatic relationship between ASEAN and China can be traced back to July 1991 when Chinese Foreign Minister Mr. Qian Qichen attended the ASEAN Ministerial Meeting (AMM) in Kuala Lumpur as a distinguished guest of the Malaysia Government, and it was the first time that the Chinese leader was present at the AMM. During the meeting, Mr. Qian expressed China's willingness to become the dialogue partner of ASEAN and showed its interest in cooperation with ASEAN, notably in science and technology.[2] In September 1993, ASEAN

---

[1] See Lee Laito, "China's Relations with ASEAN: Partners in the 21st Century?" *Pacific Review*, Vol. 13(1), February 2001 (electronic version); Ong Keng Yong (Secretary General of ASEAN), Keynote Address at ASEAN-China Forum 2004, "Developing ASEAN-China Relations: Realities and Prospects" (electronic version), Singapore, 23 June 2004, http://www.asean.org/16256.htm (accessed 5 October 2004).

[2] See ISEAS Document Service, "China-ASEAN Free Trade Area: Origins, Development and Strategic Motivations" (electronic version), http://www.bookshop.iseas.edu.sg (accessed 25 September 2003).

Secretary-General Dato' Ajit Singh paid a visit to Beijing at the invitation of the Chinese Vice Foreign Minister Mr. Tang Jiaxuan. The two sides then announced a joint press statement for the meeting to explore the establishment of the consultative relationship with the People's Republic of China.[3] But from a legal prospective, the formal cooperative relationship between ASEAN and China dated from 23 July 1994 when, in an exchange of letters, the ASEAN Secretary-General and the Chinese Foreign Minister agreed on the founding of the Joint Committee on Economic and Trade Cooperation and Joint Committee on Cooperation in Science and Technology in Bangkok.[4] The Exchange of Letters legally formalized the establishment of cooperative relationship between the two sides. At the same time, ASEAN and China began to engage in consultation on political and security issues of common concern. In July 1996, China was accorded full dialogue partner status at the 29th AMM.

### Joint Statement 1997: Programmatic Document on Partnership of Good Neighborliness and Mutual Trust

In December 1997, ASEAN and China held the informal 10 + 1 Summit for the first time and issued the Joint Statement of the Meeting of Heads of States of the Member States of ASEAN and the President of the People's Republic of China, known as the ASEAN-China Cooperation Partnership Towards the 21st Century (hereinafter referred to as the Joint Statement 1997). The Joint Statement 1997 serves as the political declaration and solemn commitment to the international society in legal form made by the leaders of both sides at the highest level, which is of vital and epic-making significance in the history of ASEAN-China relations.

In the Joint Statement 1997, the two sides confirmed that they would undertake to promote a good-neighborly and friendly relationship, increase high-level exchanges, and strengthen the mechanism of dialogue and cooperation in all areas, to enhance understanding and mutual benefits. The two sides agreed to heighten cooperation at bilateral and multilateral levels in promoting economic growth, sustainable development and social progress, based on the principles of equality, mutual benefit and shared responsibility in the interest of achieving national and

---

[3] See Joint Press Statement for the Meeting to Explore the Establishment of the Consultative Relationship with the People's Republic of China, Beijing, China, 13–14 September 1993, http://www.asean.org/5875.htm (accessed 8 October 2004).
[4] Before this event, in September 1993, the Secretary General of ASEAN, who was heading ASEAN delegation, was invited to pay a visit to Beijing. It was during this visit that the two parties agreed on the establishment of the two committees. See ASEAN-China Dialogue, Overview of ASEAN-PRC Relations, http://202.154.12.3/5847.htm (accessed 28 September 2004).

regional prosperity in the 21st century. They undertook to resolve their differences or disputes through peaceful means, without resorting to the threat or use of force to resolve their disputes, especially in settling differences and disputes in the South China Sea through friendly consultation and negotiation in accordance with universally recognized international law, including the 1982 UN Convention on the Law of the Sea. While continuing their efforts to find solutions, they agreed to explore ways for cooperation in the areas concerned. They agreed to continue to exercise self-restraint and handle relevant differences in a cool and constructive manner and not to allow existing differences to hamper the development of friendly relations and cooperation. According to the Joint Statement 1997, China appreciated and supported ASEAN's positive role in international and regional affairs. China reaffirmed that it would respect and support the efforts of ASEAN to establish a zone of peace, freedom and neutrality in Southeast Asia. In this connection, China welcomed the implementation of the Southeast Asia Nuclear Weapon-Free Zone Treaty. The two sides also welcomed the ongoing consultation between the State Parties to the Treaty and the Nuclear States to facilitate the accession by the latter to the protocol of the SEANWFZ Treaty. ASEAN member states believed that a stable, peaceful and prosperous China would constitute an important factor for the long-term peace, stability and development of the Asia-Pacific region in particular and of the world in general. ASEAN member states reaffirmed their continued adherence to the "One China" policy.[5]

*Framework Agreement on Comprehensive Economic Cooperation 2002: Symbol of Entry into Overall Substantive Cooperation in Economics and Trade*

Just as the Secretary General of ASEAN put it, even though ASEAN-China cooperation was formalized in the middle of the 1990s, substantive cooperation only picked up pace in 2001.[6] At the ASEAN-China Summit in November 2001, Zhu Rongji, the former premier of China, made a proposal to establish an ASEAN-China Free Trade Area (FTA) in ten years, and this was later accepted by ASEAN. After several rounds of consultation, the historic decision was declared by the two parties at the Summit Meeting held in 2002 and signed the Framework Agreement on Comprehensive Economic Cooperation between the Association of Southeast Asian

---

[5] See Joint Statement of the Meeting of Heads of State/Government of the Member States of ASEAN and the People's Republic of China, Kuala Lumpur, Malaysia, 16 December 1997, para. 3, 5, and 9.

[6] Ong Keng Yong (Secretary General of ASEAN), Keynote Address at ASEAN-China Forum 2004, "Developing ASEAN-China Relations: Realities and Prospects" (electronic version), Singapore, 23 June 2004, http://www.asean.org/16256.htm (accessed on 7 October 2004).

Nations and the People's Republic of China (hereinafter referred to as Framework Agreement).[7]

The Framework Agreement 2002 is an epic-making event in the history of ASEAN-China relations. It is the first formal treaty that regulates the mutual relations of the two parties. It is the most direct treaty instrument that normalizes the future comprehensive and substantive economic cooperation with CAFTA as its core. This agreement covers the trade in goods and services as well as investment and other areas of economic cooperation. It provides the guidelines, basic principles, coverage and models of the CAFTA, including the Early Harvest Program and the special and differential treatment for newer ASEAN member states, namely, Cambodia, Laos, Myanmar and Vietnam.

From 2004, the negotiations of CAFTA were formally initiated. Both sides have targeted to realize the Free Trade Area in 2010 for the six original founding member states of ASEAN (Brunei, Indonesia, Malaysia, Philippines, Singapore and Thailand) and 2015 for those less-developed member states, i.e., Cambodia, Laos, Myanmar and Vietnam. Meanwhile, in the second part of the Framework Agreement 2002, the parties agree to strengthen their cooperation in five priority areas or sections as follows: agriculture, information and communications technology, human resources development, investment, and Mekong River basin development. In addition, mutual economic cooperation shall be extended to other areas, including banking, finance, tourism, industrial cooperation, transport, telecommunications, intellectual property rights, small and medium enterprises (SMEs), environment, bio-technology, fishery, forestry and forestry products, mining, energy and sub-regional development, etc.

## *Joint Declaration on Cooperation in the Field of Non-Traditional Security Issues 2002: Legal Basis for Cooperation on New Security Issues*

In view of the increasingly serious nature of non-traditional security issues which have become important factors of uncertainty affecting regional and international security and are posing new challenges to regional and international peace and stability, and conscious of the complexity and deep-rooted background of the non-traditional security issues, the leaders of China and ASEAN recognized the need to meet the changes of the issues with an integrated approach that combines political, economic, diplomatic, legal, scientific, technological and other means.

---

[7] On 6 October 2003, China and ASEAN signed the Protocol on the Amendment to the Framework Agreement on Comprehensive Economic Cooperation, which made some changes to the articles of the 2002 Framework Agreement.

They believed that ASEAN and China are close neighbors and share extensive common interests in coping with these issues.

To this end, the two parties released the Joint Declaration of ASEAN and China on Cooperation in the Field of Non-traditional Security Issues on 4 November 2002 in Cambodia (hereinafter referred to as Joint Declaration on Non-traditional Security Cooperation 2002). This serves as the direct legal platform for China and ASEAN to launch bilateral cooperation in coping with many increasingly serious non-traditional security issues. It is one of the specific moves to fulfill the political commitments of the 1997 Joint Statement to "strengthen the mechanism of dialogue and cooperation in all areas". The Joint Declaration stipulated the objectives of such a kind of cooperation, that is, to enhance their capacity in dealing with non-traditional security issues, promote their stability and development, and maintain regional peace and security. The priorities provided in the Declaration are combating trafficking in illegal drugs, women and children smuggling, sea piracy, terrorism, arms smuggling, money laundering, international economic crime and cyber crime. The declaration also made a list of the following five forms of cooperation: to strengthen information exchange; to strengthen personal exchange and training and enhance capacity building; to strengthen practical cooperation on non-traditional security issues; to strengthen joint research on non-traditional security issues; and to explore other areas and modalities of cooperation. On these purposes, long-term and mid-term cooperation plans will be formulated as well as the plans to carry them through according to the Declaration.

Pursuant to the Joint Declaration on Non-traditional Security Cooperation 2002, the two parties, after several rounds of consultation, signed the Memorandum of Understanding between the Governments of the Member States of ASEAN and the Government of the People's Republic of China on Cooperation in the Field of Non-traditional Security Issues (hereinafter referred to as the Memorandum on Non-traditional Security Cooperation) on 10 January 2004. The Memorandum is a specific instrument that further detailed and implemented the Joint Declaration on the Non-traditional Security Cooperation 2002. The objective of the Memorandum is to develop practical strategies in accordance with their national laws and regulations to enhance the capacity of each individual country and the region as a whole in dealing with non-traditional security issues. To this end, the Memorandum respectively formulated a specific plan and its implementation of the mid- and long-term cooperation for the five forms of cooperation listed in the Joint Declaration on the Non-traditional Security Cooperation. Moreover, the Memorandum made detailed regulations on such specific items as its implementation agencies, financial arrangements, confidentiality, settlement of disputes as well as its suspension, revision, amendment, entry into force, duration and termination.

## Declaration on the Conduct of Parties in the South China Sea 2002: Solemn Commitment of Mutual Trust and Self-restraint

Due to the complexity of history and reality, there always exist territorial disputes on the islands in the South China Sea between China and some member states of ASEAN; conflicts occasionally arise as a result. Therefore, any claim over the sovereignty and jurisdiction of the islands will prejudice the bilateral relations between the related parties as well as the peace and security of Asia-Pacific area, even the global peace and security. For many years, China and member states of ASEAN concerned have been seeking for a channel that will enhance favorable conditions for a peaceful and durable solution to the differences and disputes among countries concerned. On 4 November 2002, China and ASEAN member states signed the Declaration on the Conduct of Parties in the South China Sea at the Eighth ASEAN Summit held in Cambodia. The adoption of the Declaration indicates that China and ASEAN are seeking to establish a dispute settlement mechanism on the South China Sea issues based on mutual trust, self-restraint and cooperation in a pragmatic manner.

In order to keep away from the intensification of the territorial disputes on the sea and stop or at least reduce the potential military conflicts, the parties of the Declaration reaffirmed their commitment to the purposes and principles of the Charter of the United Nations, the 1982 UN Convention on the Law of Sea, the Treaty of Amity and Cooperation in Southeast Asia, the Five Principles of Peaceful Coexistence, and other universally-recognized principles of international law; their respect for and commitment to the freedom of navigation in and over-flight above the South China Sea; their commitment to resolve their territorial disputes by peaceful means, instead of resorting to use of force; their commitment to exercise self-restraint in the conduct of activities on the island in dispute and build mutual trust and confidence; their commitment to hold exchange of views between their defense and military officials, notify, on a voluntary basis, other parties concerned of any military exercise and ensure humane treatment of all persons who are either in danger or in distress. The parties unanimously agree to undertake cooperative activities, including marine environmental protection and scientific research, safety of navigation, search and rescue operation and combating transnational crime.

## China's Entry into Treaty of Amity and Cooperation in Southeast Asia: Important Legal Behavior by a Responsible Great Power

At the beginning of October 2003, Chinese Premier Wen Jiabao took part in the signing ceremony of the Treaty of Amity and Cooperation in Southeast Asia at the

ASEAN-China Summit.[8] China became the first big country to enter into the Treaty outside Southeast Asia. The purpose of this Treaty is to "promote perpetual peace, everlasting amity and cooperation among their peoples that would contribute to their strength, solidarity and closer relationships."[9] The Treaty defines the guideline for the relations between the parties by the following fundamental principles:

1. Mutual respect for the independence, sovereignty, equality, territorial integrity and national identity of all nations;
2. The right of every state to lead its national existence, freedom from external interference, subversion or coercion;
3. Non-interference in the internal affairs of one another;
4. Settlement of differences of disputes by peaceful means;
5. Renunciation of the threat or use of force; and
6. Effective cooperation among themselves.[10]

The entry of China into the Treaty of Amity and Cooperation in Southeast Asia will exert great and positive influence for the peace and security in Southeast Asia, Asia-Pacific area and even the world.[11] The great step indicates that China is a responsible country and it will help to lessen and dispel the worries of ASEAN nations in the rise of China. It will make the "China threat theory" groundless, because China's entry into the Treaty is going on with the independent and peaceful foreign policy and it also proves that China is ready to "become a partner of all the countries in the world instead of Alliance". China's entry into the Treaty will be conducive to sustainable development of ASEAN-China relations because it serves to make a solid platform of international law for the comprehensive development of bilateral relations, showing China's solemn commitment to become ASEAN's good neighbor and partner, thus furthering the stability and continuity of ASEAN-China relations. China's entry into the Treaty will contribute to its peaceful settlement of

---

[8] The treaty was signed by the five original members of ASEAN (Singapore, Thailand, Philippine, Indonesia and Malaysia) in Indonesia on 24 February 1976. Later, other countries in the region entered the treaty. In the beginning, the treaty was only open to the countries in Southeast Asia, but after the Revisions in 1987 and 1998, countries outside the area are admissible, the first example of which is Papua New Guinea.

[9] See Article 1, Treaty of Amity and Cooperation in Southeast Asia.

[10] See Article 2, Treaty of Amity and Cooperation in Southeast Asia.

[11] See Cao Yunhua, "A Close Neighbor Means More Than a Distant Relative — The Significance of China's Accession to the Treaty of Amity and Cooperation in Southeast Asia", in *Chinese Diplomacy*, 1st edition (Information and Data Center of Renmin University of China, 2004) pp. 49–50; Zhang Xizhen, "A Great Move of Chinese Diplomacy — the establishment of ASEAN-China Strategic Partnership and Formal Accession to the Treaty of Amity and Cooperation in Southeast Asia", *People's Daily*, 9 October 2003, p. 13.

possible disputes with ASEAN, for the treaty provides the contracting parties with fundamental principles and modality for peaceful dispute settlement. In addition, China's entry into the Treaty will favor the founding of new global economic and political order and promote regional integration. Its position as a contracting party proves China's willingness to participate in the pursuit of international democracy and rule of law by its action. Meanwhile, it also shows that as a regional leading power in rising, China is ready to cooperate with small and weak nations equally and to construct a peaceful, stable, prosperous and bright future for the region.

## Joint Declarations of Strategic Partnership 2003: New Blueprint for ASEAN-China Relations

On 8 October 2003, the ASEAN-China Summit was held in Bali, Indonesia. One of its major achievements is the signing of the Joint Declaration of the Heads of State/Government of ASEAN and the People's Republic of China on Strategic Partnership for Peace and Prosperity. The signing and release of the document symbolized a new step forward for ASEAN-China relations, that is, upgrading the partnership of consultation via dialogue into a strategic one, thus enhancing the bilateral good neighborliness to a higher level.

The Joint Declaration on Strategic Partnership 2003 recalled the rapid, comprehensive and in-depth growth of the relationship between ASEAN and China in politics, economics, security and regional and international affairs since the issuance of the Joint Statement 1997. Both sides realized that ASEAN and China have become important partners of cooperation. The Declaration highlights the strategic importance of ASEAN-China relations to peace, development and cooperation in this region, and recognizes the positive contribution of such relations to world peace and development. In view of the fact that today's world is undergoing complex and profound charges, the enhanced cooperation between ASEAN and China, as two important partners in the Asia-Pacific region, will serve the immediate and long-term interests of both sides and is conducive to peace and prosperity in the region. To this end, the Declaration states solemnly: ASEAN and China establish "a strategic partnership for peace and prosperity".

The Declaration confirms that the purpose of the establishment of a strategic partnership is to foster friendly relations, mutually beneficial cooperation and good neighborliness between ASEAN and China by deepening and expanding ASEAN-China cooperative relations in a comprehensive manner in the 21st century, thereby contributing further to the region's long-term peace, development and cooperation. It also states that this strategic partnership is non-aligned, non-military and non-exclusive, and does not prevent the participants from developing their all-directional ties of friendship and cooperation with others. According to the Declaration, ASEAN-China strategic partnership is a comprehensive and forward-looking cooperation focusing on politics, economy, social affairs, security and international and regional affairs. To this end, the Declaration generally formulates the main contents and their implementation of the strategic partnership in the above-mentioned areas.

## General International Law Permeating China-ASEAN Relations

### *Reassurance of the Underlying Position of Fundamental Principles of International Law*

In line with the foregoing history, ASEAN-China relations have been developing hand-in-hand with general international law from the very beginning till today. Each important phase, and each great act and achievement of ASEAN-China relations is based on general international law in the form of international agreements. All the important legal documents signed and issued by China and ASEAN highlight the fundamental principles of international law as guidelines in their bilateral relations. As a result, all the declarations, statements and treaties between the two parties reaffirm their observance to the universally recognized norms governing international relations, by which they seek to deepen and widen their cooperation in politics, security, social affairs, international and regional affairs, and to settle the differences and disputes between them.

In the analysis of legal instruments governing ASEAN-China relations, it is easy to find a common attribute, that is, while establishing the dominant role of the fundamental principle of international law, much attention has been paid to the related titles of treaties, charters of international organizations and universally recognized norms of international law in which the principles are embodied. The 1997 Joint Statement is the very document to affirm "the UN Charter, Treaty of Amity and Cooperation in Southeast Asia, Five Principles of Peaceful Co-existence and universally accepted as the basic norms governing various international relations."[12] In the 2002 Declaration on the Cooperation in non-traditional security, the title of the foregoing treaties or principles are reiterated once more, with just a minor difference in its diction. It states that the cooperation in the field of non-traditional security issues "should be conducted on the basis of observing the Five Principles of Peaceful Co-existence and other universally recognized norms of international law, which are embodied in the Charter of the United Nations and the Treaty of

---

[12] See Joint Statement of the Meeting of Heads of State/Government of the member states of ASEAN and the People's Republic of China, Kuala Lumpur, Malaysia, 16 December 1997, para. 2.

Amity and Cooperation in Southeast Asia...".[13] There are similar statements in the Declaration on the conduct of parties in the South China Sea as "the parties reaffirm their commitment to the purposes and principles of the Charter of the United Nations, the 1982 UN Convention on the Law of the Sea, the Treaty of Amity and Cooperation in Southeast Asia, the Five Principles of Peaceful Co-existence, and other universally recognized principles of international law which shall serve as the basic norms governing state-to-state relations".[14] The 2004 Joint Declaration on Strategic Partnership similarly reiterates that "ASEAN-China cooperation will continue to take the UN Charter, the Treaty of Amity and Cooperation in Southeast Asia, the Five Principles of Peaceful Co-Existence, and other universally recognized norms governing international relations as its guidance".[15]

Through above-mentioned elaboration, some enlightening conclusions could be drawn as follows.

Firstly, with no exception, the foregoing instruments in ASEAN-China relations emphasize the UN Charter, identified as the first place in most of them. It shows that the UN Charter is the law-making treaty that shares the highest effectiveness and authority in all international legal instruments. Its seven principles constitute the basic norms which are legally binding the UN, its member states and non-member states as well;[16] all the purposes of the UN Charter are the goals to be achieved by the UN and its member states as well as by all the countries, international organizations and cooperative regimes. Meanwhile, it also indicates that, in politics, security, economy, social affairs, international and regional affairs, the cooperation between ASEAN and China shall be consistent with the purposes of the UN Charter and take its principles as the basic norms guiding their conducts.

Secondly, each of the instruments mentions the Five Principles of Peaceful Co-existence and some even list it in the first place. The fact fully shows that the Five Principles of Peaceful Co-existence, initiated by Myanmar and India in 1954, have been universally accepted as major (even the core) components of basic norms governing modern international relations, for over half a century. What attracts

---

[13] See Joint Declaration of ASEAN and China on Cooperation in the Field of Non-Traditional Security Issues, 4 November 2002, preamble.
[14] See Declaration on the Conduct of Parties in the South China Sea, para. 1.
[15] See Joint Declaration of the Heads of State/Government of ASEAN and the People's Republic of China on Strategic Partnership for Peace and Prosperity, para. 5.
[16] Article 2(6) of the UN Charter provides: "The Organization shall ensure that states that are not member of the United Nations act in accordance with these principles so far as may be necessary for the maintenance of international peace and security". There have been plenty of studies on relationship between the UN and non-member states. See bibliography listed in Benedetto Conforti, *The Law and Practice of the United Nations* (Kluwer Law International, 1996), p. 126.

much attention is that all the declarations and statements issued or signed jointly by ASEAN and China repeatedly set out "the Five Principles of Peaceful Co-Existence", which is unusual for the legal instruments between China and other countries or international organizations. As a result, the dominant position of "Five Principles of Peaceful Co-existence" in ASEAN-China relations is undoubtedly recognized by all the ASEAN member states.

Thirdly, the Treaty of Amity and Cooperation in Southeast Asia is also mentioned in all the foregoing instruments and taken as the basic instrument governing ASEAN-China relations, even in the declarations and statements issued or signed by China before its formal accession to the Treaty. The connotation of the Treaty is quite in-depth. It means that although this Treaty is not the instrument specially governing ASEAN-China cooperation, even open to the countries outside Southeast Asia, the two parties regard the basic purposes of ASEAN-China relations as the essential part of the Treaty. What's more, the principles provided in the Treaty are themselves universally taken as the major components of fundamental principles of international law. Even if this is the case, from the very beginning, China had accepted the guiding role of the Treaty in ASEAN-China relations, especially as it made such a commitment before its entry into the Treaty. This constitutes a special political stance and legal instrument in the practice of international relations, from which we could get to know China's devotion to promote ASEAN-China's durable good neighborliness and everlasting peace, stability and development of Southeast Asia.

Fourthly, other universally recognized principles of international law also develop through these instruments. Therefore, as for the fundamental principles governing ASEAN-China relations, the two parties adhere to an open policy and stand that keep up with the times. In the first place, they accept the fundamental principles of international law in the listed international instruments as well as those contained in the international customary law and unlisted instruments, even though China, ASEAN and its member states might not be the contracting parties to the latter. For instance, the ten principles of peace co-existence and friendly cooperation in the 1955 Final Communiqué of the Bandung Conference and the European Declaration on the Principles Governing the Participating Countries passed by the Conference on Security and Cooperation in Europe (CSCE) in 1975 are undoubtedly of this category. In the second place, the two parties treat the fundamental principles of international law governing ASEAN-China relations in a developing attitude, that is, both the existing fundamental principles of international law as well as those that might be formulated in the future are the basic norms guiding the mutual relations. No doubt, these "other principles" shall be "universally recognized" all over the world, either in the past or in the future.

Fifthly, it is only the Declaration on the conduct of parties in the South China Sea that treats the 1982 UN Convention on the Law of the Sea as the general international instruments which governs the issues in the South China Sea. This is certainly a special case which might be taken for granted. The 1982 UN Convention on the Law of the Sea, as the most important law-making treaty in the area of modern law of the sea, is based on the codification of customary law and treaties in contemporary and modern times. Apart from the universally recognized fundamental principles of international law, the 1982 UN Convention on the Law of the Sea shall undoubtedly be complied with, for it is the most authoritative law-making treaty in this specific area.

Then, what are the fundamental principles of international law that have been repeatedly mentioned in the legal instruments governing ASEAN-China relations? To sum up, they are listed as follows: the principle of mutual respect for independence, sovereignty, equality, integrity of the territory and nation character; the principle of no interference with domestic politics or internal affairs; denunciation/giving up of the use of force or the threat of force; the principle of peaceful settlement of differences or disputes; the principle of effective cooperation; and so on.[17]

## Stressing the Compatibility of Mutual Economic and Trade Cooperation with Fundamentals of WTO

After reviewing China-ASEAN relations, it is easy to notice that the development of economic and trade cooperation has been taking the lead in terms of speed, scale and legal framework. From international legal point of view, it is sufficient to prove the compatibility of this leading field with the WTO from the Framework Agreement of Comprehensive Economic Cooperation in 2002 and its subsequent protocol in 2003. There are two reasons for this view. On one hand, the Framework Agreement and its protocol with the establishment of the CAFTA as the core are the very basic legal sources between China and ASEAN. On the other hand, WTO is the representative universal trade organization with both China and ASEAN countries as its members and contracting parties to its various multilateral trade agreements.

---

[17] See Treaty of Amity and Cooperation in Southeast Asia, Article 2; Joint Statement of the Meeting of Heads of State/Government of the Member States of ASEAN and the President of the PRC, 16 December 1997, para. 2 and 8; Joint Declaration of ASEAN and China on Cooperation in the Field of Non-traditional Security Issues, preamble; Declaration on the Conduct of Parties in the South China Sea; Joint Declaration of the Heads of State/Government of ASEAN and the People's Republic of China on Strategic Partnership for Peace and Prosperity, para. 4 and 5.

*Consistency in respect of general purposes*

We might begin with the analysis of consistency in respect of general purposes of the CAFTA and those of the WTO.

The purposes of WTO are mainly included in the preamble of the Marrakesh Agreement Establishing the World Trade Organization (hereafter referred to as "WTO Agreement") plus preambles of GATT 1947, Havana charter and General Agreement on Trade in Services (hereinafter referred to as "GATS"). WTO's purposes could be summed up in the following points: (1) raising standards of living; (2) ensuring full employment and a large and steadily growing volume in real income and effective demand; (3) expanding the production of and trade in goods and services; (4) allowing for the optimal use of the world's resources and ensuring sustainable development; and (5) ensuring that developing countries, especially the least developed among them, secure a share in the growth in international trade commensurate with the needs of their economic development.

The purposes of China-ASEAN economic cooperation with CAFTA as the basis could be mainly identified from the Framework Agreement 2002, in addition to related documents mentioned above. The preamble of this Agreement clearly illustrates that both China and ASEAN countries desired to establish CAFTA and to make the arrangement on comprehensive economic cooperation, with the objective of forging closer economic relations in the 21st century. This is to be achieved by: minimizing barriers and deepening economic linkages between the parties; lowering costs; increasing intra-regional trade and investment; increasing economic efficiency; creating a larger market with greater opportunities and larger economies of scale for the businesses of the parties; and enhancing the attractiveness of the parties to capital and talent. Article 1 of this Agreement further clarifies the objectives as follows: (1) strengthening and enhancing economic, trade and investment cooperation; (2) progressively liberalizing and promoting trade in goods and services as well as creating a transparent, liberal and facilitative investment regime; (3) exploring new areas and developing appropriate measures for closer economic cooperation; and (4) facilitating the more effective economic integration of the newer ASEAN member states and bridging the development gap among the parties.

Compared with the objectives of both regimes, it could be easily concluded that the purposes of the China-ASEAN economic cooperation arrangement with the basis of CAFTA are fully compatible with those of the WTO in substance; the only difference is in the wording. To sum up, the objectives of WTO are global while those of CAFTA are bilateral or regional. However, the realization of both regimes' objectives is supportive of each other.

*Consistency in respect of general rules*

In addition to affirming the dominant position of those general principles of public international law in all fields of China-ASEAN relations described early in this paper, the Framework Agreement 2002 explicitly lays down fundamental rules governing mutual economic cooperation, especially for the establishment of CAFTA. All these rules and disciplines are compatible with those of the WTO either in their implication or in their equivalent expression. Some articles of the Framework Agreement even directly refer to their consistency with WTO rules and disciplines.

Principle of Most-Favored Nation Treatment (MFN) is the best illustration. There is no problem for both China and those WTO ASEAN member states to apply MFN principle in the CAFTA, since this principle is compulsory and automatic in the WTO legal regime. However, whether MFN principle is to be applied between China and those non-WTO ASEAN member states is negotiable and largely lies in the Chinese government. The latter's attitude toward this matter is rather positive. Article 9 of the Framework Agreement 2002 provides that "China shall accord Most-Favored Nation (MFN) Treatment consistent with WTO rules and disciplines to all the non-WTO ASEAN member states upon the date of signature of this Agreement".

Principle of special and differential treatment is another example in this aspect. The parties of the agreement, "recognizing the different stages of economic development among ASEAN member states and the need for flexibility, in particular the need to facilitate the increasing participation of the newer ASEAN member states in the ASEAN-China economic cooperation and the expansion of their exports, including, *inter alia*, through the strengthening of their domestic capacity, efficiency and competitiveness",[18] promised to provide "special and differential treatment and flexibility to the new ASEAN member states",[19] with view to make "closer economic cooperation" and "bridge the development gap".[20]

*Consistency in respect of measures*

According to the provisions of Article 2 of the Framework Agreement, the basic measures for establishing CAFTA and comprehensive economic cooperation between the parties could be summarized as follows: (1) progressive elimination of tariffs and non-tariff barriers in trade in goods; (2) progressive liberalization of trade in services; and (3) establishment of an open and competitive investment

---

[18] Framework Agreement 2002, preamble.
[19] Framework Agreement 2002, Article 2(d).
[20] Framework Agreement 2002, Article 1(c) and (d).

regime. In doing so, certain special and differential treatment and flexibility will be granted to the less developed ASEAN member states as mentioned above.[21] As a kind of guarantee or enhancement measures, the parties will establish effective trade and investment mechanism, such as simplification of customs procedures, development of mutual recognition arrangement and effective implementation system.

It is no doubt that all these measures are not consistent with the rationale of liberation of trade in goods and services and investment. They are even equivalent to the provisions of respective multilateral trade agreements of the WTO, such as the GATT, GATS, TRIM, etc.

Besides, some measures adopted in the CAFTA are directly referred to the WTO requirements. In the field of trade in goods, Article 2(6) of the Framework Agreement expressly provides that the commitments undertaken by the parties concerning the elimination of tariffs and non-tariff barriers including the Early Harvest which "shall fulfill the WTO requirements to eliminate tariffs on substantially all the trade between the parties". The same article also mentions that in the CAFTA safeguards, disciplines on subsidies and countervailing measures, anti-dumping measures, protection of trade-related aspects of intellectual property rights shall be based on the GATT principles, existing GATT disciplines and TRIPS. In the field of trade in services, Article 4 of the Agreement requires that in the CAFTA elimination of existing discrimination, prohibition of new or more discriminatory measures and expansion of liberalization with respect to trade in services shall be in conformity with the GATS.

Finally, in the CAFTA, any modification of commitments, safeguard actions, emergency measures and other trade remedies such as anti-dumping and subsidies and countervailing measures in the Early Harvest Program shall be conducted under the respective WTO provisions, although in the interim.[22]

*Consistency in respect of exceptions*

There are two different dimensions of exceptions. Firstly, it concerns the provisions of exceptions in the Framework Agreement related to the CAFTA, that is, the issue of their compatibility with the WTO provisions. Secondly, it concerns the compatibility of the CAFTA itself with the WTO provisions, that is, the legality or justification of the CAFTA under the WTO regime.

---

[21] Flexibility is also given to sensitive areas in the goods, services and investment sectors for all parties based on the principle of reciprocity and mutual benefits. See Framework Agreement 2002, Article 2(e).
[22] See Framework Agreement 2002, Article 6.3(d).

So far as the first dimension is concerned, it is self-evident that Article 10 of the Framework Agreement is by and large a kind of paraphrasing the provisions of Article 20 and Article 21 of the GATT, and Article 14 and Article 14bis of the GATS. There are four "general provisions" listed in Article of the Agreement, namely, (1) protection of national security; (2) protection of articles of artistic, historic and archaeological value; (3) protection of public morals; and (4) protection of human, animal or plant life and health. Obviously, the first general exception could be comparable to the "security exceptions" embodied in Article 21 of the GATT and Article 14bis of the GATT, though in a simpler way, while the residual three general exceptions are exactly the same as provisions in paragraphs (f), (a) and (b) of the GATT and paragraphs (a) and (b) of the GATS. It is worth mentioning that Article 10 of the Agreement provides an important prerequisite, which is identical to the preambles in Article 20 of the GATT and in Article 14 of the GATS, that is, such exceptional measures adopted by any party should not "constitute a means of arbitrary or unjustifiable discrimination" or "a disguised restriction on trade".[23]

Let's turn to the dimension of legality or justification of the CAFTA under the WTO. In substance, the establishment of free trade area is contrary to the MFN principle, the very basis of the multilateral trade system. However, the multilateral trade system has put it as an exception to MFN Principle from the very beginning, while at same time stipulating certain substantial and procedural conditions. The legal basis for such an exception and its conditions is laid down in Article 24 of the GATT "enabling clause" 1979 and Article 5 of the GATS. Therefore, the establishment of the CAFTA is a legitimate exception under the WTO legal system, if it meets with the requirements therein.

The requirements for the justification of the CAFTA or any other free trade area or customs union could be summed up from those provisions mentioned above and jurisprudence of the WTO Disputes Settlement Body (DSB) as follows: (1) substantial coverage of trade; (2) elimination of existing tariffs and non-tariff barriers on substantially all trade in goods and of substantially all existing discrimination on trade in services between the parties; (3) prohibition of new tariffs and non-tariff barriers and of new or more discriminatory measures between the parties; (4) prohibition of raising the overall level of barriers to trade in goods and services for non-party countries; and (5) prompt notification to the WTO.

So far as the first two requirements are concerned, the Framework Agreement has already made appropriate provisions, such as its Article 3(1) and Article 4. The former explicitly provides that the "duties and other restrictive regulations of

---

[23] A comprehensive and detailed analysis on this subject, see Chen Weidong, *Interpretation and Understanding of Exceptions of the WTO* (University of Foreign Economics and Trade Press, 2002).

commerce...should be eliminated on substantially all trade in goods between the parties". The latter similarly affirms that negotiations shall be directed to "progressive elimination of substantially all discrimination between or among the parties and/or prohibition of new or more discriminatory measures with respect to trade in services between the parties". As for the fourth condition, though there is no direct mention of it in the Agreement, there seems to be no reason to doubt consistency in this aspect. It is also obvious that the parties of the CAFTA will notify the WTO as soon as the negotiations are completed.

## ASEAN-China Relations: Legal Imperfection and Suggestions for Improvement

It is a common accord that the speedy development of China-ASEAN cooperation are both politically and economically motivated. It is also true that the establishment of the strategic partnership between China and ASEAN and its member states is not only a win-win deal for both sides, but also imperative to the stability, security and prosperity in the Asian-Pacific region and even the whole world. However, we must always keep in mind that such a far-reaching significant deal could not be carried on and completed without continuous efforts made in the legal dimension. Therefore, it is necessary to constantly evaluate and perfect the legal architecture of China-ASEAN relations.

### *Legal Imperfection*

We might proudly say that the present status of ASEAN-China relations is the best in history. But it seems to me that the legal construction of this double-win relations has left behind the political and economic dimensions. Evaluating the existing legal regulation of ASEAN-China relations, the imperfection remains chiefly in the following aspects.

1. Legal regulation is obviously uneven in all fields of the strategic partnership. Generally speaking, legal regulation is relatively faster and more intensive in the CAFTA and other sectors of economic cooperation than in the fields of political, security and social cooperation.
2. So far there are few hard-law sources regulating the China-ASEAN relations; most of the legal documents are soft-law, both in form and substance. It is true that those joint declarations, statements and memorandums are not only politically significant, but are also a kind of political commitment producing legal effect. Nevertheless, these forms of documents could not match up with treaties or agreements in terms of formation, content and legal enforcement.

Except for the Treaty of Amity and Cooperation in Southeast Asia of which China and ASEAN member states are parties, the only real treaties regulating the partnership between both sides so far are the Framework Agreement on Comprehensive Economic Cooperation 2002 and its Protocol 2003.
3. The legal level of China-ASEAN relations is largely at the preliminary stage, short of systematic implemental and operationable norms and mechanism. After checking all the existing documents between China and the ASEAN, we are deeply impressed that each document mostly set up the objectives, principles, guidelines, general disciplines and overal plans, etc., including the respective implementation measures to be adopted subsequently.
4. There is no systematic dispute settlement mechanism. So far, only general principles have been reached by the parties on this aspect, promising to resolve their differences and disputes through consultation and negotiation. Some documents even exclude other peaceful methods. For instance, Article 8 of the Memorandum of Understanding Between the Governments of the Member Countries of ASEAN and the Government of the People's Republic of China on Cooperation in the Field of Non-traditional Security Issues provides that "any dispute or differences arising out of the interpretation/implementation/application of the provisions of this Memorandum of Understanding shall be settled amicably through consultation/negotiation between the parties without reference to any third party". Although the Article 11 of the Framework Agreement on Comprehensive Economic Cooperation provides that "the parties shall, within one year after the date of entry into force of this Agreement, establish appropriate formal dispute settlement procedures and mechanism for purposes of this Agreement", such formal procedures and mechanism have not come out yet.

To sum up, at present, the overal legal regulation of China-ASEAN relations has not reached the stage of a legal regime. The subsequent endeavors have to be undertaken to achieve its perfection.

## *Suggestions for Improvement*

In order to establish a solid and systematic and operational legal regime for the strategic partnership between China and ASEAN and its member states based on the present legal framework, this author would like to make some suggestions for improvement as follows.

First of all, the departments and officials of both sides should make appropriate plans to work out implementation measures or mechanism, particularly for those legal documents that have already included such requirements. The implementation measures should be as detailed and operational as possible. Certainly, there would be tremendous amount of work and difficulty in doing this. Just take the CAFTA for example. A large number of sub-agreements have to be negotiated and concluded in the fields of trade in goods and services and investment, such as reduction and elimination schedule of tariffs and non-tariff barriers, rules of origin, customs evaluation, safeguards, emergency measures, anti-dumping, subsidies and countervailing measures, standards and conformity assessment, technical barriers, investment facilitation, etc.

Secondly, since the legal foundation is relatively more solid in economic cooperation especially for the CAFTA, it is desirable that the systematic operational implementation measure might first be concluded in this less sensitive area, then progressively expand to other fields of cooperation, such as non-traditional security, social affairs and conducts in the South China Sea. This "from-easy-to-difficult" option seems to be a sound practical choice to further the legal regulation of China-ASEAN relations.

Thirdly, it is desirable to insert the principle of national treatment in perfecting the legal regime of the CAFTA. Since national treatment, together with MFN, has long been a basic norm both in the multilateral trade system and in various forms of regional integration, there seems to be no other overwhelming reasons to exclude such a universally recognized principle in international economic transactions from the CAFTA. If it is not realistic to apply the principle of national treatment on trade in goods and services and investment at same time, it could be at least applied gradually from one field to another. It is better than to do nothing.

Fourthly, as time goes by, soft-law in certain areas of cooperation could be better upgraded to some form of hard-law so as to enhance the legal basis in respective areas. Take the Declaration on the Conduct of the Parties in the South China Sea as an example.

As mentioned early in this paper, the Declaration is definitely a seldom commitment of the parties to ensure a peaceful management of any territorial dispute and to enact an informal code of conduct based on self-restraint, the non-use of force and the freedom of navigation. However, because this seldom commitment is made in the form of declaration, some commentators regard it as an informal and non-binding code of conduct.[24] Therefore, it is strongly desirable to replace this declaration by a formal treaty or agreement.

Finally, for the long run, it is necessary to set up a dynamic dispute settlement mechanism with various political or diplomatic means and legal procedures. The

---

[24] See Ralf Emmers, "ASEAN, China and the South China Sea: An Opportunity Missed, Perspectives", Institute of Defence and Strategic Studies, Nanyang Technological University, Singapore, http://www.ntu.edu.sg/idss/Prospectives/Research-050228.htm (accessed 5 October 2004).

government of the People's Republic of China has long preferred to settle international disputes through amicable consultation and negotiation. For decades, the bilateral treaties or agreements concluded between China and foreign countries have always included this kind of provisions concerning dispute settlement. This single pattern of dispute settlement must not be too inflexible to follow the trends of international dispute settlement mechanism and to adapt to the horizontal and vertical developments of China-ASEAN relations. What if disputes could not be settled through amicable consultation and negotiation?

It is no doubt that amicable consultation and negotiation should be the first and foremost method in the forthcoming dispute settlement mechanism, but it should not be exclusive. There must be some other political or diplomatic methods to choose from, such as mediation, conciliation, fact-finding committee, etc. In addition, arbitration and some sort of quasi-judicial procedures could also be made available to the parties.

## Conclusion: Most Recent Legal Developments in ASEAN-China Relations

It might not be entirely true to state that the legal construction of the strategic partnership between China and ASEAN has been developing without any problems, but it has been steadily progressing. It is worth noting that the Tenth ASEAN Summit had been held in Vientiane on 29 and 30 November 2004. At this most recent Summit, a number of new legal documents were signed or adopted by ASEAN and China leaders.[25] All these legal documents are implementation measures for agreements, declarations, joint statements and memorandums signed and adopted in previous years. They are strong signs that China-ASEAN relations will be developing further under the direction of the rule of law.

From a legal perspective, the following points of tendency are particularly of global interest.

In the field of political and security cooperation, both sides have made further commitment to enhance the role of the Treaty of Amity and Cooperation in Southeast Asia as a code of conduct for inter-state relations in the region, and to

---

[25]Namely, (1) ASEAN-China Plan of Action to implement the Joint Declaration of the Heads of State/Government of ASEAN and China on Strategic Partnership for Peace and Prosperity; (2) Chairman's Statement at the Eighth ASEAN + China Summit; (3) Memorandum of Understanding between the Governments of the Member States of ASEAN and the Government of the People's Republic of China on Transport Cooperation; (4) Agreement on Trade in Goods of the Framework Agreement on Comprehensive Economic Cooperation between ASEAN and China; and (5) Agreement on Dispute Settlement Mechanism of the Framework Agreement on Comprehensive Economic Cooperation between ASEAN and China, etc., http://www.aseansec.org/16479.htm (accessed 10 February 2004).

cooperate in urging other ASEAN Dialogue Partners to accede to the Treaty so as to promote regional peace, security, prosperity, mutual confidence and trust. It was exciting to see that the Republic of Korea and the Russian Federation making accession to the Treaty after China and India.[26] In addition, China reaffirmed its readiness to sign the Protocol to the Southeast Asia Nuclear Weapon Free Zone Treaty at an early opportunity and to cooperate in encouraging all the nuclear weapon states to sign the Protocol.[27] On the issue of South China Sea, both sides promised to implement in an effective way the Declaration on the Conduct of the Parties in (DoC) the South China Sea, adhere to the terminologies used in the UN Convention on the Law of the Sea and other instruments of the International Maritime Organization, and to affirm the vision of the DoC state parties to work on eventual adoption of a code of conduct in the South China Sea.[28] In the area of non-traditional security, both sides agreed to jointly develop the 2005 Work Plan and subsequent annual work plans to implement the ASEAN-China Memorandum of Understanding on Cooperation in Non-traditional Security.[29]

In the field of economic cooperation, substantial breakthroughs have been made legally for the establishment of the ASEAN-China Free Trade Area in the following two aspects: the first is the signing of the Agreement on Trade in Goods which sets up a dual tracks of tariff concessions and abolishment, that is, the modality in the normal track and the sensitive track, and rules of origin;[30] the second is the signing of the Agreement on Dispute Settlement Mechanism which provides systematic principles, rules, procedures and methods to settle disputes that would arise from the implementation of the Framework Agreement on Comprehensive Economic Cooperation.[31]

It is of particular significance that this Agreement on Dispute Settlement Mechanism might be the very first time that China had signed such a special and detailed agreement with other states or international organizations. The provisions of disputes settlement methods and procedures in the Agreement might gain the

---

[26] Chairman's Statement of the Eighth ASEAN + China Summit, Vientiane, 29 November 2004, para. 4, http://www.aseansec.org/16749.htm (accessed 10 February 2005).
[27] Plan of Action to Implement the Joint Declaration on ASEAN-China Strategic Partnership for Peace and Prosperity, para. 1.4, http://www.aseansec.org/16806/htm (accessed 10 February 2005).
[28] See Plan of Action to Implement the Joint Declaration on ASEAN-China Strategic Partnership for Peace and Prosperity, para. 1.5.
[29] See Plan of Action to Implement the Joint Declaration on ASEAN-China Strategic Partnership for Peace and Prosperity, para. 1.6.
[30] Agreement on Trade in Goods, Annex I, Annex II and Annex III, http://www.aseansec.org/4979/htm (accessed 10 February 2005).
[31] Agreement on Dispute Settlement Mechanism, http://www.aseansec.org/4979/htm (accessed 10 February 2005).

applause of the legal circle, for not only persisting in the favorable consultation method, but also including third-party intervention methods, such as conciliation or mediation and arbitration. There is a lengthy and detailed provisions in the Agreement concerning the arbitration mechanism, from the appointment and composition of arbitral tribunals to their concrete functions and proceedings, including participation of third parties, implementation of arbitral decisions, compensation and suspension of concessions or benefits, etc. The far-reaching significance of this Agreement is beyond ASEAN-China relations. It is a sign of the changing attitude of the Chinese government towards international dispute settlement methods. That is, while continuing to prefer political and diplomatic methods, it is ready to adapt itself to the proliferation of dispute settlement mechanism and the important role of the legal methods such as arbitration.

Finally, it is of special political, economic and legal importance that the ASEAN leaders recently agreed to recognize China as a full-market economy and committed not to apply Sections 15 and 16 of the Protocol of Accession of China to the WTO and Paragraph 242 of the Report of the Working Party on the Accession of China to WTO in relation to trade between each of the ten ASEAN member states and China.[32] The Chinese government undoubtedly appreciates such a solemn commitment by ASEAN, since it might encourage the increasing recognition of China as a full-market economy entity by more and more states and organizations, so as to possibly avoid discrimination by WTO members taking anti-dumping, countervailing and safeguard actions against China at an early stage.

---

[32] Chairman's Statement at the Eighth ASEAN + China Summit, "Developing ASEAN-China Strategic Partnership", Vientiane, 29 November 2004, para. 15, http://www.aseansec.org/16749htm (accessed 10 February 2005).

# Part III

# China-ASEAN on Non-traditional Security Issues

# Chapter 4

# Regional Cooperation in Epidemic Prevention: China and ASEAN

*LAI Hongyi*

## Introduction

Close official cooperation between China and ASEAN dated back in the late 1990s.[1] However, China-ASEAN epidemic cooperation came about only after both parties saw highly negative consequences of non-coordinated approaches to SARS management. Only after the Chinese failed miserably to contain SARS and both sides failed to cooperate in early months did China and ASEAN agree to repair the damage and regain lost ground in this area of cooperation. However, once China-ASEAN epidemic cooperation started with SARS, it was quickly emulated in the later cases of epidemics, notably the avian flu.

This paper reviews the start and progress of cooperation in the case of SARS and then analyzes the cooperation in containing the bird flu. It identifies the main forum and channels of cooperation and gives a very tentative assessment of their effects. It suggests that even though the cooperation covers mainly basic coordination in epidemic containment, improvement in communication, and funds to support disease control, these efforts prove repeatable and probably enduring. The main cause may be that absence of the cooperation will leave each other much worse off, in terms of possible spread of disease and mutual

---

[1] For cooperation between China and ASEAN in general, refer to S.D. Muni, *China's Strategic Engagement with the New ASEAN* (Singapore: Institute of Defence and Strategic Studies, Nanyang Technological University (NTU), 2002); and Xu Jian, ed., *Guoji Huanjing yu Zhongguo he Zhanlue Jiyuqi (International Environment and Period of China's Historical Opportunities)* (Beijing: Renmin Chubanshe, 2004), pp. 292–322. For regional cooperation involving ASEAN on non-traditional issues, refer to Andrew Tan and Kenneth Boutin, eds., *Non-Traditional Security Issues in Southeast Asia* (Singapore: Institute of Defence and Strategic Studies, NTU, 2001). For ASEAN and Australian cooperation over SARS, refer to M. Curley and N. Thomas, "Human Security and Public Health in Southeast Asia: The SARS Outbreak", *Australian Journal of International Affairs*, Vol. 58, No. 1, pp. 17–32 (March 2004).

suspicion about each other's disease-control measures which might affect the other parties.

## SARS and Initial Failure of Cooperation

In January 2003, several cities in Guangdong, China experienced outbreaks of SARS. It broke out in Guangzhou, the provincial capital of Guangdong in early February 2003. Local authority swiftly denied rumors about the deadly disease, and many Chinese, including medical workers, migrants, and business people, were not aware of the highly infectious nature of the disease.

Dr. Liu Jianlun, a physician, had been treating SARS patients in Guangzhou. His nephew was getting married in Hong Kong. Although Dr. Liu felt uncomfortable before the trip, he did not detect anything unusual in his chest X-ray. In the relaxed atmosphere, he also did not take the precaution of avoiding social contact. Thus he traveled to Hong Kong on 20 February 2003. At the Metropol Hotel, he felt severely sick and infected a number of tourists who came from Canada, Singapore, Vietnam, and the United States. He also infected medical workers at a Hong Kong hospital. From then on, SARS crossed the international border and started to spread outside China.

Southeast Asia was severely hit by the SARS outbreak. In mid-April, two ASEAN countries, namely Singapore and Vietnam, were on the list of SARS-affected areas issued by the World Health Organization (WHO). On 15 April, Singapore had 162 SARS cases, Vietnam 63, Thailand 8, Malaysia 4, Philippines 1, and Indonesia 1 (see details in Table 4.1). The worst effect was that news reports about SARS cases in the region scared away many tourists and even business people. For example, Phuket, a favorite place for travelers to Southeast Asia, especially Thailand, received very few tourists in April and May, and related business was hard hit.[2] Tourism and foreign investment were significant sources of growth for Southeast Asia. Between 8 and 14 April, tourists to Singapore declined by 68% compared with the same period of the previous year.[3] The SARS outbreak dampened economic growth in the region. Morgan Stanley adjusted downward the projected GDP growth for Singapore from 2.1% to 1%.[4]

Many Southeast Asian media, people and even officials were dissatisfied with the Chinese management of SARS. They criticized the Chinese for their lack of

---

[2] Conversation with an agent of a travel agency in Phuket in August 2003.
[3] "Singapore Tourism News: Tourists to Our Country Continue to Decline", *Lianhe Zaobao (United Morning Post)*, 23 April 2003.
[4] "Morgan Stanley: Singapore May Fall Into Economic Recession Again", *Lianhe Zaobao (United Morning Post)*, 30 May 2003.

Table 4.1   SARS Statistics, 15 April 2003

| Country | Number of cases | Numbers of deaths | Number recovered |
|---|---|---|---|
| Indonesia | 1 | 0 | 0 |
| Malaysia | 4 | 1 | 0 |
| Philippines | 1 | 0 | 1 |
| Singapore | 162 | 13 | 85 |
| Thailand | 8 | 2 | 5 |
| Vietnam | 63 | 5 | 46 |

Source: WHO website.

transparency in reporting the disease and slow and ineffective containment of the virus; they blamed the exports of the virus on the Chinese ineffective management. China's early management of SARS contrasted sharply with its responsible and helpful approach to the Asian financial crisis in 1997. In 1997, unlike Japan, China refused to devalue its currency and had helped Southeast Asia to weather the Asian financial crisis. Weak management of SARS, however, undermined this positive image in Southeast Asia that China had carefully cultivated since 1997. Many officials in Southeast Asia privately complained about the Chinese mismanagement; the media were more frank in their criticism of or complaints about China.[5]

At the early stage of SARS, individual ASEAN countries and China were also making individual efforts and did not engage in region-wide cooperation. There was very limited medical information exchange regarding SARS and political consultation on both sides. Without a comprehensive cooperative framework, when China or ASEAN adopted certain measures to contain SARS, it was interpreted by the other side as unfriendly moves, and the ties between them were strained. For example, a number of countries in Southeast Asia imposed travel restrictions and health requirements on the Chinese arriving in their countries. This had an unintended consequence: the Chinese loathe to be viewed as sick people in Asia. They were called sick men in East Asia when China was invaded and exploited by the West in the late Qing Dynasty. These travel restrictions against the Chinese seemed to remind the Chinese officials and leaders of the painful past experience under Western imperialism. In response, the Chinese Bureau of Tourism urged

---

[5] For a summary of criticisms by Southeast Asian officials and experts on China's SARS management, refer to Joseph Kahn, "China's Image Tarnished by Lack of Candor on SARS", *The New York Times*, 17 April 2003; Wei Li, "Atypical Pneumonia Syndrome of China's Image", *Lianhe Zaobao (United Morning Post)*, 3 April 2003; Lai Hongyi, "SARS Storm and a Responsible Great Power", *Lianhe Zaobao (United Morning Post)*, 14 April 2003.

Chinese nationals not to visit Thailand, Malaysia, and Singapore with the reason that these places were declared as affected areas by the WHO. This might be a likely Chinese attempt to retaliate against these countries. Only with the persuasion of the Ministry of Trade and Industry in Singapore did China consider softening its ban.[6]

## Launch and Deepening of Cooperation over SARS

Diplomatic skirmishes between China and ASEAN died down after China reversed its SARS management and adopted a more transparent, responsible, and effective approach. China was feeling the pain at home and abroad for following a secretive and ineffectual approach to SARS — the disease spread from one province to another across the country; mainland China was criticized heavily for mismanaging the epidemic by other countries, including Southeast Asian nations.[7] After mid-April, China's new leaders headed by President Hu Jintao and Premier Wen Jiabao decisively shifted to a more open and effective approach to tackling SARS. On 17 April, the Politburo Standing Committee convened a meeting on SARS. It stressed that containing SARS was a task of top priority and that all officials should report the SARS situation honestly without delay. On 20 April, two senior officials, the Minister of Public Health and the Beijing Mayor were relieved of their posts for their failure to contain the epidemic within the country and the national capital, respectively. With intense pressure from above, officials at all levels started to act swiftly to cope with the disease, released epidemic information, and educated the mass on preventive measures.

On 26 April 2003, Health Ministers from ASEAN except Vietnam, China (including Hong Kong), Japan, and South Korea held an ASEAN+3 health meeting in Kuala Lumpur. The attendants discussed ways to contain the spread of SARS and return the lives of millions of people in their countries to normalcy. This marked the first and high-profiled official cooperation between ASEAN and other Asian countries, especially China, the epicenter of SARS, in fighting the infectious epidemic.

In a joint declaration, the Health Ministers called for stringent pre-departure screening of passengers at border entry points and agreed that all ASEAN countries

---

[6]"China Expected to Revoke Ban on Group Travel to Singapore", *Lianhe Zaobao (United Morning Post)*, 18 April 2003.

[7]For a summary of criticisms on China's early SARS management by Southeast Asian officials and experts, refer to Lai Hongyi, "SARS Storm and a Responsible Great Power", *Lianhe Zaobao (United Morning Post)*, 14 April 2003; Joseph Kahn, "China's Image Tarnished by Lack of Candor on SARS", *The New York Times*, 17 April 2003.

should bar people with SARS symptoms from traveling abroad. The Ministers agreed to make it mandatory for travelers from affected countries to fill SARS health declaration form and to put in place a surveillance mechanism to ensure prompt exchange of relevant information on SARS cases and people who have come into contact with them. They also agreed to set up a contact point for routine information sharing and install a "hotline" in every country to facilitate communication in an emergency. The Ministers also agreed to urge their heads of government to provide adequate resources to respond effectively to the epidemic and to establish a national multi-sectoral task force with real power of enforcement.[8] This meeting also laid the foundation for the coming ASEAN and China summit on SARS in Bangkok, as well as future health ministerial meetings.

In late April, Singapore proposed to have an ASEAN summit meeting on SARS in Bangkok and publicized the fruitful cooperation between Singapore and Malaysia in containing the spread of SARS across their common border. The proposal was accepted by Hun Sen, Cambodia's Prime Minister who was chairing the ASEAN Standing Committee, and Thai Prime Minister Thaksin. At the invitation of Mr. Thaksin, Premier Wen attended the second-round joint summit meeting in Bangkok on 29 April. He acknowledged China's inadequate management of the epidemic and pledged a swift and effective approach. He told journalists and others attending the summit that he came to the meeting in a spirit of candor and expressed his thanks to ASEAN for its understanding of China's situation. He declared that China was a responsible nation and that his administration was responsible and would take care of people's health.[9] During his 48-hours stopover in Bangkok, Premier Wen also engaged in private leadership diplomacy by initiating meetings with leaders of individual ASEAN members, in order to repair damaged ties due to the spread of SARS.

Specifically, Singapore pushed for the implementation of screening at the border checkpoint and sharing of information on contact-tracing of those infected with SARS. These measures would be effective in containing the virus while keeping the border open. The summit took up a proposal from Singapore to keep the border open and trade and investment flowing, while acting to contain the disease. ASEAN leaders also successfully obtained China's support for these measures.[10]

---

[8]"ASEAN+3 Health Ministers Discuss Ways to Stop SARS Spread", http://english.people.com.cn, 27 April 2003 (accessed 12 October 2004).

[9]"Wen Jiabao: Adversities Reveal True Friendship", *Lianhe Zaobao (United Morning Post)*, 30 April 2003; Ellen Nakashima and John Pomfret, "Chinese Premier Vows to Cooperate — Asian Summit Sets Initiatives on Virus", *Washington Post*, 30 April 2003, p. A10.

[10]"Asean and China — United in Adversity", *Far Eastern Economic Review*, 8 May 2003.

At the summit, Premier Wen proposed the following cooperative measures:

> "To establish a reporting mechanism with respect to epidemic and disease prevention and treatment; to carry out exchanges of experience and joint SARS research; to accelerate bilateral health cooperation process; to coordinate border exit and entry control measures; and to do everything possible to minimize the adverse effects of SARS, including the establishment of a China-ASEAN workshop to look into possible counter-measures."

At the summit, both China and ASEAN supported a number of measures, some of which had been incorporated into the declaration of the ASEAN+3 health meeting in Kuala Lumpur. In addition to screening at border checkpoints and continued efforts to promote trade and tourism, the following measures were also pledged:

1. information sharing and setting up a reporting mechanism regarding the epidemics and remedies for SARS;
2. cooperation in medical research and training on treating SARS patients;
3. convening meetings of immigration and health officials on border check and health measures to fight SARS, and
4. holding a meeting to discuss the political and economic effects of SARS on the region.

A check suggested that measures 1, 2 and 4 described above were incorporated into the joint declaration and proposed by Premier Wen.[11] Premier Wen also proposed to establish the China-ASEAN Special Fund for SARS research and study, and offered US$1.2 million for the fund. Thai Prime Minister Thaksin offered US$250,000, and Mr. Hun Sen also pledged US$100,000 for the fund. However, other ASEAN members, led by Singapore's Prime Minister Goh Chok Tong, were not enthusiastic about the proposed fund.[12] Nevertheless, the China-ASEAN joint declaration still reiterated China's pledge for the fund.

The Philippines had organized the "Aviation Forum on the Prevention and Containment of SARS" from 15 to 16 May 2003 in Clark Special Economic Zone,

---

[11] "China Calls for Close Cooperation with ASEAN in Fight against SARS: Wen Jiabao at China-ASEAN Meeting on SARS", http://english.people.com.cn (accessed 19 October 2004).

[12] "Eight Countries Including Singapore Openly Rejected Wen Jiabao's Proposal: China's Diplomacy across the South China Sea Hits the Rock", http://bbs.fmprc.gov.cn/detail.jsp?id=58436 (accessed 19 October 2004).

Pampanga, Philippines. The forum discussed measures to prevent and contain the spread of SARS, such as standardized airport procedures for passenger screening.

On 31 May and 1 June, Beijing hosted a China-ASEAN Border Quarantine Conference. A joint action plan passed on 1 June announced measures to make traffic and passengers safer during travel. The plan required exit-entry travelers to undergo temperature screening checks and if necessary medical checks. The plan also required passengers to fill out a health declaration card. Passengers deemed a potential SARS risk would be referred to health or quarantine authorities for isolation and treatment, instead of being denied entry. It also required planes, ships, trains and buses boarded by SARS patients or suspected cases to be disinfected, and ships with suspected cases were required to contact the nearest port authority and dock at the nearest harbor.[13] By early June, ASEAN member countries had established their national multi-sectoral task forces, appointed their contact points, and set up a "hotline" for the exchange of information on SARS, as pledged in the Bangkok declaration.

On 10 and 11 June 2003, ASEAN, China, Japan and South Korea held an ASEAN+3 Health Ministers' special meeting on SARS in Siem Reap, Cambodia. The meeting pointed to the ASEAN success in fighting SARS and preventing new SARS cases for 30 days by 11 May. It declared the region SARS-free and urged concerned countries to withdraw travel advisories to ASEAN countries. The conference congratulated the government of China for its very strong political commitment and efforts in containing SARS. It suggested that China could make further improvement in obtaining information on the dates when cases were isolated, and how the patients became infected. The Ministers adopted the ASEAN+3 Action Plan on Prevention and Control of SARS and other Infectious Diseases, and stressed the guidelines for international travel, ASEAN SARS containment information network, capacity building for outbreak alert and response; and public education and information.

The most important achievement at the meeting was the effort to institutionalize the cooperation among the ASEAN+3 countries. First, the Ministers requested the ASEAN Secretariat to prepare, in consultation with ASEAN+3 senior health officials, a detailed Action Plan. The plan would include institutions for coordination, strategies for resource mobilization, timelines for implementation, and a list of meetings and activities. They asked the detailed Action Plan to be circulated to member countries for comment by August 2003. Secondly, the Ministers agreed that follow-up action on the implementation of the Action Plan should conform

---

[13]"China, ASEAN Formalize Control on Travellers", *China Daily*, 2 June 2003.

to relevant international rules and regulations and should not discriminate against any nation or individual.[14]

## Cooperation Beyond SARS: Tackling the Bird Flu

China-ASEAN anti-epidemic cooperation did not stop with the end of the SARS rampage. It stays on and extends to other diseases. On 8 October 2003, at the ASEAN+3 meeting in Bali, Indonesia, ASEAN and China issued a joint statement. They pledged to cooperate in containing and avoiding of SARS, HIV/AIDS and other communicable diseases. In addition, they agreed to further exchanges in science and technology, education, human resource development and culture, as well as personnel exchanges and improve the mechanisms for cooperation in these areas.[15] They also promised to implement the consensus reached at the Bangkok summit in April 2003 over cooperation in public health by setting up a ASEAN+1 public health cooperation fund and activating ASEAN+1 Health Ministers' meeting mechanism.[16]

The outbreak of the highly pathogenic avian influenza (HPAI) in a number of countries in December 2003 provided both parties new opportunities to utilize their cooperation mechanism and work together to fight epidemics. In early March, the bird flu, known as H5N1, had spread to Cambodia, Indonesia, Japan, Laos, South Korea, Thailand and Vietnam. Weaker cases were also reported in parts of the United States and Canada, as well as in Taiwan and Pakistan. In China, all but two of the 49 infected areas were declared safe. By 2 March 2004, the bird flu had killed 41 people in Vietnam and 14 in Thailand. In addition, poultry production in Thailand, Vietnam, China, and other Asian countries were affected by slaughters of millions of chickens and ducks, and consumers' fear of infection through eating poultry meat. According to the UN's Food and Agriculture Organization, more than 100 million birds worldwide had died or been culled in the wake of the disease, including five million in China.[17]

According to statistics from the UN Food and Agriculture Organization (FAO), Asia is one of the world's major bird markets, and produces 20 million tons of

---

[14] Joint Statement of the Special ASEAN+3 Health Ministers' Meeting on Severe Acute Respiratory Syndrome (SARS), Siem Reap, Cambodia, 10–11 June 2003.
[15] Press Statement by the Chairperson of the ASEAN+China Summit, 8 October 2003, Bali, http://www.aseansec.org/15287.htm (accessed 12 October 2004).
[16] Joint Statement by the People's Republic of China and Leaders of ASEAN, 10 October 2003, http://www.fmprc.gov.cn (accessed 12 October 2004).
[17] "Waning Bird Flu Still Hurts China: Recovery Slow for Small Farms", *The Atlanta Journal — Constitution*, 11 March 2004.

chicken products or 27% of the world's total each year. China and Thailand are two major poultry exporters, together accounting for 15% of the world's market share. China is the world's second-largest producer of chicken meat after the United States.[18]

After the HPAI outbreak, however, many major poultry-consuming countries have banned imports from the two countries.[19] The outbreak of HPAI in many countries in Asia hurt the region's agriculture, poultry industry and foreign trade, as well as human health.

In order to cope with this outbreak of the relatively new epidemic, a series of international and regional conferences were convened to share information and deliberate on prevention measures. They included a Ministerial-level conference on the HPAI situation in Bangkok on 28 January, an emergency conference on HPAI in Rome on 3 February, an Emergency Regional Meeting on Avian Influenza Control in Animals in Asia in Bangkok, Thailand on 26 February 2004, and the China-ASEAN Special Meeting on HPAI Control in Beijing, China on 2 March 2004.

Agriculture and Health Ministers and senior officials of ASEAN+3 (China, Japan, and South Korea), the United States and the European Commission met in Bangkok, Thailand on 28 January 2004. This might well be the first official regional meeting on the HPAI. Participants exchanged views and experiences on measures to fight HPAI, and discussed the current situation of the poultry disease as well as its effects on the poultry population, human health and the economy. They recognized that the outbreaks of the disease constituted a potential threat to human health.

The participants realized that containment of the disease required closer cooperation among governments, communities and businesses through regional and international organizations, and other necessary mechanisms. They called for careful monitoring and investigation to prevent any possible human-to-human H5N1 transmission. They also reiterated the usefulness and need for measures that had been introduced by affected countries. These measures included "rapid diagnosis and confirmation, rapid killing of infected and susceptible poultry populations, vaccination of poultry, quarantine of infected areas, intensified surveillance, movement control, epidemiological investigation and hospitalization and monitoring of affected patients."

Participating countries decided to adopt more stringent surveillance and effective response systems, strengthen research and development capabilities, share

---

[18]"Waning Bird Flu Still Hurts China: Recovery Slow for Small Farms", *The Atlanta Journal — Constitution*, 11 March 2004.
[19]"Bird Bug Brings Lessons", *Beijing Review*, 11 April 2004.

information and technology, and carry out domestic measures to control avian influenza that were recommended by the World Organization for Animal Health (OIE), World Health Organization (WHO) and the Food and Agriculture Organization (FAO). They also decided to form effective outreach and communication strategies to foster transparency and better public awareness of the extent and nature of the disease. They also declared their commitment to assist with biosecurity developments of the poultry sector for both small-scaled and commercial production.

Participants at the meeting reached a significant and new consensus for intensifying national, regional and international efforts to tackle the outbreak of this disease and similar threats in future. Three areas were identified. Firstly, they pledged to strengthen and develop reporting and surveillance systems. The Ministers pledged to work closely with OIE to strengthen guidelines for the system. They also declared to promote exchanges of scientific information in order to provide early warning of potential outbreaks. They also stressed the need to raise public awareness of the extent and nature of the disease. More importantly, they considered creating a regional veterinary surveillance network and connecting it with the existing health surveillance mechanisms, including the APEC Task Force on Health, ASEAN Ministers' Health Special Meeting in Kuala Lumpur, and ASEAN Heads of States Meeting on SARS in Bangkok, respectively.

Secondly, these countries decided to strengthen their cooperation with regional and international organizations on research and development initiatives, to reduce the hazards of epizootic outbreaks to human health by sharing effective practices, devising measures, and developing effective and low-cost diagnostic kits, vaccination and drugs. Thirdly, the meeting called for assistance and exchange of expertise to help affected countries to exchange their epidemiological and laboratory capacity for prompt detection, monitoring, surveillance and control of the disease.[20]

On 2 March, a China-ASEAN Special Meeting on HPAI Control was held in Beijing. Participants included the Deputy Prime Minister of Thailand, the Minister of Animal Husbandry and Fisheries of Myanmar, Vice-Ministers and senior officials for agriculture and health, and experts from China and ten ASEAN countries, representatives from Hong Kong and Macao Special Administrative Regions of China as members of the Chinese delegation, as well as officials from ASEAN Secretariat, UN Food and Agriculture Organization, World Health Organization and World Organization for Animal Health.

---

[20] Joint Ministerial Statement on the Current Poultry Disease Situation, Bangkok, Thailand, 28 January 2004, http://www.thaiembassy.se (accessed 22 October 2004).

Remarks made by Du Qinglin, Chinese Minister of Agriculture, at the opening ceremony signified a drastic shift in China's view on anti-epidemic regional cooperation since the SARS outbreak. He declared: "The trans-regional spread of HPAI is a challenge confronting the international community, particularly our region. The disease will be successfully defeated only through regional and international cooperation."

The participating countries acknowledged the cross-boundary nature of the disease, its threat to public health and economy, and the need for China and ASEAN to cooperate in order to minimize losses. Participants recognized that strong leadership, political commitment, and inter-agency cooperation and partnership at both national and regional levels were needed to contain the disease. The meeting concluded that participating countries and areas should take the following measures to address the challenge of HPAI. The first measure referred to a surveillance network that had been evolving in the past year. The Ministers pledged to exchange information and experience on HPAI status and control through the ASEAN disease surveillance network, the ASEAN+3 SARS focal point network and the proposed regional veterinary network. The aim was to create an early warning system for epidemic recognition and control. Suwit Khunkitti, Deputy Prime Minister of Thailand, told a China magazine that the exchange of information and network of data were very important for every country. The above declaration thus devoted a great deal of attention to the issue by relying on evolving mechanism. In addition, the meeting also declared to set up a China-ASEAN cooperation mechanism for public health, through regular meetings of agriculture or health Ministers and their senior officials, and joint meetings of the health and agriculture Ministers.

As before, the Ministers decided to strengthen extensive cooperation and exchanges with other countries, regional and international organizations, such as WHO, FAO, OIE on HPAI prevention and control. They also wanted to step up cooperation among their inspection and quarantine agencies for border control. Financial, technical and medical cooperation was also stressed. China and ASEAN declared to mutually provide bilateral financial, material and technical assistance to countries in the region hit by or at risk of being infected by HPAI. The Ministers also pledged to exchange HPAI expert teams and organize joint technical training courses on HPAI-related technologies, including laboratory management, diagnosis and testing, emergency response measures and quality of vaccines in compliance with OIE international standards. The National Reference Laboratory of China pledged to share its experience and offer technical cooperation with ASEAN countries in terms of diagnostic technology. Finally, the Ministers decided to use the "China-ASEAN Fund for Public Health" to finance cooperation.[21] For reasons to be explained below, in the case of bird flu, China's financial support proved to be helpful, instead of being purely symbolic.

According to the FAO, killing birds in HPAI-hit areas can effectively eradicate the disease. In order to make up for the lost income of the farmers, governments needed to compensate the losses of all the affected farmers. However, most of the countries hit by the bird influenza were developing countries and lacked the resources for carrying it out. International organizations did offer valuable help for affected poultry farmers. On 13 February, the World Bank decided to provide the Vietnamese government $10 million in donations and long-term low-interest loans for compensating local farmers' losses caused by bird flu. In late February, the FAO gave $5.5 million to help peasants and individuals in Asia to fight the disease. China also provided additional assistance, by offering over $450,000 in financial help to Thailand, Pakistan, Vietnam and Cambodia in fighting the epidemic in February.[22] With painstaking efforts and international cooperation, Vietnam, where the bird flu first broke, announced that it successfully eradicated HPAI in early March. Meanwhile, quarantine was declared over for most of the affected regions in China by mid-March 2003.

China-ASEAN cooperation in health issues is not restricted to SARS and the bird flu. In fact, both sides have exchanges in other health issues before the SARS outbreak. A key area is HIV/AIDS. As early as 24–26 April 1999, in the wake of the ASEAN Workshop on Population Movement and HIV Vulnerability held in Chiang Rai, Thailand in November 1998, a consultation meeting of the Cambodia/China (Guangxi and Yunnan provinces)/Lao PDR/Myanmar/Vietnam cluster was held, with Vietnam as the cluster coordinator.[23] The ASEAN Task Force on AIDS met with China as early as 2000.[24]

Since 2000, HIV/AIDS and related-narcotic drugs have become a new area for cooperation between China and ASEAN in non-traditional security. In October 2000, the two sides established a cooperation mechanism on fighting illegal drugs.

---

[21] Joint Press Statement of China-ASEAN Special Meeting on HPAI Control, 2 March 2004, Beijing, China, http://www.aseansec.org (accessed 10 October 2004).

[22] "Bird Bug Brings Lessons", *Beijing Review*, 11 April 2004; "Isolation Ends in Last Two Bird Flu-Hit Areas", http://www.asean-disease-surveillance.net (accessed 10 October 2004).

[23] Lee-Nah Hsu, "Update News Letter — Mobility and HIV Vulnerability, April 2000", http://www.hiv-development.org/publications/sea_pub_update_april00.asp (accessed 1 March 2005).

[24] Press statement by the Chairman of the Seventh ASEAN Summit and the Fifth ASEAN+3 Summit November 6, 2001, Bandar Seri Begawan, http://www.dfa.gov.ph/about/offices/7thasean.htm (accessed 1 March 2005).

In August 2001, China, Laos, Myanmar and Thailand held a ministerial meeting on fighting illegal drugs in Beijing and issued the Beijing Declaration. China has thereafter been taking an active part in the AIDS control and prevention program in the Mekong River Basin.[25] On 6 November 2001, at the Seventh ASEAN Summit and the Fifth ASEAN+3 Summit in Bandar Seri Begawan, ASEAN and China agreed to cooperate to fight HIV/AIDS and combat drugabuse through ASEAN-China Cooperative Operations in Response to Dangerous Drugs. In 2002, China proposed the Project of Establishing an ASEAN AIDS Laboratory Network. It was accepted by ASEAN. China hosted an AIDS Lab Networking Construction training course for ASEAN countries.[26]

Cooperation over HIV/AIDS has been helped by proactive exchanges and initiatives in the wake of the SARS outbreak. At the Ninth ASEAN Summit and the Seventh ASEAN+3 Summit in October 2003 in Bali, ASEAN and China stated their commitment to "further intensify cooperation in the area of public health, including the prevention and control of infectious diseases, such as HIV/AIDS, SARS and avian flu". Less than one year later, at the First ASEAN+3 Health Ministers Meeting on 23 April 2004 held in Penang, Malaysia, ASEAN also expressed its commitment to its individual dialogue partnerships with China, Japan and the ROK for addressing shared health concerns. ASEAN welcomed China's interest to make proposals for activities on communicable diseases (HIV/AIDS, disease surveillance) and capacity-building.[27] The mechanisms for health cooperation between ASEAN and China discussed earlier and later, serve to facilitate cooperation over HIV/AIDS.

## Conclusion: Catastrophe Prompts Strides in Regional Health Cooperation

For decades, China had been emphasizing bilateral ties and distrusted multilateral efforts. It saw the former manageable and more productive. It dismissed the latter as difficult and futile even if driven by good intention. China's distrust of multilateral initiatives and organizations were rooted in its recent bitter experience with them. For the Chinese, the most effective multilateral efforts and organizations were NATO and the Warsaw Pact in the Cold War, both of which China disliked for their active role in the Cold War confrontation. For the Chinese, these

---

[25]Speech by Foreign Minister Tang Jiaxuan at the ASEAN-China Dialogue, http://un.fmprc.gov.cn/eng/33782.html (accessed 1 March 2005).

[26]Non-paper China's Follow-up Actions Since the Fifth 10+3 and 10+1 Summits, http://un.fmprc.gov.cn/eng/33784.html (accessed 1 March 2005).

[27]Co-Chairs' Statement, Seventh ASEAN Health Ministers' Meeting, 22 April 2004; First ASEAN+3 Health Ministers' Meeting, 23 April 2004, Penang, Malaysia, http://www.aseansec.org/16092.htm (accessed 1 March 2005).

multilateral institutions oriented toward traditional security, dominated by the superpowers, and were used as a ploy against other nations. In the last decade, China has significantly moved away from bilateralism in its diplomacy and come to accept and practise multilateralism. This also occurs in non-traditional security areas, including epidemic prevention.

In particular, China viewed the Association of Southeast Asian Nations (ASEAN) with more receptiveness for two reasons. ASEAN has distanced itself from the superpower policy and has tried hard to assert its political independence. China also seeks a breakthrough in its good neighbor policy. ASEAN seems to the best place for China to try and succeed in establishing smooth and lasting ties with its neighbors.

China's cooperation with ASEAN over epidemic prevention had been very limited prior to April 2003. Since then, both parties have endeavored to develop and expand their cooperation. Numerous initiatives and institutions were launched. These rapid advances can be attributed to urges, willingness and receptiveness on both sides. China wanted desperately to mend its ties with ASEAN that were damaged by its inadequate management of SARS in the early 2003. ASEAN, on the other hand, learned from its painful lesson during the early 2003 that when China did not manage epidemics in a cooperative manner, Southeast Asia would inadvertently import epidemics and experience economic and social disruptions. Thus, ASEAN has also been receptive to China's multilateral efforts to fight epidemics.

ASEAN+3 health collaboration continues to progress after March 2004. On 23 April 2004, ASEAN+3 health Ministers convened in Penang, Malaysia. They declared their satisfaction with the progress made in establishing collaborative partnerships in the ASEAN+3 countries to effectively respond to emerging infectious diseases. "ASEAN+3 collaboration in responding to diseases has steadily gained momentum since the first time ASEAN+3 countries came together to address the spread of Severe Acute Respiratory Syndrome (SARS) in April 2003, and more recently the avian flu." Around the time of this conference, health agencies and experts of these countries were preparing a Phase I Work Plan for the Framework ASEAN+3 Action Plan on Prevention and Control of SARS and Other Infectious Diseases adopted by the Special ASEAN+3 Health Ministers Meeting on SARS in June 2003, which was later known as the ASEAN+3 Emerging Infectious Diseases (EID) Program. The implementation of the Phase I Work Plan will contribute to increasing the effectiveness of regional surveillance, early warning and response to emerging infectious diseases. The work plan would provide ASEAN+3 with greater opportunities to share and exchange information, experience and expertise. These countries also agreed to prepare against future threats of diseases, including bioterrorism in the ASEAN+3 EID Program with the participation of animal health experts, and also to facilitate partnerships among networks in the region on public and animal health.[28] Therefore, regional cooperation has moved beyond SARS, the first epidemic it tackled, into other epidemics and bio-threats that could harm public health.

---

[28] Co-Chairs' Statements of the Seventh ASEAN Health Ministers' Meeting on 22 April 2004 and of the First ASEAN+3 Health Ministers' Meeting on 23 April 2004, Penang, Malaysia, http://www.aseansec.org (accessed 10 October 2004).

# Chapter 5

# Anti-Piracy Cooperation Dilemma: ASEAN and China

*XU Ke*

## Introduction

Maritime piracy is one of the non-traditional security issues regarding Southeast Asian waters. Since the last decade, piracy incidents have dramatically increased and received increasing amount of media and political attention. After the event of 9/11, piracy in Southeast Asian waters became a greater concern. The possibility of the conflation of piracy and terrorism in Southeast Asia has been reflected in the mass media as well as in academic journals (Young and Valencia, 2003, p. 269). Since piracy in Southeast Asian waters is a trans-national crime, international anti-piracy cooperation becomes a hot issue under discussion.

This paper focuses on the anti-piracy cooperation among the three ASEAN coastal countries, namely Indonesia, Malaysia and Singapore, and China. The first section reviews the piracy and anti-piracy cooperation in Southeast Asian history. The second section introduces the anti-piracy agencies of ASEAN coastal countries and China. The third section describes the anti-piracy cooperation between ASEAN and China. The fourth section examines the dilemma in anti-piracy cooperation.

## Piracy in Southeast Asia: A Brief Historical Review

### Piracy in "Malay World"

Maritime piracy in Southeast Asian waters has a long history. Piracy dated back to as early as the fifth century. Shih Fa-Hsien, the Buddhist monk from Ceylon (now Sri Lanka), recorded cases of sea raiding in the waters of the Straits of Malacca and South China Sea in 414 AD (Moorhead, 1957, p. 34).

In the late 7th–12th century, the Srivijaya Kingdom in South Sumatra suffered from piracy. The sea nomads in its water engaged in piracy and preyed upon passing merchant ships. The Srivijayan monarchs, unable to suppress these pirates, had to surrender a portion of the post's revenue for their agreement not to raid the ships at sea (Tarling, 1992, p. 202).

At the end of the 13th century, the Malay empire of Malacca was established on the west coast of the Malay Peninsula, where the Malacca town now stands. It developed rich trading links with China, India and other small kingdoms in this region. As trade increased, so did the number of pirates. The southern end of the Straits of Malacca became a piracy-prone area, for it was a transportation gateway for the junks from China, Bugis boats from the Celebes and craft from Siam and Java (Miller, 1970, p. 14).

In the first decade of the 15th century, the Chinese imperial embassies established direct contact with north Sumatran pepper sources. The Ming Dynasty (1405–1430) became concerned about security in the Strait. It was recorded that China sent out a fleet under the admiral Cheng Ho (Zheng He) to suppress a Chinese pirate nest at Palembang in 1405 for the purpose of keeping peace and bringing order to the region (Tarling, 1992, p. 227).

After the Portuguese sailor, Vaseo Da Gama, went round the Cape of Hope to Calcutta, India in 1498, the ships of Portugal, Holland, England, France and Spain sailed on one after another to the Far East. Each of these countries created spheres of influence and dominated particular regions there — the Dutch in Java; the English in India, Singapore and the Malay peninsula; the French in small areas of India; and the Spanish in the Philippines.

The trade that the West brought to the Far East gave a fresh impetus to piracy. From 18th to 19th centuries, the Dutch and then the British tried to monopolize their own control over the lucrative East-West trade routes through the Straits of Malacca and Singapore. British East India Company resorted to using Malay "privateers" for commerce raiding against its Dutch counterpart in the "Malay world". In order to gain more power and wealth, the local sultans also supported pirates for sharing the booty and slaves. The most feared privateers were the Iranun and Balangingi of the Southern Philippines. Iranun and Balangingi squadrons plundered the shores of the Philippines, and sailed around Borneo, the Celebes and the Moluccas, Java, the east coast of the Malay Peninsula, the Gulf of Siam, Riau and up the Straits of Malacca to the Mergui Archipelago of Burma for the booty and slaves (Warren, 2002, pp. 3–4).

In the beginning of 19th century, these Malay privateers were too powerful to be controlled, and they became professional pirates. European colonizers eventually realized that piracy in the Malay world was their common threat. In 1819, the British founded Singapore and consolidated British control over the island by signing a Anglo-Dutch Treaty with the Dutch on 17 March 1824. In Article 5 of the Anglo-Dutch treaty, both parties agreed to cooperate in anti-piracy operations (Keppel, 1968, Appendix I).

After the Anglo-Dutch Treaty, there were many attempts by the British to suppress piracy. Between 1845 and 1848, the British attacked the principal Iranun base in Marudu and other Iranun bases in Tempasuk, Pandasan and Tanku (Trocki, 1979, pp. 208–211). By around 1843, white Raja Brooke and Captain Henry Keppel destroyed many pirates strongholds, i.e., Saribas stronghold at Padu, Paku and Rembas, and the Sekrang stronghold at Patusan and Undap (Rutter, 1970, pp. 99–126).

In 1848, with the aid of steamboats, the Spanish destroyed the Balangingi's stronghold at Sipac and deported hundreds of Balanginigi people to the distant mountain valleys of north central Luzon and eventually eradicated the Balangingi pirates (Warren, 2002, pp. 267–309).

Another dimension of this anti-piracy campaign was the elimination of slavery to destroy the markets of the maritime marauders. In 1836, the British destroyed the slave marts at Galang and Endau. With the founding of a naval base and commercial port in Labuan in 1846, the maritime marauders could no longer sell their booty there anymore. Jolo, the foremost slave mart, was no longer accessible after increased Spanish naval presence there (Tarling, 1963, pp. 17–20).

## Piracy in the South China Sea

Pirates had been a disturbing factor in the South China Sea for centuries, since the South China Sea was the gateway for trade between Western maritime powers and China. In 1516, the Portuguese sailed into Canton. At that time, there were many pirates threatening the river mouth of Canton. The Portuguese promised to help the Chinese suppress piracy, and the Chinese officials gave the Portuguese the right to found a settlement at Macao as a reward for their anti-piracy effort.

In the chaotic interim period (1640–1646) between the Ming and Qing dynasties, the Chinese pirate head, Cheng Chih-lung (*Zheng Zilong*), with a powerful fleet and thousands of followers, seized many southern coastal provinces of China and controlled trade with India, the Portuguese, the Spanish and the Dutch. In 1646, he was lured to Peking and killed by Manchus (Zheng, 1998, p. 268).

However, his powerful fleet remained under the command of his son, Cheng Cheng-kung (*Zheng Chenggong*), also known as Koxinga. Koxinga seized Formosa (Taiwan) from the Dutch in 1661 and other coastal provinces for his base. He supported the declining Ming dynasty, and fought against southward Manchu. After the Manchu's Qing dynasty stabilized its control of the mainland, Koxinga retreated to Formosa, and established a kingdom. In 1683, the grandson of Koxinga surrendered Formosa to the Manchus (Fox, 1940, p. 78).

The next famous pirate figure was Ching Yih (*Zheng Yi*), who had a confederation of between 50,000 and 70,000 pirates by 1805 and controlled the trade and fishing industry along the coast of Guangdong province. Even Europeans were forced to negotiate with them for their safety. Murray's famous monograph (1987) gives details of the pirate confederation.

After 1834, the year the British government abolished the East India Company's monopoly of the Anglo-China trade, the number and strength of pirates increased with the growth of foreign trade and the civil war in China. However, their fleets never approached the magnitude or power of the great pirates of the 17th and early 19th centuries. Pirates interfered both directly and indirectly with foreign commerce. They attacked and plundered some European ships, and they captured an incalculable number of native cargoes. After several years of official indifference towards the marauders, resulting from strained relations with the Chinese authorities, the British government ordered its navy to suppress piracy in the China seas and to assist the Chinese officials "within reasonable bounds" to give security to the coastal trade of the Empire. Britain's suppression of piracy between 1834 and 1869 promoted anti-piracy cooperation with the Chinese officials even during periods of Anglo-China hostility, and introduced a limited scheme of international co-operation in the China seas (Fox, 1940, pp. 187–191).

From the end of 19th century onwards, with the establishment of strong centralized state power in Southeast Asia through European colonial rule, piracy declined to a low level. However, throughout the colonial period, colonial authorities were never able to eradicate piracy completely.

### Modern Piracy in Southeast Asia

The political situation in Southeast Asia has changed dramatically. The new national states of Indonesia, Malaysia and Singapore are in place of the old colonial domains in the 20th century. At same time, piracy recrudesced in Southeast Asian waters. There were many records of Indonesian sea raiders or privateers attacking Commonwealth naval forces during the Malayan Emergency (1948–1960). In the subsequent "Confrontation" period (1963–1966), many piratical incidences against the newly created Malaysian Federation were recorded (Ellen, 1986).

During the 1970s–1980s, hundreds of attacks against Vietnamese boat-people fleeing the Communist regime were documented, with many of the assaults carried out in the Gulf of Thailand by Malaysian and Thai raiders (Chalk 2000, p. 57). In 1981 alone, when 452 boats arrived in Thailand carrying 15,479 refugees,

349 boats were attacked by pirates at an average of three times each; 578 women had been raped; 228 women had been abducted; and 881 people were dead or missing (UNHCR, 2000, p. 87).

Since the end of Cold War, piracy has made a spectacular comeback. According to IMB Piracy reports, the number of piratical attacks rose form 88 in 1991 to 124 in 1996. In 2000, the number of acts of piracy and armed robbery against ships reached 242, an increase of 81 over the 1999 figure and nearly three times that of 1991. A significant portion of these incidents occurred in Southeast Asian waters. Indonesian waters alone accounted for 119 out of the 469 incidents reported worldwide in 2000, 91 out of 335 incidents in 2001, and 103 out of 370 in 2002 (IMB, 1991–2003).

In 1992, in response to the escalating number of piracy incidents, the International Maritime Bureau (IMB) Piracy Reporting Center was established in Kuala Lumpur. IMB Piracy Reporting Centre, financed by voluntary contributions from ship-owners' associations and insurance companies, is a division of a non-government organization, International Chamber of Commerce (ICC). The IMB Piracy Reporting Centre performs a number of essential services to help victims of piracy, and provides weekly updates on pirate activities via the Internet and publishes comprehensive quarterly and annual reports detailing piracy statistics. IMB Piracy reports are widely used as the reliable source for piracy research (IMB, 1992).

In addition to the problem of increasing number of acts of piracy and armed robbery reported above, there is another very sophisticated issue of definition and categories of contemporary maritime piracy, which aggravated the anti-piracy cooperation situation.

## Definition

There are two definitions of piracy now. The first adopted definition is according to the United Nations Convention on the Law of the Sea (UNCLOS), Article 101. This definition applies only to any of the described illegal acts committed either on the high seas or outside the jurisdiction of any state. Thus, technically any of violent and illegal acts against ships or property and people on board ships, taking place in ports or territorial waters do not fall under the definition of Article 101. The illegal acts, taking place in ports or territorial waters, are defined as "armed robbery against ships" by IMO.

The second is the IMB definition. IMB defines piracy as "an act of boarding any vessel with the intent to commit theft or any other crime and with the intent or capability to use force in the furtherance of that act" (IMB, 1999, p. 2). This definition is much wider than UNCLOS Article 101. Given that few acts of piracy in Southeast Asia take place on high seas, and that most of the attacks occur in territorial waters instead, this paper uses the IMB definition as the working definition.

## Typology

The IMB classifies the piratical cases currently taking place in Southeast Asian waters into three categories: Low Level Armed Robbery; Medium Level Armed Assault and Robbery; and Major Criminal Hijack.

Low Level Armed Robbery refers to the assaults committed at anchorage or near harbour that target cash or portable high-value items, with an average theft of between US$5,000 and US$15,000. The pirates are generally armed with rudimentary weapons, such as knives, pistols and machetes, and operate from small, manoeuvrable dinghies or other similar light crafts. This type of piracy largely reflects the relatively lax security measures in many small and medium-sized ports in this region.

Medium Level Armed Assault and Robbery target ships traveling on the high seas or in territorial waters. It is a more serious form of piracy; the pirates use the powerful guns, fast boats and advanced communication equipments. More violence is involved and causes the casualty of the crew. Furthermore, once it is carried out in narrow sea lanes, such as the Straits of Malacca, it has the potential to disrupt maritime navigation, particularly in cases where vessels are out of control.

Major Criminal Hijack assaults are the most serious piracy type, which involve the hijack of ships and their subsequent re-registration under other names for the purpose of illegal trading. The vessels are seized and their cargoes are then offloaded into lighters at sea. These cargoes either being kept by the pirates themselves or sold off to private bidders. The ships are then re-registered, generally under lax shipping registration states, such as Panama, Liberia or Honduras, and issued with false documents to enable them to take on board a fresh payload. Known as the "phantom ship" phenomenon, the new cargoes are never delivered to its intended destination, but transferred to pre-arranged buyers.

## Anti-Piracy Agencies of ASEAN Countries and China

### The ASEAN Countries

#### Indonesia

Indonesian territorial water is the most serious piracy-prone area. In Indonesia, there are many agencies involved in law enforcement activities at sea. The Indonesian Navy (TNI-AL) and Indonesian police are the two main agencies in combating piracy. The TNI-AL is poorly equipped, lacking fast patrol boats and sophisticated weaponry. The *Bakorkamla* coordinates these security agencies at sea.

The *Bakorkamla* is being criticized for lack of efficiency and effectiveness, especially after the separation of the police from the armed forces (Djala, 2002, p. 108).

*Malaysia*

There are several agencies authorized to enforce the various maritime legislations, namely the Royal Malaysian Marine Police (RMMP), the Fishery Department, the Royal Navy, the Department of Environment and the Immigration Department. In 1985, Malaysia established a Maritime Enforcement Coordinating Centre (MECC) under the National Security Division of the Prime Minister's Department to enhance inter-agency cooperation in managing maritime security, especially in surveillance and enforcement functions.

After the event of 9/11, the RMMP has increased its patrols against piracy and the maritime terrorism in the Straits (Agence France-Presse, 31 October 2002). A new inter-agency Malaysian Maritime Enforcement Agency (MMEA) is scheduled to operate in November 2005. The MMEA will start operating with 72 patrol vessels and more than 4,000 personnel. The new agency is a paramilitary force that will absorb personnel from the navy, police and other existing government agencies. Besides tackling piracy, the agency would also be responsible for enforcing all Malaysian maritime laws, fighting coastal smuggling, and monitoring pollution in the Straits (The New Straits Times, 15 May 2005).

*Singapore*

The Singapore Police Coast Guard is in charge of combating piracy in its territorial waters. The Police Coast Guard is made up of six patrol squadrons and a fleet of more than 80 vessels. The Police Coast Guard works in collaboration with Republic of Singapore Navy (RSN). The Coastal Command of RSN is the agency directly involved in deterring and preventing acts of piracy and maritime terrorism in Singapore's territorial waters. It is composed of patrol vessels, in-shore fast boats and mine counter-measure ships. The Coastal Command collaborates closely with the Police Coast Guard, the Maritime and Port Authority and the Port of Singapore Authority to combat piracy and maritime terrorism. Besides, the Republic of Singapore Air Force provides maritime air surveillance supports. Singapore has continued to strengthen its coastal patrol capabilities over the last few years by purchasing new equipment and vessels (Ministry of Defence, 2000).

*China*

Since the mid-1990s, the South China Sea has become a popular region among pirates. More specifically, the Hong Kong-Luzon-Hainan Island triangle and waters off the Philippines have become the focal point of piracy. China was under great international pressure for either not taking effective legal action against pirates or providing a safe haven for pirates and their hijacked vessels.

One of the complaints against China was that it deports pirates, rather than prosecute them. On 16 April 1998, MV *Petro Ranger* sailed from Singapore with a cargo of gas, oil and kerosene. The vessel was boarded by armed pirates and hijacked. The crew members were threatened with death and kept in custody for ten days. The vessel sailed to Hainan Island in south China. At Hainan, the vessel was seized, the cargo confiscated and the legitimate crew of the vessel were detained and questioned for over two weeks. According to the Chinese authorities, the vessel was engaged in smuggling operations. The Indonesian pirates were sent back to Indonesia without being prosecuted. However, after that case, China has taken a very effective measure to suppress piracy. China successfully prosecuted many piratical cases, such as the "Tenyu", "Cheung Son" and "Marine Fortuner" cases. In the "Cheung Son" case, which was conducted in January 2000, 13 pirates were sentenced to death for the killing of 23 crew members in December 1998. China has gained an international reputation for its anti-piracy efforts. The piratical activities in China territorial waters have been effectively reduced (Zou, 2000).

In China, there are two government agencies directly in charge of anti-piracy operation: the Ministry of Public Security and the Ministry of Communications. The Ministry of Public Security is the law enforcement agency. Its two divisions, the Criminal Investigation Division and Coastal Police Division, take control of the related law enforcement operations. Criminal Investigation Division investigates the piracy cases while Coastal Police Division patrols, monitors and combats piracy in territorial waters.

The Maritime Security Administration of China Maritime Bureau, a division of the Ministry of Communications, is in charge of maritime information gathering and communication with the IMB Piracy Reporting Center.

## ASEAN and China's Anti-Piracy Cooperation

### ASEAN

There have been a number of bilateral and multilateral cooperation among the ASEAN countries.

In 1992, parallel bilateral agreements among Singapore, Indonesia and Malaysia coordinated naval patrols and conducted periodic anti-piracy exercises in the Malacca and Singapore Straits which resulted in a significant reduction of piracy attacks in the Straits for several years.

In 1997, ASEAN adopted a Plan of Action to combat trans-national crimes, and constituted a plan for combating piracy in the South-East Asia region. In the two ASEAN Regional Forum (ARF) Experts Group Meetings on Trans-national Crime held in October 2000 and April 2001, the experts recognized that piracy is an increasingly serious trans-national crime with regional security implications. The ARF adopted the Statement on Cooperation against Piracy and Other Threats to Maritime Security at its ministerial meeting in Phnom Penh in June 2003. The ARF participants agreed to undertake a series of measures to tackle the piracy problem, i.e., information sharing, the provision of technical assistance and capacity building to the states in need of equipment, and called for the adherence to "the Rome Convention on Suppression of Unlawful Acts against the Safety of Maritime Navigation (SUA Convention)" to combat piracy.

However, the developments of anti-piracy cooperation were placed on hold until the 9/11 terrorist attacks on New York and Washington, DC. Since then, the patrols in the Straits of Malacca were stepped up and piratical attacks decreased significantly between 2000 and 2001.

In 2002, Malaysia and the Philippine navies conducted a six-day joint military exercise, while Indonesia, Malaysia and the Philippines signed a trilateral security pact to bolster cooperation in trans-national crime issues (Agence France-Presse, 2002).

In February 2003, Indonesia and the Philippines conducted anti-piracy drills with Japan similar to those held between Malaysia and the Philippines (Associated Press, 2003).

On 20 July 2004, Malaysia, Singapore and Indonesia launched new coordinated patrols, i.e., "MALSINDO" in the Straits of Malacca, after US marines and Special Forces claimed that they would help ASEAN patrol it. The new coordinated patrols would involve year-round patrols with ships from the littoral states (Jakarta Post, 30 June 2004).

On 27 May 2005, the Republic of Singapore Navy (RSN) and the Indonesian Navy (TNI-AL) officially launched Project SURPIC (Surface Picture) in Batam. This is a sea surveillance system, which allows the RSN and TNI-AL to share common real-time information on the sea situation in the Singapore Strait. It will enable the two navies to monitor this busy waterway, exchange information, and deploy their patrol vessels in the area more effectively. The Project SURPIC is a significant step forward in enhancing security and the safety of shipping in the Singapore Strait. It strengthens the already close working relationship between the RSN and TNI-AL, and enhances the interoperability of the two navies to combat piracy and sea robberies (The Straits Times, 28 May 2005).

## China

China has supported and actively participated in cooperation with ASEAN countries on maritime security issue. In 2002, China-ASEAN reached consensus on the Declaration on the Conduct of Parties in the South China Sea, which not only reaffirmed both sides' commitment to maintain peace and stability of the region, but also the willingness to enhance cooperation in maritime environment protection, maritime transport and navigation safety, and fight against trans-national crimes at sea.

In accordance with the Joint Declaration of China and ASEAN on Cooperation in the Field of Non-traditional Security Issues, which was released at the 2002 China-ASEAN Summit, China and ASEAN signed a Memorandum of Understanding on Cooperation (MOU) on Non-traditional Security Issues in 2004. The objective of the MOU is clearly stated: "To develop practical strategies in accordance with their national laws and regulations to enhance the capacity of each individual country and the region as a whole in dealing with such non-traditional security issues as trafficking in illegal drugs, people smuggling including trafficking in women and children, sea piracy, terrorism, arms smuggling, money laundering, international economic crime and cyber crime" (ASEAN Secretariat, 2004).

In 2004, the China-ASEAN Prosecutors-General Conference was held in China, and both sides agreed to work together in the fight against crimes, including maritime trans-national crimes.

## Dilemma in Anti-Piracy Cooperation

Although so many efforts have been made to combat piracy, the IMB annual Piracy and Armed Robbery against Ships report for 2003 shows that piratical incidents continued to rise. Compared to 370 in 2002, the total number of piratical incidents increased to 445 in 2003, the second highest number of attacks since the IMB Piracy Reporting Centre commenced compiling statistics in 1991. Indonesia continued to record the highest number of attacks with 121 reported incidents in 2003, the figures of 2003 show a rise to 28 incidents in the Straits of Malacca. In 2004, there was an increase in the number of incidents to 60 in the Malacca Straits, compared with 36 in 2003 (IMB, 2005).

There are many reasons for the rising figures. However, the ineffectiveness of anti-piracy cooperation in Southeast Asia is the main reason. Many difficulties have not yet been solved in anti-piracy cooperation.

## Divergence of Perception, Incentive and Priority

### Perception

Most of the Indonesian scholars and officials regard piracy in the Straits of Malacca and Southeast Asian waters as petty theft rather than a serious issue. The Indonesian chief of navy, Admiral Bernard Kent Sondakh, argued that the piracy situation in the Straits of Malacca had been deliberately exaggerated, and that it was part of an international strategy to justify foreign intervention in Indonesia by portraying the country as weak and incapable of looking after its own waters (Sondakh, 2004, pp. 5–6).

Malaysia was inclined to view piracy as nothing more than a criminal act which does not pose a significant threat to the state system. As its Defence Minister and Deputy Prime Minister Najib Abdul Razak put it, "The Malaysian position was that the piracy in the Straits had been exaggerated." (Mak, 2004, p. 19).

Singapore's perception toward piracy is different from these two countries. Since Singapore's economy depends on international trade and international transportation, the maritime security in the Straits of Malacca is crucial to its national security. Piracy in the Straits of Malacca can directly affect its national security and economic prosperity. After the event of 9/11, the Singapore authorities are convinced that there exists the possible piracy and terrorism nexus. In December 2001, Singapore arrested 13 members of the *Jemaah Islamiyah* terrorist group, and uncovered that it had made preliminary plans to conduct suicide attacks on US military vessels entering Singapore harbor. Singapore thus expressed its great concern on the piracy issue (Emmers, 2004, p. 41).

China's economy has been booming for the last two decades. The sea lanes of communication and sea-borne commerce have become increasingly important for the country. More than 80% of China's energy imports pass through the Straits of Malacca. The maritime security in Straits of Malacca and South China Sea is becoming more and more important for Chinese economic and national security.

### Incentive

As the main Indonesian seaports are not along the Straits of Malacca, compared to those of Singapore and Malaysia, Indonesia does not benefit as much from it. However, Indonesia has to take the responsibility of safeguarding maritime security in the Malacca Straits. At present, maritime users of this waterway appear unwilling to acknowledge that they should contribute to its security. In the eyes of Indonesian scholars and policy makers, foreign ships passing through the Straits of Malacca are "free riders" who do not pay the cost of maintaining safety in the Straits. Even the Indonesian government knows that to combat maritime piracy will contribute towards the long-term interests of the country's maritime economy, but the immediate cost of establishing and maintaining an effective anti-piracy task force is still too high. It was estimated that it would cost Indonesia about US$38.5 million to implement anti-piracy measures between the Strait of Singapore and Jakarta (Djala, 2004, p. 424). Thus Indonesian government is reluctant to spend more money on anti-piracy operations.

Another probable reason for Indonesia's lack of incentive to spend money on safeguarding the Straits of Malacca is that, if it were closed, Indonesia will benefit from the use of two alternative Indonesian waterways for international shipping — the Sunda Strait, and the Lombok and Makassar Straits (Richardson, 2004, p. 38). Both straits are deep-water channels and totally controlled by Indonesia, but using the route through the Lombok and Makassar Straits and Sulu Sea would raise the cost of transportation by over 6% for east-bound ships (Naidu, 1997, p. 35).

Malaysia's economy relied heavily on trade and the sea lines of communications in the Straits of Malacca and the South China Sea. Malaysia's maritime industries include offshore exploitation of resources and sea-based tourism industries, which together with the manufacturing sector contribute significantly to the well-being of Malaysia's economy. From the perspective of internal trade, Malaysia's two geographically separated land masses require the safe passage through the South China Sea. Safeguarding the security in the Straits of Malacca and Southeast China Sea is an important task for the Malaysian government.

The Port of Singapore (PSA) is the regional premier container port. On average, Singapore attracts some 140,000 vessel calls annually. PSA is the world's single largest container terminal owner-operator. It handles about one tenth of the world's container business (Chia *et al.*, 2003, p. 32). Maritime security is therefore critical to Singapore's survival, and it is very enthusiastic about anti-piracy cooperation.

The demand of oil import is increasing substantially with China's economic development, as mentioned before. With a high percentage of its oil exports being transported through the Straits of Malacca and South China Sea, China is thus actively participating in anti-piracy cooperation with ASEAN countries.

*Priority*

Combating piracy is not on Indonesia's priority list. Its priorities are to prevent conflict between the provinces and between the districts; to suppress the Free Aceh Movement (GAM) separatist's movement in Aceh; and to prevent illegal fishing, which is causing the Indonesian government to lose an estimated US$5.0 billion a year (Djala, 2004, p. 423).

Malaysia's primary concern is state sovereignty. This issue will be discussed later in this paper.

For Singapore, as a vulnerable city-state, anti-piracy is high on its priority list. It is convinced that terrorists will attack the heavy traffic congesting the entrance to Singapore port. Thus Singapore had proposed that US Marines and Special Forces patrol the Straits of Malacca.

China's priority is on its economic development and the national unity. Though anti-piracy is not on its priority list, piracy in the Straits of Malacca and South China Sea is a potential threat to its oil imports and its shipping fleet. Furthermore, if there were war across the Taiwan Straits in the near future, the Straits of Malacca will be the crucial strategic sea line for victory. Thus maritime security has been attracting more and more attention of Chinese policy makers.

## Capabilities of Law Enforcement Agencies

Indonesia is a large maritime and archipelagic state, which controls about three million square kilometers of archipelagic waters and territorial seas, and nearly three million square kilometers of EEZ and continental shelf. Several important sea lanes for the communications between the Pacific and the Indian Oceans pass through its territorial waters. It is estimated that Indonesia needs more than 300 vessels to protect its maritime space, port facilities and natural resources. However, Indonesia now only has about 115 vessels, and out of these, only about 25 can be operated at sea at the same time. To make matter worse, half of the Indonesian air force is grounded due to lack of funding, which further undermines Indonesia's capability to ensure its maritime security and anti-piracy operations (Emmers, 2004, p. 46).

The low defence budget is one of the reasons for its weak law enforcement capabilities. Indonesia's defence spending is among the lowest in the region, either in terms of dollar value or in relation to the GDP. For example, Singapore spent US$4.2 billion for military spending in 1999 or roughly 24.9% of the total government spending at that time, while Indonesia spent only US$1.5 billion in the same period or only 5.9% of the total government spending. Indonesia does not have any extra fund for anti-piracy purpose, for it is already accumulating billions of dollars in foreign debt (Djalal, 2004, p. 424).

Furthermore, rivalry over jurisdiction between the Indonesian police and navy also undermines the efforts to prevent piracy attacks. The police have full jurisdiction over piracy incidents in its territorial waters, while the navy has more enforcement capabilities but lacks the power of arrest. The navy wants to have a share of the jurisdiction. Kusanato Anggoro, a military analyst at the Centre for Strategic and International Studies (CSIS) in Jakarta, points out that "jurisdiction is territory, territory means authority and authority means resources and money" (The Strait Times, 23 July 2002). It was reported that many corrupt officials in the law enforcement agencies were involved in piracy activities.

*National Sovereignty Concern*

Bilateral and multilateral anti-piracy cooperation is hampered by the consideration on national sovereignty. Indonesia and Malaysia are worried that anti-piracy cooperation will erode their sovereignty. Malaysia states that anti-piracy cooperation should not be used as an excuse to compromise the country's sovereignty (Mak, 2004). Mohamed Nazri Abul Aziz, a Minister in the Prime Minster's Department, said, "The safety of the Straits of Malacca is important. If it is not guarded properly, foreign powers may be prone to intervene in its management and this will pose a threat to Malaysia's sovereignty." (Ramachandran, 16 June 2004).

The notion is reflected in bilateral and multilateral anti-piracy patrols among Indonesia, Malaysia and Singapore. These are coordinated patrols, not joint patrols. Coordinated patrol means that the law enforcement agencies can only patrol in their own territorial waters. It does not allow for "hot pursuits" into a neighbor's territorial waters. The Malaysian MMEC declared, "Under no circumstances would we intrude into each other's territory. If we chase a ship and it runs into the other side, we let the authorities there handle it." (Chalk, 1997).

The concept of coordinated patrols was criticized as "a potential cause for confusion, inefficiency and misallocation of resources" (Gatsiounis, 26 July 2004). IMB described the system as "ridiculous". Under this system, even if Malaysia and Singapore have done their best to combat piracy, they still cannot help Indonesia to solve its insufficient capability problem. The pirates take advantage of jurisdictional limitations and often commit their crimes in territorial waters of one state, then flee into another country's territorial waters. In a nutshell, anti-piracy cooperation in Southeast Asia cannot be effective in this kind of situation.

The SUA Convention, which was endorsed in 1988 in response to the 1985 *Achille Lauro* terrorist incident, provides a legal structure for inter-state cooperation against maritime terrorism and piracy. It empowers signatory states to extradite or prosecute pirates arrested in their territorial waters for crimes committed under the jurisdiction of the other countries. It aims to remove the issue of jurisdiction that has often stopped states from prosecuting pirates who found sanctuary in their territories. Indonesia and Malaysia have not ratified the convention so far.

China is the signatory state of the SUA Convention, and is ready to cooperate with ASEAN countries to combat piracy and build an enduring and stable regional maritime security environment. However, some ASEAN countries are suspicious of China's active participation.

## Conclusion

Maritime piracy is a trans-national crime in contemporary Southeast Asia waters. Unilateral anti-piracy operation is not sufficient to suppress piracy, and it requires consistent and comprehensive multilateral cooperation at the government level.

However, finding ways to bridge the divergence of perception, incentive and priority toward piracy, improve the law enforcement capabilities and deal with the national sovereignty issue, is still an intractable mission testing the political will and wisdom of the leaders of ASEAN countries and China.

## References

ASEAN Secretariat, "Memorandum of Understanding between the Governments of the Member Countries of the Association of Southeast Asian Nations (ASEAN) and the Government of the People's Republic of China on Cooperation in the Field of Non-Traditional Security Issues", http://www.aseansec.org/15647.htm (accessed 23 November 2004).

Chalk, Peter, "Contemporary Maritime Piracy in Southeast Asia", *Studies in Conflict and Terrorism*, Vol. 21. No. 2, pp. 87–112 (16 March 1997).

Chia, Lin Sien, et al., *Southeast Asian Regional Port Development: A Comparative Analysis* (Singapore: Institute of Southeast Asian Studies, 2003).

Djala, Hasjim, "Piracy and Challenges of Cooperative Security and Enforcement Policy", *The Indonesian Quarterly*, Vol. XXX. No. 3, pp. 106–116 (2002).

———, "Piracy In Southeast Asia: Indonesian and Regional Responses", *Indonesian Journal of International Law*, Vol. 1. No. 3, pp. 419–440 (April 2004).

Ellen, Eric, ed., *Violence at Sea: A Review of Terrorism, Acts of War and Piracy, and Countermeasures to Prevent Terrorism* (Paris: ICC Publishing SA, 1986).

———, *Piracy at Sea* (Paris: ICC Publishing SA, 1989).

Emmers, Ralf, *Non-Traditional Security in Asia-Pacific: The Dynamics of Securitisation* (Singapore: Marshall Cavendish International, 2004).

Fox, Grace, *British Admirals and Chinese Pirates: 1832–1869* (London: Kegan Paul, Trench, Trubner & Co. Ltd, 1940).

Gatsiounis, Ioannis, "Strait: Target for Terror", *Asia Times Online*, http://www.atimes.com/ptatimes/Southeast_Asia/FH11Ae02.html (accessed 12 October 2004).

IMB, *Piracy and Armed Robbery Against Ships: Annual Report* (UK: ICC International Maritime Bureau, 1991–2005).

"Intelligence Centre to Curb Sea Piracy", *The Straits Times* (29 April 2005).

Keppel, Henry, *The Expedition to Borneo of HMS Dido for the Suppression of Piracy: With Extracts from the Journal of James Brooke* (London: Cass, 1968; original work published in 1846 by London: Chapman and Hall).

Mak, Joon Num, "Unilateralism and Regionalism: Working Together and Alone in the Malacca Straits", paper presented at Workshop on Maritime Security, Maritime Terrorism and Piracy in Asia Singapore (September 2004).

"Malaysia, RP Begin Maritime Exercises", *Jakarta Post* (13 May 2002).

"Malaysia Ships Up Patrols in the Straits Against Terror Attacks", *Agence France-Presse* (31 October 2002).

"Maritime Agency to Have 97 Vessels", *The New Straits Times* (15 May 2005).

Miller, H., *Pirates of the Far East* (London: Hale, 1970).

Ministry of Defence, *Defending Singapore in the 21st Century* (Singapore: 2000).

Moorhead, F. J., *A History of Malaya and Her Neighbours* (Kuala Lumpur: Longmans of Malaysia, 1957).

Murray, Dian H., *Pirates of the South China Coast 1790–1810* (California: Stanford University Press, 1987).

Naidu, G., "The Straits of Malacca in the Malaysian Economy", in A. Hamzah, ed., *The Straits of Malacca: International Cooperation in Trade, Funding and Navigational Safety* (Petaling Jaya: Maritime Institute of Malaysia and Pelanduk Publications, 1997).

Ramachandran, Sudha, "Divisions over Terror Threat in Malacca Straits", *Asia Times Online*, http://www.atimes.com/atimes/Southeast_Asia/FF16Ae01.html (accessed 14 October 2004).

Richardson, Michael, *A Time Bomb for Global Trade: Maritime-related Terrorism in an Age of Weapons of Mass Destruction* (Singapore: Institute of Southeast Asian Studies, 2004).

Rutter, Owen, *The Pirate Wind: Tales of the Sea-Robbers of Malaya* (Oxford University, 1986).

Sondakh, Bernard Kent, "National Sovereignty and Security in the Straits of Malacca", paper presented at the Conference on Straits of Malacca: Building a Comprehensive Security Environment, Kuala Lumpur, Malaysia (October 2004).

Tarling, Nicholas, *Piracy and Politics in the Malay World* (Singapore: Donald Moore Gallery, 1963).

———, *The Cambridge history of Southeast Asia Vol. 1* (UK: Cambridge University Press, 1992).

UNHCR, *The State of World's Refugees 2000: 50 Years of Humanitarian Action* (Oxford University Press, 2000).

Warren, James Francis, *Iranun and Balangingi: Globalisation, Maritime Raiding and the Birth of Ethnicity* (Singapore: Singapore University Press, 2002).

Young, A. J. and Valencia, M. J., "Conflation of Piracy and Terrorism in Southeast Asia: Rectitude and Utility", *Contemporary Southeast Asia*, Vol. 25. No. 2. pp. 269–283 (August 2003).

Zheng, Guangnan, *The History of Chinese Piracy* (China: Huatong Ligong University Press, 1998).

Zou, Keyuan, "Piracy at Sea and China's Response", *Lloyd's Maritime and Commercial Law Quarterly*, pp. 364–382 (August 2000).

# Part IV

## Towards a Free Trade Area

## Chapter 6

# WTO Rules and China-ASEAN FTA Agreement

*ZENG Huaqun*

The Framework Agreement on Comprehensive Economic Co-operation (FACEC) between the Association of Southeast Asian Nations (ASEAN) and the People's Republic of China (China) was concluded on 4 November 2002 and ratified on 1 July 2003.[1] The agreement would lead to the creation of China-ASEAN Free Trade Area (FTA) within ten years[2] and pave the way towards closer cooperation in the elimination of tariffs and non-tariff barriers on trade in goods and services, create an open and competitive investment regime and encourage the simplification and standardization of customs procedures.[3] As China and seven out of the ten member states of ASEAN were full members of World Trade Organization (WTO) at that time, and the provisions of FACEC involve liberalization of trade in goods and services as well as investments (the same subjects as those regulated by WTO rules) among the parties, the parties to FACEC have to deal with the issues on the application of WTO rules in the agreement. This paper deals with the issues on characteristics of the parties to FACEC, the WTO responsibilities of parties to

---

[1] The Framework Agreement on Comprehensive Economic Cooperation between the Association of Southeast Asian Nations and the People's Republic of China. The extracts of the Agreement are available from *China: An International Journal*, 1(1) pp. 170–178 (2003). For the brief history of FACEC, see Qingjiang Kong, "China's WTO Accession and the ASEAN China Free Trade Area: the Perspective of a Chinese Lawyer", *Journal of International Economic Law*, 7(4), pp. 802–803 (2004).

[2] The timeframes of the negotiations on the Agreement are: for trade in goods, negotiations on the Agreement for tariff reduction/elimination shall commence in early 2003 and concluded by 30 June 2004 in order to establish the FTA for trade in goods by 2010 for China and ASEAN6 and by 2015 for newer ASEAN member states; for trade in services and investments, negotiations shall commence in 2003 and conclude as expeditiously as possible in accordance to the timeframes to be agreed; and for other areas of cooperation, the parties shall continue to build on existing programs, develop new programs and conclude agreements expeditiously for each implementation. See Briefing Paper on the Establishment of the ASEAN-China Free Trade Area (FTA), 18 February 2003, at http://www.tariffcommission.gov.ph/briefing_paper_fta.htm12/23/2003.

[3] ASEAN Secretariat, News Release, 30 October 2002, ASEAN-China Free Trade Area Negotiations to Start Next Year, http://www.aseansec.org/13125.htm.

FACEC and provisions of FACEC within or beyond WTO regime, and then tries to explore and indicate the relationship between WTO rules and the development of FTA agreement between China and ASEAN.

## Parties to FACEC: Facts and Differences

According to the general principle of international law, parties to a treaty are only bound by the treaty they have entered into. Then, in the context of FACEC, the important questions are why and how the parties to FACEC relate to WTO rules? Do they bear any WTO obligations when they conclude and/or implement FACEC?

It is clear that the legal status of the two parties to FACEC is different in WTO regime. China is a WTO full member while ASEAN is not. For ASEAN, it therefore has no obligation to observe WTO rules in general.[4] Considering the complexity of ASEAN as an international economic organization and the unique position of China in the WTO regime, the characteristics of the two parties to FACEC are the first and most important issues for this study. The characteristics of these two parties might be summarized as follows.

### One ASEAN, Two Groups of Member States

For ASEAN member states, they are divided into two groups based on two criteria provided in the text of FACEC: one group comprises WTO members and non-WTO members by the criterion of WTO membership; and the other group consists of "ASEAN6" and the newer ASEAN member states by the criterion of level of development to a large extent.

### WTO members and non-WTO members

First, we shall distinguish between the WTO members and non-WTO members of ASEAN member states in the context of FACEC. Among the ten ASEAN member states, seven were WTO members and three were not at the point of concluding FACEC. The seven WTO members were Brunei, Indonesia, Malaysia, Myanmar, Philippines, Singapore and Thailand. Obviously they have their commitment to WTO, though ASEAN itself does not bear any obligation to WTO.

The three non-WTO members were Cambodia, Laos and Vietnam. Cambodia became the eighth WTO member among member states of ASEAN on

---

[4]It is possible that ASEAN could become a *de facto* or even a *de jure* member of WTO after all its ten member states have joined WTO. EC is a pioneer in this regard in GATT/WTO regime.

13 October 2004.[5] Both Vietnam and Laos are currently observers of WTO and are in the process of accession to WTO. Vietnam has made great progress in the process[6] while Laos' process is relatively slow.[7] For them, currently there are no other obligations and responsibilities to WTO except those they are already committed to. However, as they are applying for accession to WTO, they have potential obligations and responsibilities to WTO.

In FACEC, there are different provisions for WTO members and non-WTO members of ASEAN member states.

According to Article 3(b) of FACEC, "applied MFN tariff rates" shall: "(i) in the case of ASEAN member states (which are WTO members as of 1 July 2003) and China, refer to their respective applied rates as of 1 July 2003; and (ii) in the case of ASEAN member states (which are non-WTO members as of 1 July 2003), refer to the rates as applied to China as of 1 July 2003."[8] It is noticeable that for the ASEAN member states (which are non-WTO members as of 1 July 2003), the rates as applied to China as of 1 July 2003 are not multilateral "MFN tariff rates" since they are not bound by WTO treaties.

The results of the above provisions are that for WTO members in ASEAN, they enjoy the benefits from the regional trade agreement (RTA); for non-WTO members in ASEAN, they also enjoy the treatment of WTO members, as China shall accord MFN treatment consistent with WTO rules and disciplines to them.[9]

## "ASEAN6" and the newer ASEAN member states

The formation of "Great ASEAN" marked the important development of the RTA in the region. At the same time, ASEAN members are divided into developed

---

[5] http://www.wto.org/english/thewto_e/countries_e/cambodia_e.htm2004/10/19.
[6] The Working Party on the accession of Vietnam was established on 31 January 1995. Bilateral market access contacts have been initiated. Topics under discussion in the Working Party include: agriculture, customs system, import licensing, national treatment, SPS and TBT, state trading, trading rights and TRIPS. The sixth meeting of the Working Party took place on 12 May 2003, and the seventh meeting was held on 12 December 2003. The 63 WTO members negotiating a membership package with Vietnam praised the Southeast Asian country on 15 June 2004 for considerably improving its market access for goods and services, and its programme for applying WTO agreements. See http://www.wto.org/english/thewto_e/acc_e/a1_vietnam_e.htm2004/10/19.
[7] The Lao People's Democratic Republic's Working Party was established on 19 February 1998. The Memorandum on the Foreign Trade Regime was circulated in March 2001 and replies to questions from Members made in October 2003. The Working Party has not yet met. See http://www.wto.org/english/thewto_e/acc_e/a1_laos_e.htm2004/10/19.
[8] There is a same provision in Article 6 Early Harvest of FACEC.
[9] Article 9 of FACEC.

members and less developed members.[10] "ASEAN6" refers to the six original ASEAN member states of Brunei, Indonesia, Malaysia, Philippines, Singapore and Thailand. The newer ASEAN member states refer to the four less developed member states of Cambodia, Laos, Myanmar and Vietnam.[11] Three of the four newer ASEAN member states, namely Cambodia, Laos and Myanmar, are on the list of 48 least developed countries.[12] The latter will be offered special and differential treatment and flexibility in implementation of FACEC.[13] The memorandum of understanding on agricultural cooperation was signed separately. It will be especially helpful to the newer ASEAN member states.

There are different requirements for "ASEAN6" and the newer ASEAN member states. Considering the internal differences, liberalization will be done by different tracks and different timetables. In fact, FACEC will serve as the fulcrum for establishing China-ASEAN FTA by 2010 for China and "ASEAN6", and by 2015 for China and the four newer ASEAN member states.[14]

It should be further pointed out that the ambiguous legal personality[15] and division of ASEAN members into two groups do not refute the bilateral nature of FACEC. In fact, as the treaty deals with comprehensive economic cooperation between ASEAN and China, it is clear that one contracting party is China, while the other parties are the ten ASEAN members taken as a whole. However, the fact that all the names of ASEAN members and China appear in FACEC as the contracting parties except ASEAN, and the provision that "the parties undertake to complete their internal procedures for the entry into force of this Agreement"[16] indicate

---

[10] For the formation of Great ASEAN and its related impacts, see Chen Ning, "The Formation of Great ASEAN and the Problems Confronting us in the Future" (in Chinese), *Southeast Asian Studies*, 3, pp. 22–24 (1998).

[11] Briefing Paper on the Establishment of the ASEAN-China Free Trade Area (FTA), 18 February 2003, http://www.tariffcommission.gov.ph/briefing_paper_fta.html12/23/2003.

[12] http://www.wto.org/english/thewto_e/whatis_e/tif_e/org7_e.htm2004/10/19.

[13] ASEAN Secretariat, News Release, 30 October 2002, ASEAN-China Free Trade Area Negotiations to Start Next Year, http://www.aseansec.org/13125.htm.

[14] Article 3(4)(a) of FACEC deals with the "Normal Track" and stipulates that products listed in the Normal Track by a party on its own accord shall have their respective applied MFN tariff rates gradually reduced or eliminated in accordance with specified schedules and rates (to be mutually agreed by the parties) over a period from 1 January 2005 to 2010 for ASEAN6 and China, and in the case of the newer ASEAN member states, the period shall be from 1 January 2005 to 2015 with higher starting tariff rates and different staging.

[15] Some scholars challenged the legal personality and treaty-making power of ASEAN and concluded that FACEC is a multilateral treaty among 11 countries rather than a bilateral treaty between China and ASEAN. See Jiangyu Wang, "Two International Legal Issues in the China-ASEAN Free Trade Agreement", paper for Symposium on China's Relations with ASEAN: New Dimensions, 3–4 December 2004.

[16] Article 16 of FACEC.

that the legal obligations of complying with the treaty may be borne by up to ten ASEAN members collectively or individually.[17]

## One China, Four WTO Memberships

In general, China's membership in WTO has greatly extended the horizon of China's regional economic cooperation and brought ASEAN members together within the multilateral trading system.[18]

It is very important to clarify and emphasize that China in FACEC context only refers to the Chinese mainland, not including Hong Kong Special Administrative Region, Macao Special Administrative Region and Chinese Taipei.

On 11 December 2001 and 1 January 2002, China and Separate Customs Territory of Taiwan, Penghu, Kinmen and Matsu (Chinese Taipei) became members of WTO one after another. Together with Hong Kong and Macao, the two original WTO members since 1 January 1995, the "one China, four WTO memberships" situation has emerged, and this is the new phenomenon in the General Agreement on Tariffs and Trade (GATT)/WTO history.[19]

To ASEAN and its member states, this phenomenon has great significance. As equal members under WTO regime, China, Hong Kong, and Macao may separately establish a customs union (CU) or FTA or negotiate and conclude an agreement necessary for establishing a CU of FTA with ASEAN and/or one or more than one of its member states in accordance with Article 24 of GATT.

Recently, the authorities of Taiwan tried to conclude FTA agreements with other countries like US, Japan, Singapore, New Zealand and Central American countries in the name of "Taiwan" instead of its formal name of "Separate Customs Territory of Taiwan, Penghu, Kinmen and Matsu (Chinese Taipei)" in WTO.[20] It also proposed establishing a Free Trade Agreement with the 10-member ASEAN nations.[21] It reflects the "creeping independence" or "*de facto* independence" policy of Taiwan's authorities. It should be emphasized that since 1949, China's position has been that any state or international organization, which wishes to deal with China, should recognize the government of PRC as the sole legitimate government

---

[17] This is a new form of bilateral treaty, better than EU's mixed treaty.
[18] See Qingjiang Kong, "China's WTO Accession and the ASEAN China Free Trade Area: the Perspective of a Chinese Lawyer", *Journal of International Economic Law*, 7(4), p. 810 (2004).
[19] Zeng Huaqun, "On the One Country Four Memberships in WTO Regime" (in Chinese), *Xiamen University Journal*, 5, pp. 5–14 (2002).
[20] Zhu Lei, "How Taiwan Concludes 'Free Trade Agreements'", *Kai Fang Chao*, 7, pp. 24–25 (2002).
[21] On 12 November 2002, Taiwan's "Economic Minister" Lin Yifu made the suggestion in Malaysia. Yow Cheun Hoe, "China-ASEAN Relations in 2002: Chronology of Events", *China: An International Journal*, 1(1), p. 168 (2003).

of China and Taiwan as an inalienable part of China. This "one China" principle has been generally accepted and recognized by the international community. The Taiwan issue, therefore, is completely an internal affair of China and will be settled by Chinese themselves. There is no legal ground for any foreign country, international organization or other entity to engage in the Taiwan issue in accordance with international law or international practice.[22] The WTO membership of Chinese Taipei does not bring any change in the legal status of Taiwan. ASEAN and its member states shall adhere to the "one China" principle consistently and shall not set up any official relations with Taiwan.

At the same time, China, Hong Kong, Macao and/or Chinese Taipei may also establish a CU or FTA mutually, tripartitely or quadruply. The significance of a CU or FTA among China, Hong Kong, Macao and/or Chinese Taipei under WTO regime is to deal with the special problems in the application of the WTO non-discrimination principle, setting up a necessary legal basis for closer economic relations among the four regions. Considering China's "one country two systems" policy, there are big differences among China, Hong Kong, Macao, or Chinese Taipei in political, economic and legal systems. At the economic level, establishing a FTA, rather than a CU, is more suitable for them. Before setting up a FTA, they have to negotiate and conclude an agreement necessary for it. This arrangement may also have impact on ASEAN and its member states. In fact, both of the Closer Economic Partnership Arrangements (CEPAs) between the Chinese mainland and Hong Kong and between Chinese mainland and Macao came into effect from 1 January 2004. The potential CEPAs among the Chinese mainland, Hong Kong, Macao and even Chinese Taipei leading to so-called "great China FTA" certainly provide opportunities for further expansion of FACEC. It is probably logical that the great China FTA will be established before the completion of the China-ASEAN FTA.

## WTO Responsibilities of Parties to FACEC

For identifying the WTO responsibilities of parties to FACEC, we shall first study WTO rules on RTAs and the differences between RTAs and WTO rules.

### *WTO Rules on RTAs*

The non-discrimination principle, one of the basic principles of WTO, stipulates that trade among WTO members should be conducted without any discrimination,

---

[22] However, it is worth noting that there are some comments on the topic in Western countries from time to time. Most recently, Jonathan I. Charney and J. R. V. Prescott, "Resolving Cross-Strait Relations between China and Taiwan", *American Journal of International Law*, 94, pp. 452–477 (2000).

requiring all members to extend the "most-favored nation (MFN)" treatment and "national treatment" to other members, except in certain circumstances detailed otherwise in GATT/WTO.

Establishing a RTA among some WTO members is likely conflict with the non-discrimination principle of WTO, as the treatment for members (including WTO members) of the RTA might be more favorable than that of outsiders (including WTO members). Therefore, WTO members who wish to establish a RTA have to seek exception for the non-discrimination principle, so as to avoid violation of this principle.

As the provisions of GATT "shall not prevent, as between the territories of contracting parties, the formation of a customs union or of a free-trade area or the adoption of an interim agreement necessary for the formation of a customs union or a free-trade area",[23] a CU and a FTA, as typical models of RTAs, become exceptions of the non-discrimination principle with certain conditions. Following the GATT practice, WTO allows the existence of its members' CU and FTA with some detailed rules. The key provisions on RTAs in GATT/WTO regime are as follows:

*Article 24 of GATT and understanding of Article 24*

Article 24(8) of GATT makes clear definitions of a CU and a FTA.[24] It seems that the key difference between these two is that members of a CU have to apply substantially the same duties and restrictive regulations to the trade of territories not included in the union, while these requirements are not applicable for members of a FTA. In fact, a CU, a FTA and an interim agreement necessary for the formation of a CU or a FTA[25] are different types or levels of RTAs. Among them, a CU is at the top level, a FTA comes second, while an interim agreement necessary for the formation of a CU or a FTA is at the primary stage of RTAs.

---

[23] Article 24(5) of GATT.
[24] According to the provisions, "A customs union shall be understood to mean the substitution of a single customs territory for two or more customs territories, so that (i) duties and other restrictive regulations of commerce (except, where necessary, those permitted under Articles XI, XII, XIII, XIV, XV and XX) are eliminated with respect to substantially all the trade between the constituent territories of the union or at least with respect to substantially all the trade in products originating in such territories", and "substantially the same duties and other regulations of commence are applied by each of the members of the union to the trade of territories not included in the union". "A free-trade area shall be understood to mean a group of two or more customs territories in which the duties and other restrictive regulations of commence (except, where necessary, those permitted under Articles XI, XII, XIII, XIV, XV and XX) are eliminated on substantially all the trade between the constituent territories in products originating in such territories".
[25] A comment indicates that the interim agreement is a form or a level for RTA. See Zhao Weitian, *The Legal Systems of WTO* (in Chinese) (Jilin People Press, 2000), pp. 84–85.

The Understanding on the Interpretation of Article 24 of the General Agreement on Tariffs and Trade 1994 (Understanding on Article 24) was formulated as part of the Final Act of the Uruguay Round. It clarifies some rules of Article 24 of GATT. However, the Understanding has not resolved the more difficult issues, and it was hoped that further clarification would come from the Committee on Regional Trade Agreements (CRTA),[26] which also has in its mandate the responsibility "to consider the systemic implications of such agreements and regional initiatives for multilateral trading system and the relationship between them, and make appropriate recommendations to the General Council".[27]

For RTAs, two major principles are laid down in Understanding on Article 24: one is that a RTA must cover substantially all trade; the other is that other countries must not be worse off as a result of a RTA.[28] In addition, according to Article 24 and Understanding on Article 24, the principles and rules on a FTA and/or an interim agreement necessary for the formation of a FTA are mainly as follows:

1. The purpose of a FTA should be to facilitate trade between the constituent territories and not to raise barriers to the trade of other contracting parties with such territories.[29]
2. The duties and other regulations of commerce maintained in each of the constituent territories and applicable to the formation of such FTA, or the adoption of such interim agreement to the trade of contracting parties not included in such area or not parties to such agreement, shall not be higher or more restrictive than the corresponding duties and other regulations of commerce existing

---

[26] The CRTA was established in February 1996. The mandate of the Committee is to carry out the examination of agreements referred to it by the Council for Trade in Goods, the Council for Trade in Services and the Committee on Trade and Development. The CRTA is also charged to make recommendations on the reporting requirements for each type of agreement and to develop procedures to facilitate and improve the examination process.

[27] Jo-Ann Crawford and Sam Laird, "Regional Trade Agreements and the WTO", http://www.tradeobservatory.org/library12/23/2003.

[28] The "substantially all trade" requirement ensures that RTA is not abused as a cover for narrow sectoral discriminatory arrangement. The same requirement is also laid down in the Article 5 of GATS which requires a "substantially sectoral coverage" in services. Unfortunately, there is not agreement on the meaning of these wordings among WTO members, and in fact some RTAs omit from their coverage major and sensitive sectors such as agriculture and textile. The "not worse off" requirement is to ensure that non-participating members are not harmed by the RTAs. See Qingjiang Kong, "China's WTO Accession and the ASEAN China Free Trade Area: the Perspective of a Chinese Lawyer", *Journal of International Economic Law*, 7(4), pp. 811–812 (2004).

[29] Article 24(4) of GATT.

in the same constituent territories prior to the formation of the FTA, or interim agreement as the case may be.[30]
3. Any interim agreement leading to the formation of a FTA shall include a plan and schedule for the formation of such a FTA within a reasonable length of time.[31] The "reasonable length of time" should exceed ten years only in exceptional cases.[32]
4. Any contracting party deciding to enter into a FTA or an interim agreement leading to the formation of a FTA shall promptly notify the other contracting parties and make available to them such information regarding the proposed FTA as will enable them to make such reports and recommendations to contracting parties as they may deem appropriate.[33] All such notifications shall be examined by a working committee, in the light of the relevant provisions of GATT and of paragraph 1 of Understanding on Article 24, that will submit a report to the Council for Trade in Goods on its findings in this regard.[34]
5. The provisions of Articles 22 and 23 of GATT as elaborated and applied by the Dispute Settlement Understanding may be invoked with respect to any matters arising from the application of those provisions of Article 24 relating to FTAs or interim agreements leading to the formation of a FTA.[35]

The relevant practice shows that the systems of pre-examination, supervision and dispute settlement on RTAs have been established and operated properly, though there are still some defects.[36]

*Article 5 of GATS*

The provisions of Article 5 of General Agreement on Trade in Services (GATS) comform with the principles of Article 24 of GATT. It adopts the concept of "economic integration", and provides that "this Agreement shall not prevent any of its members from being a party to or entering into an agreement liberalizing trade in services between or among the parties to such an agreement", provided that such an agreement meet certain conditions. WTO Rules on RTAs in the field of trade in goods therefore extend to the field of trade in service.

---

[30] Article 24(5)(b) of GATT.
[31] Article 24(5)(c) of GATT.
[32] Para. 3 of Understanding on Article 24.
[33] Article 24(7)(a) of GATT.
[34] Para. 7 of Understanding on Article 24.
[35] Para. 12 of Understanding on Article 24.
[36] Zhong Liguo, "On the Regulation of WTO on the Regional Trade Agreements on its Completion" (in Chinese), *Jurists' Review*, 4, pp. 156–157 (2003).

Most recently, these have attempted to identify the issues arising from the interpretation and application of individual provisions of Article 5 of GATS and the possible linkages between Article 5 of GATS and Article 24 of GATT, particularly with regard to the scope of "substantial sectoral coverage" and "substantial all discrimination".[37]

*The enabling clause*

The GATT contracting parties adopted a decision in 1979, allowing derogations to the most-favored nation treatment in favor of developing countries in trade in goods. This includes differential and more favorable treatment, reciprocity and fuller participation of developing countries, and provisions related to the establishment of preferential trade agreements among developing countries, granting the possibilities of not meeting the substantially-all-trade requirements.[38]

From the contents of FACEC, it is obvious that FACEC in nature is an interim agreement necessary for the formation of a FTA and meets the requirement of Article 24 of GATT, Understanding on Article 24 and Article 5 of GATS. As China and member states of ASEAN are all developing countries, the enabling clause is also applicable to them.

*WTO Rules Versus RTAs*

Even for a WTO member, it only bears obligations within the ambit of WTO. It seems that for a WTO member who is at the same time a member of RTA, the key questions are: what are WTO members' obligations within the ambit of WTO; what is the relationship between RTAs and WTO rules?

The objects of WTO rules and RTAs are different. They are not identical or overlapping. The main reasons are that firstly, the WTO and RTA systems have been developing concurrently and have not been subordinate to each other. Secondly, WTO is a trade organization, regulating the trade relations among its members, and a RTA is an economic integration arrangement, regulating the economic (including trade) relations among its parties. Although the fields governed by WTO rules have been extended greatly after the Uruguay Round, WTO rules certainly cannot cover all economic sectors of its members. In fact, the fields governed by RTAs might be

---

[37] Jo-Ann Crawford and Sam Laird, "Regional Trade Agreements and the WTO", http://www.tradeobservatory.org/library12/23/2003.

[38] Sung-Hoon Park, "The Current Status and Future Prospects of Regionalism and Multilateralism in the World Economy: A Case Study of Economic Relations Between EU and APEC in the WTO Era", pp. 4–5, http://www.ecsanet.org/conferences/ecsaworld3/park.htm12/23/2003.

wider than those governed by WTO rules.³⁹ Thirdly, even in the same fields covered by WTO rules and RTAs,⁴⁰ the WTO rules could not replace RTAs, due to inherent defects of a multilateral framework. Fourthly, a membership of WTO is not a prerequisite for parties to a RTA.⁴¹ Some parties of a RTA who are not WTO members have not any obligations to WTO. Finally, many RTAs have been established before GATT or WTO. To some extent, the rules on RTAs or GATT/WTO are recognition or compromise to the reality of existing RTAs. Therefore a RTA concluded by WTO members is not necessarily required to be consistent fully with WTO rules or only limited in the fields governed by WTO rules. The only requirement for WTO members is that those subject to WTO rules in a RTA shall be consistent with WTO rules.⁴² In other words, when a WTO member engages in a RTA, it should fulfill its WTO obligations.

To a member of WTO, it is very important to distinguish between the concepts of "within the ambit of WTO" and "beyond the ambit of WTO". Within the ambit of WTO, it is doubtless that all members have to observe WTO rules and fulfill their commitments to WTO seriously. At the same time, as all the duties and obligations of all WTO members are limited to their commitments to WTO, beyond the ambit of WTO, members may regulate their trade relations through bilateral or RTAs in accordance with general international law.

There is some other Legal Regime for FACEC. Article 12A of Protocol to Amend FACEC, titled as "Agreements Outside this Agreement" provides: "Nothing in this Agreement shall prevent or prohibit any individual ASEAN member state from entering into any bilateral or pluralateral agreement with China and/or the rest of the ASEAN member states relating to trade in goods, trade in services, investment, and/or other areas of economic cooperation outside the ambit of this Agreement.

---

³⁹For example, the development of EU from FTA to the Union of Economics and Currency, the investment rules made by ASEAN, NAFTA, etc.

⁴⁰WTO rules are inherently deficient. For example, the consensus system on which WTO is based has shortcomings such that many issues end up without any measurable results. In contrast, countries in a region generally have more commonality, so they are more likely to diffuse differences and reach RTAs. See Qingjiang Kong, "China's WTO Accession and the ASEAN China Free Trade Area: the Perspective of a Chinese Lawyer", *Journal of International Economic Law*, 7(4), p. 804 (2004).

⁴¹For example, the membership of International Monetary Fund is the prerequisite of applying for membership of International Bank for Reconstruction and Development.

⁴²In fact, the full conformity of RTAs with GATT provisions is rare. In the 46 years of GATT up to the end of 1994, a total of 98 RTAs had been notified under Article 24 of GATT, most of which were examined in individual working parties. But, consensus on the conformity of these RTAs with GATT provisions was reached in only one case: the Czech-Slovak customs union. Jo-Ann Crawford and Sam Laird, "Regional Trade Agreements and the WTO", http://www.tradeobservatory.org/library12/23/2003.

The provisions of this Agreement shall not apply to any such bilateral or plurilateral agreement." The provision shows there might be an overlap in international relations between an ASEAN member state and China or among some of ASEAN member states and China. FACEC is not a closed agreement as it provides a more flexible way for China-ASEAN member states relations.

*Responsibilities of Parties to FACEC under WTO Regime*

The general WTO rules on RTAs is that where the WTO deems a RTA to be inconsistent with WTO provisions, the WTO members concerned shall modify it with a recommendation by CRTA. If they are not prepared to modify it, they "shall not maintain or put into force" such an agreement. If the WTO members concerned fail to implement the recommendation, any provision in a RTA is likely to be subject to WTO dispute settlement procedures. Each member undertakes to accord sympathetic consideration to and to afford adequate opportunity for consultation on the RTA disputes before panel procedure.[43] As a matter of fact, the WTO Dispute Settlement Body (DSB) once faulted Turkey's move to impose quotas on non-EU textile imports as part of a customs union agreement with EU.[44] When the DSB ruled that a provision of GATT has not been observed, the responsible member shall take such reasonable measures as may be available to it to ensure its observance. Failure to do that would mean the provisions relating to compensation and suspension of concessions or other obligations would apply.[45]

Although ASEAN itself and two ASEAN member states are currently not WTO members, China and eight of the ten ASEAN member states are WTO members. It is reasonably safe for WTO members to formulate a RTA in accordance with WTO rules.[46] Considering that ASEAN itself might become a *de facto* WTO member through its further integration and the two non-WTO members of ASEAN might become WTO members in the near future, the two parties to FACEC definitely cannot ignore WTO rules when they conclude and implement FACEC and negotiate further FTA agreement in future.

## WTO Rules in FACEC: Within Coverage or Beyond

As China and eight member states of ASEAN are WTO members and they concluded FACEC after WTO agreements, some Articles of FACEC refer directly or

---

[43] Para. 15 of Understanding on Article 24.
[44] See Qingjiang Kong, "China's WTO Accession and the ASEAN China Free Trade Area: the Perspective of a Chinese Lawyer", *Journal of International Economic Law*, 7(4), p. 813 and note 28 (2004).
[45] Para. 14 of Understanding on Article 24.
[46] Kong Oingjiang, "Closer Economic Partnership Agreement between China and Hong Kong", *China: An International Journal*, 1(1), pp. 136–137 (2003).

indirectly to WTO rules. This is unique in the history of RTAs, indicating the consensus of the WTO members in FACEC to observe WTO rules.

There are two categories of provisions in FACEC that are related to WTO rules.

*Provisions Related to WTO Rules*

In FACEC, there are some articles related to WTO rules, dealing with the traditional fields of GATT, namely the trade in goods and new fields of the Uruguay Round, including trade in service and trade-related intellectual property rights (TRIPS).

Article 2 of FACEC deals with the measures for comprehensive economic cooperation. It states that the measures include (1) progressive elimination of tariffs and non-tariff barriers in substantially all trade in goods; and (2) progressive liberalization of trade in services with substantial sectoral coverage. The provisions are similar to the provisions of Article 24(8)(b) of GATT and Article 5(1) of GATS. Some terms like "in substantially all trade in goods" and "trade in services with substantial sectoral coverage" are even exactly the same as those in WTO rules.[47]

Article 6(3)(d) of FACEC titled as "Application of WTO provisions" clearly indicates the general application of WTO rules, providing that WTO rules governing modification of commitments, safeguard actions, emergency measures and other trade remedies, including anti-dumping and subsidies and countervailing measures, shall, in the interim, be applicable to the products covered under the Early Harvest Program.[48]

Furthermore, some provisions of FACEC refer directly to specific WTO rules. For example, Article 3 titled as "Trade in Goods" refers to the exception for RTAs in Article 24 of GATT: "except, where necessary, those permitted under Article XXIV (8)(b) of the WTO General Agreement on Tariffs and Trade".

Some provisions even require the parties to FACEC to fulfill the requirements of WTO rules on RTAs in spite of non-WTO member positions of ASEAN and its three members at that time. Article 3(6) of FACEC states that the commitments undertaken by the parties under this Article and Article 6 of this Agreement shall fulfill the WTO requirements to eliminate tariffs on substantially all the trade between the parties.

---

[47] For the discussion of the coverage of RTAs, see Jo-Ann Crawford and Sam Laird, "Regional Trade Agreements and the WTO", http://www.tradeobservatory.org/library12/23/2003.

[48] The "Early Harvest" program involves the quick reduction of tariffs on a number of goods that will take effect within a targeted date of three years from July 1 next, year when the Agreement comes into force. ASEAN Secretariat, News Release, 30 October, 2002, ASEAN-China Free Trade Area Negotiations to Start Next Year, http://www.aseansec.org/13125.htm; Qingjiang Kong, "China's WTO Accession and the ASEAN China Free Trade Area: the Perspective of a Chinese Lawyer", *Journal of International Economic Law*, 7(4), p. 815 (2004).

Moreover, the parties to FACEC also agree to conduct their negotiations to establish the ASEAN-China FTA in accordance with the relevant WTO rules. According to Article 3(8) of FACEC, the negotiations between the parties to establish the ASEAN-China FTA covering trade in goods shall also include but not be limited to the following, *inter se*: (1) modification of a party's commitments under the agreement on trade in goods based on Article 28 of GATT;[49] (2) safeguards based on the GATT principles, including but not limited to the following elements: transparency, coverage, objective criteria for action, including the concept of serious injury or threat thereof, and temporary nature;[50] (3) disciplines on subsidies and countervailing measures and anti-dumping measures based on the existing GATT disciplines;[51] and (4) facilitation and promotion of effective and adequate protection of trade-related aspects of intellectual property rights based on existing WTO, World Intellectual Property Organization (WIPO) and other relevant disciplines.[52]

As to the field of trade in services, Article 4(a) of FACEC provides that progressive elimination of substantially all discrimination between or among the parties and/or prohibition of new or more discriminatory measures with respect to trade in services between the parties, except for measures permitted under Article 5(1)(b) of GATS.

From the above-mentioned provisions of FACEC, it is safe to conclude that there is a close link and relationship between some provisions of FACEC and WTO rules. To a certain extent, some provisions of FACEC are based on or governed by WTO rules.

### Provisions Beyond the Coverage of WTO Rules

The important features of modern RTAs are that they have more extensive product coverage than in earlier agreements, and many go beyond the traditional tariff-cutting exercises. They now may cover services, investment, intellectual property, technical barriers to trade, dispute settlement, super-national institutional arrangements, etc.[53] There are some provisions beyond the coverage of WTO rules in FACEC.

---

[49] Artcile 3(8)(d) of FACEC.
[50] Article 3(8)(f) of FACEC.
[51] Article 3(8)(g) of FACEC.
[52] Article 3(8)(h) of FACEC.
[53] Jo-Ann Crawford and Sam Laird, "Regional Trade Agreements and the WTO", http://www.tradeobservatory.org/library12/23/2003.

The objectives of FACEC are: (1) strengthen and enhance economic, trade and investment cooperation between the parties; (2) progressively liberalize and promote trade in goods and services as well as create a transparent, liberal and facilitative investment regime; and (3) explore new areas and develop appropriate measures for closer economic cooperation between the parties.[54]

It is noticeable from the above provisions that firstly, "investment cooperation" and "investment regime" show the importance of investment in FACEC.[55] Article 5 of FACEC further deals solely with the investment.[56] Secondly, "new areas" means the areas in addition to the traditional trade in goods and investment. The parties to FACEC adopt a gradual approach in the negotiation for the FTA — comprehensive, but easy first, trade goods first, service and investment later. Moreover, they aim at not just a FTA, but a closer partnership based on comprehensive cooperation. As for areas of cooperation or "new areas", five priority sectors for cooperation in FACEC are agriculture, information and communications technology, human resources development, investment, and Mekong River basin development. Other areas would include banking, finance, tourism, industrial cooperation, transport, IPR, SMEs, environment, biotechnology, fishery, forestry, mining, energy and sub-regional development.

The measures for comprehensive economic cooperation provided in FACEC also indicate the areas beyond the coverage of WTO rules. The measures include: (1) establishment of an open and competitive investment regime that facilitates and promotes investment within the ASEAN-China FTA;[57] (2) establishment of effective trade and investment facilitation measures, including but not limited to simplification of customs procedures and development of mutual recognition arrangements;[58] and (3) expansion of economic cooperation in areas as may be mutually agreed between the parties and formulation of action plans and programs

---

[54] Article 1 of FACEC.
[55] On 20 March 2004, Swiss President Joesph Deiss concluded an official visit to Thailand after launching an ASEAN-China Investment Fund aiming at financing small- and medium-sized enterprises in Southeast Asia and China with US$125 million. Switzerland was one of the initial investors in the fund, along with the Asian Development Bank, French financial development institution Proparco, the Japan Asia Investment Company and the United Overseas Bank of Singapore. See ASEAN/China investment fund launched, http://news3.xinhuanet.com/english/2004-03/20/content_1376020.htm.
[56] It provides that: to promote investments and to create a liberal, facilitative, transparent and competitive investment regime, the parties agree to: (a) enter into negotiations in order to progressively liberalize the investment regime; (b) strengthen cooperation in investment, facilitate investment and improve transparency of investment rules and regulations; and (c) provide for the protection of investments.
[57] Article 2(c) of FACEC.
[58] Article 2(f) of FACEC.

in order to implement the agreed sectors/areas of cooperation.[59] Again, "new areas" are open to negotiation and might be agreed between ASEAN and China.

Even in the fields within the coverage of WTO rules, it seems that the parties to FACEC wish to go further than their commitments in WTO regime. For example, Article 4 of FACEC deals with trade in services, providing that: expansion in the depth and scope of liberalization of trade in service beyond those undertaken by ASEAN member states and China under GATS;[60] and enhanced cooperation in services between the parties in order to improve efficiency and competitiveness, as well as to diversify the supply and distribution of services of the respective service suppliers of the parties.[61]

In summary, the parties to FACEC might develop traditional areas and "new areas" for economic cooperation and integration to meet their needs in substantial aspect and the regional dispute settlement regime in procedural aspect.

## Conclusions

FACEC is a milestone for the development of China-ASEAN comprehensive economic cooperation and the groundwork for establishment of China-ASEAN FTA. It has both economic and political significance.[62] It also sets up important principles for the China-ASEAN FTA process: mutual benefit; compromise; and consolidation.

At the same time, FACEC is a RTA with its partial contents regulated by WTO rules. WTO rules provide the legal basis for and have close link to the development of FACEC and the China-ASEAN FTA process. After the two non-WTO members in ASEAN have completed their accession to WTO, WTO rules might be more applicable in general. Though there are different and even opposite views on the influence of RTAs to multilateral trade systems,[63] the close link between WTO rules

---

[59] Article 2(g) of FACEC.
[60] Article 4(b) of FACEC.
[61] Article 4(c) of FACEC.
[62] The future of China-ASEAN FTA is encouraging. ASEAN is currently China's fifth largest trade partner, and China the sixth of ASEAN. The planned China-ASEAN FTA is projected to boost Chinese and ASEAN exports by 50% while adding 1% growth of ASEAN's GDP and 0.3% of China's economy. The decision to set up the China-ASEAN FTA is a demonstration of the Chinese and ASEAN leaders' commitment to "enhancing friendship and cooperation in the face of economic globalization". See Qingjiang Kong, "China's WTO Accession and the ASEAN China Free Trade Area: the Perspective of a Chinese Lawyer", *Journal of International Economic Law*, 7(4), pp. 805–806 (2004).
[63] Jo-Ann Crawford and Sam Laird, "Regional Trade Agreements and the WTO", http://www.tradeobservatory.org/library12/23/2003.

and FACEC indicates the possible complementarity between WTO and RTAs on the road to economic liberalization.

It should be further pointed out that ASEAN is composed of ten developing countries, and China is also a developing country. FACEC is therefore an arrangement of South-South cooperation with win-win results. As developing countries, China and ASEAN members share much in common in their history, economic development, international participation, etc. The common goal of the parties to FACEC is to facilitate their economic development through regional economic integration. Moreover, the establishment of China-ASEAN FTA may provide a forum for regional issues, including the development of South China Sea.

## Chapter 7

# The International Legal Personality of ASEAN and the Legal Nature of the China-ASEAN FTA

*Jiangyu WANG*

In the past decade, the proliferation of regional (free) trade agreements — often known as "trade regionalism", has posed serious challenges to the multilateral trading system and raised significant questions in international law as well. In the East Asia area, one of the most salient developments in trade regionalism is the negotiation on a free trade agreement between the People's Republic of China (China) and the Association of Southeast Asian Nations (ASEAN), titled as the China-ASEAN Free Trade Agreement (CAFTA).[1] This paper is an effort to examine CAFTA in the context of international law and discuss two major international legal issues relating to CAFTA. It is organized as follows. The first section provides a brief introduction of the development of the CAFTA proposal and the current negotiating process, as well as the results achieved. The second section considers an international law issue, i.e., on the part of ASEAN, who is the party to CAFTA? Is it that ASEAN as a group is, *vis-à-vis* China, a single contracting party to CAFTA, or are individual ASEAN countries, in their own rights, contracting parties to CAFTA? The third section examines the compatibility of CAFTA with the major pillar of contemporary international trade law, namely the legal rules of the World Trade Organization (WTO).

## Trade Regionalism in East Asia and the Emergence of the CAFTA

Trade regionalism is not a new thing. Jagdish Bhagwati observes that there have been two waves or phrases of regionalism in the post-War period.[2] The

---

[1] Jiangyu Wang, "China's Regional Trade Agreement Approach: The Law, the Geopolitics, and the Impact on the Multilateral Trading System", in *Singapore Yearbook of International Law* (2004).
[2] Jagdish Bhagwati, "Regionalism and Multilateralism: An Overview", in Jagdish Bhagwati *et al.*, eds., *Trading Blocs: Alternative Approaches to Analyzing Preferential Trade Agreements* (Cambridge: The MIT Press, 1999), pp. 3, 9–10.

"First Regionalism", emerging in the 1950s and spreading to both developed and developing countries, demised by the end of the 1960s with only two exceptions in Europe.[3] The "Second Regionalism", signaling the revival of regionalism, has been back since the 1980s and demonstrated strong characteristics of survival and even prosperity.[4]

The second wave of regionalism is not only back — it is actually taking over the central stage in terms of inter-governmental arrangements on trade. Over the past decade, the global trading system has been dominated by the drive towards the conclusion of regional trade agreements (RTAs), also known as free trade agreements (FTAs).[5] By the end of 2003, a total of 193 RTAs in force had been notified to the WTO.[6] Given the short history of the WTO, which was founded in 1995, this is a sharp contrast with the 124 notifications of RTAs received by the WTO's predecessor, the General Agreement on Tariffs and Trade (GATT).[7] The year 2003 along saw 18 RTAs notified to the WTO.[8]

A latecomer to this trend, East Asia shows strong enthusiasm for participation. In 2000, the only RTA in this region was the proposed ASEAN Free Trade Area (AFTA), which is still in the process of negotiation.[9] In the beginning of the 21st century, East Asia has seen intense RTA activities. Singapore has been leading the way in accelerating trade regionalization in East Asia through the signing of FTAs with Japan, New Zealand, the United States, Australia, the European Free Trade Association, and Jordan.[10] Other countries quickly followed suit, leading to a number of recent RTA initiatives and negotiations, as summarized in the

---

[3] Regional economic integration through the formation of the European Community and the European Free Trade Area (EFTA) is, however, a success. See Bhagwati, *supra* note 2, p. 10.
[4] Bhagwati, *supra* note 2, p. 13 (noting the "Second Regionalism" will endure as "it shows many signs of strength and few points of vulnerability").
[5] RTA(s) and FTA(s) are used interchangeably in this paper.
[6] WTO, Annual Report 2004 (Geneva: The World Trade Organization, 2004), p. 68.
[7] WTO, Regional Trade Agreements: Facts and Figures, online at http://www.wto.org/english/tratop_e/region_e/regfac_e.htm.
[8] WTO Report 2004, *supra* note 5, p. 68.
[9] AFTA was started in 1992 through the conclusion of the Framework Agreement on Enhancing ASEAN Economic Cooperation and the Agreement on the Common Effective Preferential Tariff (CEPT) Scheme for the ASEAN Trade Area in 28 January 1992. The two agreements aim to establish an ASEAN Free Trade Area within 15 years following 1992. AFTA, however, is not yet an established RTA. A 1997 declaration by the Heads of State/Government of ASEAN countries, ASEAN Vision 2020, called for efforts to "fully implement the ASEAN Free Trade Area and accelerate liberalization of trade in services, realize the ASEAN Investment Area by 2010 and free flow of investments by 2020." ASEAN Vision 2020, Kuala Lumpur, 13 December 1997, http://www.aseansec.org (accessed 10 November 2004).
[10] For more information on Singapore's FTAs, see http://www.fta.gov.sg.

WTO Annual Report 2004:

> "In Asia Pacific [in 2003] ... negotiations are ongoing on a number of FTAs and several more are at a proposal/study phase. Japan is negotiating an FTA with Mexico and announced in October that negotiations will soon be launched on an FTA with the Republic of Korea. In view of strengthening its ties with ASEAN, a number of working groups have also been established to study the feasibility of FTAs between Japan and ASEAN as a whole, and with individual countries. An additional group is studying the feasibility of an East Asia Free Trade Area (EAFTA) comprising ASEAN plus China, Japan and the Republic of Korea. China for its part has concluded FTAs with Hong Kong, China; and Macao, China; and is engaged in negotiations with ASEAN. Korea has concluded an FTA with Chile and it is considering FTAs with Japan and Singapore, while Thailand has concluded an FTA with Australia and announced its intention in October 2003 to begin negotiations with the United States. In October 2003, India signed a Framework Agreement on Comprehensive Economic Partnership with ASEAN, it has concluded an FTA with Thailand, a framework agreement with MERCOSUR, and is engaged in negotiations with Singapore.[11]"

China jumped onto the bandwagon of regionalism quite recently. In addition to the above mentioned FTAs with Hong Kong and Macao and the negotiations with ASEAN, China has also reached out to a number of countries for RTA talks, including Australia, New Zealand, Middle Asia countries, MERCOSURE countries, and the Gulf Cooperation Council (GCC) countries.[12]

## Economic and Political Significance of CAFTA

The initiative of CAFTA is characterized as a "surprising" movement by some commentators.[13] At the ASEAN-China Summit in November 2000, Chinese Premier Zhu Rongji put forward a basket of proposals on strengthening cooperation in East Asia, including, "seen in the long run", one that China and ASEAN

---

[11] WTO Annual Report 2004, *supra* note 5, p. 69.
[12] Wang, *supra* note 1.
[13] See e.g., China as an Emerging Regional and Technology Power: Implications for US Economic and Security Interest: Hearing before the US-China Economic and Security Review Commission (USCC), 108th Cong., February 12–13, 2004 (testimony of Richard Feinberg and Stephan Haggard), http://www.uscc.gov.

should explore the possibility on the formation of an FTA.[14] At his suggestion, a China-ASEAN Experts' Group on Economic Cooperation (Experts' Group) was established to look into the possibility of establishing a free trade area between the two sides.[15] In its final report issued in October 2001, the group suggested the construction of a "WTO-consistent ASEAN-China FTA within ten years".[16] It noted the profound implication of CAFTA as follows:

> "[T]he establishment of a FTA between ASEAN and China will create an economic region with 1.7 billion consumers, a regional GDP of about US$ 2 trillion and total trade estimated at US$ 1.23 trillion... [T]he removal of trade barriers between ASEAN and China will lower costs, increase intra-regional trade and increase economic efficiency. The establishment of an ASEAN-China FTA will create a sense of community among ASEAN members and China. It will provide another important mechanism for supporting economic stability in East Asia and allow both ASEAN and China to have a larger voice in international trade affairs on issues of common interest." [17]

The report also recommended that China and ASEAN adopt a comprehensive and forward-looking framework of economic cooperation to forge closer economic relations in the 21st century. In November 2001, the Seventh China-ASEAN Summit endorsed the ideas envisaged by the Experts' Group and initiated the negotiation process.[18]

At the Eighth China-ASEAN Summit in Phnom Penh, Cambodia in November 2002, ASEAN and Chinese leaders signed the "Framework Agreement on the Comprehensive Economic Cooperation between ASEAN and China" (FA) which, coming into force on 1 July 2003, provides the groundwork for the eventual formation of the CAFTA by 2010 for the six original ASEAN members and 2015 for the newly admitted members (Cambodia, the Lao PDR, Myanmar and

---

[14] "Zhu Rongji chuxi di sici Zhongguo-Dongmeng lingdaoren huiwu" [Zhu Rongji Attends the Fourth China-ASEAN Summit], *Xinhua News Agency*, 25 November 2000.
[15] ASEAN Secretariat, "ASEAN Annual Report 2000–2001", p. 110, http://www.aseansec.org.
[16] "Forging Closer ASEAN-China Economic Relations in the 21st Century", a report submitted by the ASEAN-China Experts Group on Economic Cooperation, October 2001, p. 36, http://www.aseansec.org.
[17] *Id*, at 2.
[18] ASEAN Secretariat, "ASEAN Annual Report 2001–2002", pp. 123–124, http://www.aseansec.org.

Vietnam).[19] The FA was amended by a protocol signed on 6 October 2003 by China and ASEAN at their 2003 annual summit in Bali.[20] The FA represents the first FTA initiative of both ASEAN (as a group) and China (outside the Greater China Area[21]).

As the WTO Secretariat notes, the formation of RTAs is inevitably driven by both economic and political considerations, which might include geopolitical concerns such as maintaining peace and regional security, acquiring greater bargaining power in multilateral negotiations, pursuing good governance and durable democracy, and maintaining the position of bureaucratic teams involved in RTA negotiations.[22] In the case of CAFTA, apart from the perceived economic benefits, the politics underlying the initiatives should never be underestimated. As one commentator points out, the strategic and political considerations are weighted more heavily at this stage in China's and ASEAN's approach to the free trade agreement.[23] For China, CAFTA is a strategic movement to woo ASEAN for regional security and influence, to compete with Japan and even the United States for regional leadership, and to allay the "China threat" concern — both economically and militarily — of its smaller neighbors by offering a number of "sweeteners" to ASEAN countries.[24] On the part of ASEAN, it has no alternative but to engage a rising China which is widely recognized as marching towards an economic and political superpower, "locking in" China's role as a promoter of regional stability.[25] Furthermore, there are few reasons for ASEAN not to accept the economic benefits offered by the fast-growing Chinese markets through preferential trade arrangement.

## Scope and Coverage of CAFTA

### Objectives and General Coverage for Economic Cooperation Under the FA

With a view to establishing the CAFTA before 2010, the parties to the FA agree to strengthen cooperation and to "progressively liberalize and promote trade in

---

[19] Framework Agreement on Comprehensive Economic Co-Operation between the Association of South East Asian Nations and the People's Republic of China, Phnom Penh, 5 November 2002, http://www.aseansec.org [hereinafter the FA].
[20] The Protocol to Amend the Framework Agreement on Comprehensive Economic Co-operation Between the Association of South East Asian Nations and the People's Republic of China, Bali, 6 October 2003, http://www.aseansec.org [hereinafter the Protocol].
[21] The Greater China Area includes Mainland China, Hong Kong, Macao, and Taiwan.
[22] WTO Secretariat, World Trade Report 2003, p. 49 (Geneva: the World Trade Organization, 2003), http://www.wto.org.
[23] See Wang, *supra* note 1.
[24] *Id.*
[25] *Id.*

goods and services as well as create a transparent, liberal and facilitative investment regime."[26] This suggests that the proposed CAFTA will cover trade in goods and services as well as trade and investment facilitation. In addition, the FA opens the door for the parties to "explore new areas and develop appropriate measures for closer economic cooperation."[27] Specific measures towards the realization of CAFTA, which will be implemented progressively in the coming years, include the following:[28]

1. elimination of tariffs and non-tariff barriers in substantially all trade in goods;
2. liberalization of trade in services with substantial sectoral coverage;
3. establishment of an open and competitive investment regime that facilitates and promotes investment within CAFTA;
4. special and differential treatment and flexibility to the newer ASEAN member states, including Cambodia, the Lao PDR, Myanmar and Vietnam;
5. flexible measures to allow the parties in CAFTA negotiations to address their sensitive areas in the goods, services and investment sectors with such flexibility to be negotiated and mutually agreed based on the principle of reciprocity and mutual benefit;
6. trade and investment facilitation measures, such as simplification of customs procedures and the development of mutual recognition arrangements;
7. an open attitude toward further liberalization in new areas/sectors; and
8. the establishment of appropriate mechanisms for the effective implementation of the FA.

*Early-Harvest Program (EHP) for Trade in Goods Under the FA*

In addition to the obligations to enter into negotiations for the ultimate free trade pact, the FA also establishes an Early Harvest Program (EHP), implemented as of 1 January 2004, which is aimed to reap the immediate concessions offered by the parties, mainly China. The EHP allows the reduction of tariffs on certain products before the onset of CAFTA. Initially, it aims to implement tariff reduction on these products over three years: to 10% before 2004, to 5% before 2005, and to zero tariffs no later than 1 January 2006.[29] A distinctive feature of the EHP is that China has also given unilateral concessions to ASEAN members who feel they would not benefit as

---

[26] FA, *supra* note 34, Article 1(b).
[27] *Id.*, Article 1(c).
[28] *Id.*, Article 2.
[29] Annex 3 of the FA, Part B. A slightly different schedule is used for the newer ASEAN members.

much from the EHP, covering over 130 agricultural and manufacturing products.[30] In essence, it "allows ASEAN products to be exported to China at a significant concessionary rate so that ASEAN countries can actually benefit from the free trade agreement even before the agreement itself is finalized."[31] In return, the ASEAN countries agree to give tariff concessions to China under the Harmonized System (HS) on tariffs for agricultural products, including meat, fish, fruits, vegetables and milk.[32] In total, the EHP has targeted a host of some 600 products listed in Chapters 1–8 of the HS, mostly agricultural products who are to be unilaterally liberalized by China.[33] In addition, China agrees to grant WTO benefits (mainly MFN treatment) to those ASEAN members who are not yet official WTO members.[34]

Initially the FA took a multilateral approach to tariff reduction, namely — the tariff concessions under the HS approach shall be multilateralized to all parties (i.e., all ASEAN members and China) provided that the same products are included in their EHP. But because the Philippines and China failed to establish an EHP scheme, other countries in the region were to subsequently consider that this would therefore allow the Philippines to a "free ride". In addition, fear that some ASEAN countries like Thailand have more efficient farm sectors that could suppress the growth of the agricultural sector in others has also deterred ASEAN members from implementing a multilateral approach under the EHP. Malaysia was amongst the first that negotiated a clause in 2003, allowing it to offer lower agriculture tariffs *only* to China in return for the latter's concessions under the EHP.[35] This practice has since been consolidated in the 2003 Protocol, which replaces the original Article 6(3)(b)(i) of the FA with a new provision. The new provision recognizes that a party may accelerate its tariff reduction and/or elimination under the EHP to the rest of the parties "on a unilateral basis."[36] Meanwhile, one or more ASEAN members are still allowed to conduct negotiations and enter into acceleration arrangements with China to fast-track their tariff reduction or elimination, which shall be done, however, only on "bilateral or plurilateral" basis. In other

---

[30] Annex 2 of the FA.
[31] "ASEAN, China Launch First Stage of Free-Trade Plan," *AFP*, 7 October 2003 (on file).
[32] *Id.*
[33] China, however, has excluded from EHP some agricultural products such as rice and palm oil, which are said to be major exports from ASEAN countries. These products are to be negotiated in the coming years.
[34] FA, Article 9. Currently Cambodia, Vietnam and the Lao PDR are not WTO members.
[35] "Malaysia Early Harvest Clause a Way out for RP," *The Manila Times*, 7 April 2003, http://www.manilatimes.net/national/2003/apr/07/business/ 20030407bus4.html.
[36] The Protocol, *supra* note 35, Article 2.

words, no conditional or unconditional MFN status is granted under the EHP except to Brunei and Singapore.[37]

A good example of a tariff acceleration arrangement is the China-Thailand Agreement on Accelerated Tariff Elimination under the Early Harvest Program,[38] which prescribes that China and Thailand shall eliminate tariffs on all vegetable and fruit products no later than 1 October 2003 (ahead of the EHP effective date of 1 January 2004, as well as the targeted zero tariff date of 1 January 2006).[39]

## Negotiation Agenda for Goods, Services, Investment and Other Areas to Complete CAFTA and the Latest Development in Negotiations

Goods not covered by the EHP are subject to further negotiation within specified timeframes. The FA categorizes these goods into two tracks. The first, the "Normal Track", contains products tariff rates which shall be gradually reduced or eliminated in accordance with specified schedules (to be mutually agreed by the parties) over a period from 1 January 2005 to 2010 by the ASEAN 6 and China, and from 1 January 2005 to 2015 for the newer ASEAN members.[40] The "Sensitive Track" represents a second category of products specified by an individual party on its own accord in respect of which a timeframe is not imposed on further liberalization. Parties are merely required to reduce (or eliminate when applicable) tariffs on Sensitive Track products "in accordance with the mutually agreed end rates and end dates."[41] In addition to tariffs, the negotiations shall also include, among others, rules of origin, out-quota-rates, renegotiation of concession schedules, non-tariff measures, trade remedy laws, trade facilitation measures, as well as trade-related intellectual property protection.[42] In terms of the timeframe, the FA prescribes that the goods negotiations shall commence in early 2003 and be concluded by 30 June 2004.[43] However, an agreement on trade in goods was not concluded until

---

[37] *Id.* Only Brunei and Singapore will, subject to conformity with specified requirements, automatically become parties to any arrangements that have been agreed on or will be agreed to between China and any other ASEAN state under the EHP. See *id.*, Annex 2. Singapore has already become a party to the China-Thailand EHP arrangement.

[38] Text of the agreement is available on China's Ministry of Commerce website at http://www.mofcom.gov.cn.

[39] *Id.*, Article 1.

[40] The FA, Article 3(4)(a)(i).

[41] *Id.*, Article 3(4)(b).

[42] *Id.*, Article 3(8).

[43] *Id.*, Article 1(1).

November 2004, when China and ASEAN signed an instrument to start tariff-reduction program from 2005 and completely remove tariffs on merchandise goods by 2010. In addition, China and ASEAN have also signed agreement on the dispute settlement mechanism in CAFTA.[44]

The FA, however, does not have a mandatory timeframe for the negotiations on services and investment, except that negotiations shall commence in 2003 and be concluded "as expeditiously as possible for implementation in accordance with the timeframes to be mutually agreed."[45] Substantive negotiations in these two areas have already started for the conclusion of the framework texts and the first packages of commitments were set for the second half of 2004.[46] The negotiations on services seek to progressively eliminate substantially all discrimination and to prohibit new discriminatory measures with respect to trade in services between the parties.[47] The FA also seeks to expand the depth and scope of liberalization in trade in services beyond those undertaken by China and ASEAN members under the GATS.[48] This could result in an acceleration of China's WTO commitments on services for ASEAN, not unlike what China has done for Hong Kong and Macao. As for investment, the FA aims, through negotiations, "to obtain commitments on liberalizing the investment regime, increasing market access as well as commitments on protection of investment in the China market."[49]

Besides trade and investment, China and ASEAN have also identified five priority sectors for strengthened cooperation, including agriculture, information and communications technology, human resources development, investment, and the Mekong River basin development, as well as 11 other activities, including standard mutual recognition and harmonization, electronic commerce, technology transfer, and specific projects such as the acceleration of the railway project linking Singapore and China's southern city of Kunming.[50] All these efforts indicate that the FA is indeed a herald for an all-embracing pact of economic cooperation.

---

[44] *Xinhua News*, 28 November, 2004.
[45] *Id.*, Article 8(3).
[46] ASEAN-China FTA Factsheet, Ministry of Trade and Industry (MTI) of Singapore, http://www.mti.gov.sg (accessed 16 January 2004).
[47] The FA, Article 4(a).
[48] *Id.*, Article 4(b).
[49] MTI of Singapore, *supra* note 62.
[50] The FA, Annex 4.

## Legal Personality of ASEAN as an International Organization

### International Law on Legal Personalities and Treaty-Making Powers

If signed, CAFTA will be a treaty based on international law. Article 2 of the Vienna Convention on the Law of Treaties of 1986 defines a "treaty" as "an international agreement concluded between states in written form and governed by international law, whether embodied in a single instrument or in two or more related instruments and whatever its particular designation." Thus, to be a treaty, an agreement also has to be "an international agreement" which is "concluded between states", "in written form", and "governed by international law". The phrase "governed by international law" entails an "intention to create obligations under international law" according to the International Law Commission's Commentary.[51] The Vienna Convention on the Law of Treaties between States and International Organizations or between International Organizations of 1986, in its Article 2:1(a), modifies this definition by recognizing treaty "between one or more states and one or more international organizations; or between international organizations." As such, both states and international organizations can conclude treaties or agreements and thus create binding international legal obligations.

The treaty-making power of an international organization is closely associated with whether it possesses objective international legal personality. As a leading text on international law observes:

> "Whether an organization possesses personality in international law will hinge upon its constitutional status, its actual powers and practice. Significant factors in this context will include *the capacity to enter into relations with states and other organizations and conclude treaties with them*, and the status it has been given under municipal law. Such elements are known in international law as the indicia of personality.[52]"

States, as the original and major subjects of international law, derives their personality from the very nature and structure of the international system which recognizes independence (or sovereign) and equality as the fundamental characteristics of states.[53] For international organizations, their role in the world order centres on their possession of international legal personality as well as the scope

---

[51] See Fourth Report on the Law of Treaties, *Year Book of the International Law Commission* (ILC), 1965, p. 12.
[52] Malcolm N. Shaw, *International Law*, 5th Ed. (Cambridge: Cambridge University Press, 2003).
[53] *Id.*, pp. 189–193, 242.

of powers derived from this personality. The acquisition of personality of international organizations in the first instance depends primarily upon the terms of the instrument establishing the organization. "If states wish the organization to be endowed specifically with international personality, this will appear in the constituent treaty and will be determinative of the issue."[54] The International Court of Justice, in the "Legality of the Use by a State of Nuclear Weapons in Armed Conflict" case, opined that:

> "[t]he constituent instruments of international organizations are also treaties of a particular type; their object is to create new subjects of law endowed with a certain autonomy, to which the parties entrust the task of realizing common goals. Such treaties can raise specific problems of interpretation owing, *inter alia*, to their character which is conventional and at the same time institutional; the very nature of the organization created, the objectives which have been assigned to it by its founder, the imperatives associated with the effective performance of its functions, as well as its own practice, are all elements which may deserve special attention when the time comes to interpret these constituent treaties."[55]

This line of reasoning also implies the capacity of international organizations to conclude international treaties, which also depends primarily upon the constituent instrument. Having independent legal personality is important, but "the existence of legal personality is, on its own, probably insufficient to ground the competence to enter into international agreements."[56] Thus, Article 6 of the Vienna Convention on the Law of Treaties between States and International Organizations or between International Organizations stipulates that "the capacity of an international organization to conclude treaties in governed by the rules of that organization."[57] As will be discussed later, the provision has a wider meaning than reliance solely on the constituent instrument.[58] However, the constituent instrument is without doubt the most important source of treaty-making power of international organizations. In this regard, vivid evidence is the Treaty Establishing the European Community (EC). Article 281 of the treaty provides that "the Community shall have legal personality". Furthermore, a number of provisions confer on the Community express

---

[54] *Id.*, pp. 1187–1188.
[55] ICJ Reports, 1996, pp. 66, 74–75, quoted in Shaw, *supra* note 52, pp. 1193–1194.
[56] Shaw, *supra* note 52, p. 1197.
[57] Vienna Convention on the Law of Treaties between States and International Organizations or between International Organizations, Article 6.
[58] See *infra* note.

competence across a range of policy issues in reaching agreements with foreign states or international organizations.[59]

As a principle of international law, however, the absence of provisions concerning legal personalities and treaty-making powers does not necessarily lead to a lack of capacity in this regard. In many cases, the legal personality of international organizations may be inferred from the powers or purposes of the organization and its practice. The International Court of Justice (ICJ), in a now famous case called the "Reparation for Injuries Suffered in the Services of the United Nation", declared that, although the Charter of the United Nations does not explicitly stipulate the legal personality of the UN, the UN nevertheless had international legal personality because this was indispensable in order to achieve the purposes and principles provided in the Charter, and this was also the intention of the UN members who created it. In the words of the ICJ:

> "It must be acknowledged that [the UN] members, by entrusting certain functions to it, with the attendant duties and responsibilities, have clothed it with the competence required to enable those functions to be effectively discharged."[60]

To support the conclusion that the UN did have objective legal personality, the ICJ examined not only the United Nations Charter and subsequent relevant treaties, but also the UN practice, including the obligations of member states toward the UN, its ability to conclude international agreements, as well as the underlying meaning of certain provisions of the UN Charter.[61] With regard to the powers of international organizations, the ICJ held that:

> "[u]nder international law, the organization must be deemed to have those powers which, though not expressly provided in the charter, are conferred upon it by necessary implication as being essential to the performance of its duties."[62]

It can be submitted that, despite the absence of provisions explicitly granting legal personality and treaty-making power in an international organization's constituent instrument, such legal characteristics can still be established by other rules or even

---

[59] See e.g., Article 181 EC (on development policy), Article 174 EC (on environment policy), Article 170 EC (on research and technology), and Article 181a EC (on economic, financial, and technical cooperation with foreign States).

[60] "Reparation for Injuries Suffered in the Services of the United Nations Case", Advisory Opinion, ICJ Reports 1949, pp. 174, 179 [hereinafter Reparation].

[61] Shaw, *supra* note 52, p. 1188.

[62] This is quoted in Shaw, *supra* note 52, p. 1195.

practice. It has been argued that Article 6 of the Vienna Convention on the Law of Treaties between States and International Organizations or between International Organizations[63] has "a wider formulation than reliance solely upon the constituent instrument and permits resource to issues of implied powers, interpretation and subsequent practice."[64] This view was already supported in the commentary of the International Law Commission, which noted that the phrase "the rules of the organization" in Article 6 meant the constituent instruments plus relevant decisions and resolutions and the *established practice* of the organization.[65] As such:

> "[d]emonstration of treaty-making capacity will revolve around the competences of the organization as demonstrated in each particular case by reference to the constituent instruments, evidenced implied powers and subsequent practices."[66]

## Legal Status of ASEAN

Popular view has it that CAFTA is a bilateral agreement between ASEAN and China. This view, however, has to be called into question after a serious examination based on international law and ASEAN's own legal and policy instruments.

The assertion that ASEAN cannot be a party to an international agreement such as the CAFTA FA is caused by ASEAN's loose institutional structure and the unsettled question of ASEAN's legal personality, as well as the institutional powers existing thereunder. Since ASEAN is an inter-governmental organization, it is largely up to its constituent instruments to determine its legal personality and powers. ASEAN was established by the ASEAN Declaration reached by the five original member states on 8 August 1967.[67] Also known as the Bangkok Declaration, the ASEAN Declaration 1967 is actually the founding instrument establishing ASEAN as a regional international organization. The Declaration contains five major clauses, laying down some very general principles of cooperation. It has been argued that, although lacking in detail, the Declaration "goes beyond 'a mere statement of intent for cooperation', to enumerate the aims and purposes of the Association and to establish a machinery to carry out the aims set forth in the

---

[63] See *supra* note, p. 57.
[64] Shaw, *supra* note 52, p. 1197.
[65] *Id.*, p. 1198.
[66] *Id.*
[67] The legal text of the ASEAN Declaration 1967 is available online at http://www.aseansec.org/1629.htm. The five states are, respectively, Indonesia, Malaysia, Philippines, Singapore and Thailand.

Declaration."⁶⁸ Arguably, the parties' subsequent conducts, including the conclusion of a number of political accords for the gradual strengthening of ASEAN's institutional character, are indicative that ASEAN is an international institution with some power.⁶⁹ However, none of the instruments gives explicit legal personality to ASEAN, nor do they confer treaty-making power to the organization.

As analyzed above (i.e., the ICJ's analysis in "Reparation"), the legal personality and powers of an international institution do not necessarily have to come from its constituent instruments. Without a provision in the constituent instrument, these legal characteristics can also be evidenced by other agreements, established practices, direct or implicit intention of the members, or an understanding that they are indispensable for the fulfilment of the organization's purposes.⁷⁰

Rodolfo C. Severino, the formal Secretary-General of ASEAN, summarized the organization's mission as follows:

> "ASEAN's founders in 1967 intended ASEAN to be an association of all the states of Southeast Asia cooperating voluntarily for the common good, with peace and economic, social and cultural development as its primary purposes."⁷¹

However, he further noted that, achieving these purposes does not entail the need for a strong ASEAN institution:

> "[ASEAN] is not and was not meant to be a supranational entity acting independently of its members. It has no regional parliament or council of ministers with law-making powers, no power of enforcement, no judicial system."⁷²

---

⁶⁸ Paul J. Davidson, *ASEAN — The Evolving Legal Framework for Economic Cooperation* (Singapore: Times Academic Press, 2002).

⁶⁹ Major political accords after the ASEAN Declaration 1967 includes: the Zone of Peace, Freedom and Neutrality Declaration, Kular Lumpur, 27 November 1971; Declaration of ASEAN Accord, Bali, 24 February 1976; Treaty of Amity and Cooperation in Southeast Asia, Bali, 24 February 1976; ASEAN Declaration on the South China Sea, Manila, 22 July 1992; Treaty on the Southeast Asia Nuclear Weapon-Free Zone, Bangkok, 15 December 1997; ASEAN Vision 2020 I, Kuala Lumpur, 15 December 1997; and Declaration of ASEAN Accord II, Bali, 7 October 2003.

⁷⁰ See *supra* note, pp. 56–66 and the accompanying text.

⁷¹ "Asia Policy Lecture: What ASEAN is and What It Stands for", by Rodolfo C. Severino, Secretary-General of ASEAN, at The Research Institute for Asia and the Pacific, University of Sydney, Australia, 22 October 1998, http://www.aseansec.org/3399.htm (accessed 10 November 2004).

⁷² *Id.*

One commentator quite rightly observed the basic shortcomings suffered by the present ASEAN machinery and *operandi*:[73]

- It lacks an integrated decision-making structure. ASEAN basically serves as a forum for talks and there is virtually no central decision-making body in the organization.
- Application of the consensus method is applied to all issues and all levels, which has considerably reduced the effectiveness of intra-ASEAN cooperation.
- The principle of rotation, heavily emphasized in ASEAN, has had a debilitating effect on the role of the ASEAN Secretariat, which is largely marginalized by the system of national secretariats.
- The ASEAN structure reflects the dominant emphasis on national interests and national representation; in contrast, there is no provision for representation of the ASEAN "community interest".
- ASEAN relies almost entirely on a policy regime and does not have a legal regime.

In short, ASEAN is deliberately designed to have a loose structure, which provides opportunities and room for behaviors that conform to the cultural environment in this part of the world, i.e., "face saving" which is considered vital for regional solidarity and cohesion.[74] More significantly, ASEAN was not designed as a sovereign body, but as a regional grouping of sovereign nations. At this stage, ASEAN is more an instrument of cooperation than integration.[75] The common desire of the member states here is to promote regional and national peace, progress and security, and these goals are perceived to be better achieved by a social community, not a legal community.[76] In conclusion, it is yet too early to decide that ASEAN has an independent, objective international legal personality and treaty-making power.

Insofar as legal personality and power of an international organization can also be acquired through practices and recognition, this cannot be applied to the CAFTA FA as it has already been rebuffed by ASEAN's external relations practices. In recent years, ASEAN has developed "dialogue" relations with major nation-states and regions, including China, Japan, Korea, India, EU, among others. Increasingly, foreign countries and regions have shown an interest in dealing with ASEAN as a collectivity.[77] ASEAN, however, hasn't taken this opportunity to make use of the "recognition" principle in international law. Instead, it seems to be developing

---

[73] Muthiah Alagappa, "Institutional Framework: Recommendations for Change", in *Second ASEAN Reader* (Singapore: Institute of Southeast Asian Studies, 2004) pp. 22–24.
[74] *Id.*, p. 22.
[75] Davidson, *supra* note 69, p. 29.
[76] *Id.*
[77] *Id.*, p. 37.

a "selective exercise of legal personality" practice. In recent years, for important agreements with foreign countries, ASEAN's ten Heads of States have affixed their signatures without representation of ASEAN as an organization. Examples in this regard include the three "Framework" agreements on comprehensive economic cooperation/partnership with China, India and Japan, as well as the Instruments of Extension to the Treaty of Amity and Cooperation in Southeast Asia, which admit the accession of China, Japan and India, respectively, to the Treaty of Amity and Cooperation in Southeast Asia. In contrast, agreements concerning issues in a special area — which is also probably considered less important — can be signed by the Secretary-General of ASEAN. For example, the Memorandum of Understanding between the Association of Southeast Asian Nations (ASEAN) Secretariat and the Ministry of Agriculture of the People's Republic of China, concluded on 2 November 2002, was signed by ASEAN's then Secretary-General Rodolfo C. Severino and the Vice Minister of Agriculture Qi Jingfa. A more representative agreement is the Memorandum of Understanding between the Governments of the Member Countries of the Association of Southeast Asian Nations (ASEAN) and the Government of the People's Republic of China on Cooperation in the Field of Non-Traditional Security Issues, which was signed by ASEAN's incumbent Secretary-General Ong Keng Yong and the representative of China on 10 January 2004.

This "selective personality" practice, as I call it, is not yet sufficient to establish that ASEAN as an institution has international legal personality and treaty-making powers. One can easily see from the title of the two MOUs mentioned above that the ASEAN Secretary-General's representation was based more likely on an *ad hoc* basis rather than a regular exercise of authority. Furthermore, assuming "selective personality" can be established through this practice, it actually strengthens the view that those "important agreements" between ASEAN states and foreign countries, including the Framework Agreements with China, India and Japan, are definitely not agreements between those individual countries and ASEAN as a collectivity. In any case, the CAFTA FA is not a bilateral agreement between China and ASEAN as a independent entity.

## Legal Nature of CAFTA and its Profound Implication on the Further Negotiation and Implementation of CAFTA

### Legal Nature of CAFTA

It is yet too early to tell the nature of this agreement as it is in the negotiating process. Its current form looks more like a multilateral agreement, with its parties being China and individual ASEAN countries. However, a more precise — and thus more acid — view would be that it is largely a collection or combination of bilateral agreements between China and individual ASEAN members.

With certain exceptions, this view is confirmed by the language and rights/obligation structure of the FA. The provisions in the FA and its various Annexes support this view: they show that China's obligations — e.g., those under the Early Harvest Programme — are towards individual countries, and vice versa, although sometimes "ASEAN member states" as a whole was referred for purpose of convenience. This is first evidenced by the language of the FA. In the Preamble of the CAFTA FA,[78] the Parties declared that the FA was signed by:

> "WE, the Heads of Government/State of Brunei Darussalam, the Kingdom of Cambodia, the Republic of Indonesia, the Lao People's Democratic Republic (Lao PDR), Malaysia, the Union of Myanmar, the Republic of the Philippines, the Republic of Singapore, the Kingdom of Thailand and the Socialist Republic of Vietnam, member states of the Association of South East Asian Nations (collectively, ASEAN or ASEAN member states, or individually, ASEAN member state), and the People's Republic of China (China) . . ."[79]

It is noted that all the names of the ten member states of ASEAN as well as the name of China appear in the FA as the contracting parties. Although it is also stated in the Preamble that the ten Southeast Asian states should be "collectively" called "ASEAN" or "ASEAN member states", or "individually" as "ASEAN member state", in the end of the FA, only the Heads of the 11 nation-states (including ASEAN-10 and China) affixed their signatures to the agreement, while no ASEAN's own representative was called to sign the FA. Paragraph three of the Preamble, stating the parties' desire to adopt a Framework Agreement on Comprehensive Economic Cooperation *between ASEAN and China*, interprets the phrase "ASEAN and China" as "collectively, 'the Parties', or individually referring to an ASEAN member state or to China as a 'party'".[80] All these expressions, with little doubt, explicate the Contracting Parties' intention to make the 11 sovereign nations parties to the FA.

The core of the FA is its Article 6 which prescribes the Early Harvest Program (EHP). Unlike other provisions which aim to establish a framework for conducting negotiations, Article 6 is a provision concerning substantive rights and obligations. In international law, it is considered as imposing binding obligations upon the contracting parties. As noted previously, it allows ASEAN members states to export

---

[78] See *supra* note 17.
[79] CAFTA FA, the Preamble, para. 1, *supra* note 19.
[80] CAFTA FA, the Preamble, para. 3, *supra* note 19.

to China at preferential tariff rates for all the goods covered by Chapters 1–8 of the Harmonized System. China's unilateral concessions to the member states of ASEAN are embodied in the stipulation that ASEAN countries are allowed to "carve out" those products for which they have difficulties to grant market access.[81] In short, every ASEAN country can put forward an Exclusion List for agricultural products, denying market access to imports. In addition, China agreed to grant concessions on 130 categories of manufactured goods to individual ASEAN countries, provided that those countries can put forward an "inclusion" list of products. For goods included in the list, China will extend special and preferential tariff treatment to the ASEAN country. The ASEAN country concerned has to render some concessions to China, which however will be offered on less than full reciprocity basis.[82]

In terms of mutual grant of preferential benefits, the FA initially followed a "multilateralism" approach, which at once made the FA like a multilateral agreement. The essence was that, under the EHP, for agricultural products, except for those placed in the Exclusion List as prescribed in Article 6(a)(i) of the FA, each ASEAN country should not only extend preferential treatment to China, but also give the same privilege to other ASEAN members. As the Philippines was eventually not able to put forward an exclusion list, Malaysia had managed to secure a clause allowing it to offer EHP benefits only to China.[83] This practice was subsequently incorporated in the FA through the 2003 Protocol, changing effectively the direction of CAFTA in terms of agricultural products from multilateralism to bilateralism.[84] The 2003 Protocol to amend the FA consolidated this practice by adding the following amendments, among others, to Article 6(3)(b)(i) of the FA:

1. A party may accelerate its tariff reduction and/or elimination under this Article in relation to the rest of the parties on a unilateral basis; and
2. One or more ASEAN member states may also conduct negotiations and enter into a bilateral or plurilateral acceleration arrangement with China to accelerate their tariff reduction and/or elimination under this Article.

Half-hearted multilateralism is also embodied in Article 6(a)(iii) of the FA, which prescribes that:

> "The specific products set out in Annex 2 of this Agreement shall be covered by the Early Harvest Program and the tariff concessions shall

---

[81] See *supra* note 29–34 and the accompanying text.
[82] The lists of goods put forward by ASEAN countries are contained in Annex 2 of the FA.
[83] See *supra* note 35 and the accompanying text.
[84] See *supra* note 35 and the accompanying text.

apply only to the parties indicated in Annex 2. These parties must have extended the tariff concessions on these products to each others."

So far, only Indonesia, Malaysia and Thailand have submitted their "inclusion lists" under Article 6(a)(iii). These countries, plus Brunei and Singapore which are allowed to "be parties to any arrangements that have been agreed on or will be agreed to between China and any other party pursuant to Article 6(a)(iii)",[85] constitute only half of the ASEAN states which can participate in the EHP.

The FA only set up principles for conducting negotiations toward the final FTA, the CAFTA. The nature of CAFTA itself is thus yet undisclosed. However, the essential question in this regard actually depends on the institutional characteristics of ASEAN. If ASEAN continues to be an organization without legal personality, which can be established either explicitly through legal provisions in the constituent instruments, or implicitly through practice, CAFTA cannot be a bilateral FTA deal between ASEAN as a group and China.

## Implications of CAFTA Being Not a Bilateral Agreement Between ASEAN as an Entity and China

The recognition that CAFTA is not a bilateral agreement between China and ASEAN as an entity will have impact on a number of significant issues, including the organization's and the member states' respective responsibilities and liabilities to a third state (in this case China), especially when disputes arise out of the agreement.

International legal personality brings forth responsibility and liability. However, it is still an unsolved international law issue — whether a treaty concluded by an international organization has binding effect upon the member states of the organization. In the negotiating process of the 1986 Vienna Convention on the Law of Treaties between States and International Organizations or between International Organizations, the International Law Commission put forward a draft clause on this issue, providing that member states of an international organization shall be bound by a treaty if (1) the member states have agreed to be bound by virtue of the constituent instrument of the organization; and (2) the assent of the member states to be bound has been duly brought to the knowledge of the negotiating states and negotiating organizations.[86] Such arrangement has particular value to international trade deals, such as a tariff agreement concluded by a closed regional organization, such as the European Community, and another state/organization.

---

[85] Annex 2 of the FA.
[86] Shaw, *supra* note 52, p. 859.

As one commentator observes, "such agreement would be of little value if they were not to be immediately binding on member states."[87] However, despite the strong support from the European Community, the draft clause was rejected at the conference adopting the 1986 Convention, and was replaced by Article 74(3) of the Convention, which stipulates:

> "The provisions of the present Convention shall not prejudge any question that may arise in regard to the establishment of obligations and rights for member states of an international organization under a treaty to which that organization is a party."

It has now become a general principle of international law that the question of the legal effect of a treaty concluded by an international organization on its member states should be "resolved on the basis of the consent of the states concerned in the specific circumstances and on a case-by-case basis".[88] The European Community, having objective personality and implementing common policy in social, commercial and other arenas, has a closed organizational structure compared with any other major international institutions in the world. In the international arena, the EC often acts as a representative of member states with explicitly authorized powers from its constituent instrument.[89] The EC is an independent member of the WTO, the major international economic institution in the world. It has become a widely recognized state practice that, in the WTO's dispute settlement system, the EC acts as one party, bearing rights and obligations collectively.[90]

ASEAN has not been like the EC. It is an international organization with ambiguous legal personality. It does not have a central decision-making structure or enforcement regime. It is more like a forum for political dialogue. In all the formal agreements with outside countries, all the sovereign member states, rather than the organization itself, sign them. Accordingly, it falls directly on individual members of ASEAN to implement the treaty obligations. The language and rights and obligations structure of the CAFTA FA further conform this: the rights and obligations of the treaty are directly borne by individual members; the rights and obligations are directly between individual ASEAN members and China or, as the case may be, between individual ASEAN members. This will have a profound implication on enforcement of the obligations and the dispute settlement relating

---

[87] *Id.*
[88] Shaw, *supra* note 52, p. 860.
[89] *Supra* note 59 and the accompanying text.
[90] See e.g., WTO Appellate Body report, "European Communities — Regime for the Importation, Sale and Distribution of Bananas", WT/DS27/AB/R, 9 September 1997.

to CAFTA. The parties, namely the 11 nation-states, are directly responsible to each other in terms of performing the obligations. In this sense, CAFTA is more like the WTO which is a multilateral agreement but is indeed bilateral in nature. CAFTA, however, will probably contains mainly the bilateral obligations and rights between individual ASEAN members and China, with something on ASEAN nations' obligations and rights to each other. Among ASEAN countries, if the reciprocal benefits offered by CAFTA are not better than those in the Asean Free Trade Agreement (AFTA), CAFTA will be of little value to them in terms of the relationships among ASEAN countries.

This bilateral nature of the obligations will be reflected in the dispute settlement mechanism. Basically, it will be a mechanism for resolving disputes between China and the individual ASEAN countries, and disputes between two or more ASEAN nations, if the obligations between them are breached. However, in no case can be there a dispute resolution system between China and ASEAN as a collectivity. For China, the trouble will be that it will have to pursue every ASEAN member to enforce the trade privileges accorded to it. ASEAN as an organization will offer political, goodwill help, but will not give legal assistance by commanding the individual country concerned to perform its obligations. Given its current structure, ASEAN is not in the legal position to call upon the responsible member to remedy its default. ASEAN, the institution, will however bear no responsibility and liability towards China.

## Conclusion

This paper discusses the international legal personality of ASEAN and its impact on the negotiation and implementation of the China-ASEAN Free Trade Agreement. Although CAFTA is popularly regarded as a bilateral agreement between China and ASEAN, it is actually not the case in international law. Instead, it is more like a multilateral agreement signed by 11 nation-states. In its nature, it contains mainly bilateral obligations between China and individual ASEAN countries. This situation is mainly caused by the ambiguous legal personality and loose structure of ASEAN. This situation will also have profound impact on enforcement of obligations and dispute settlement relating to the implementation of CAFTA.

# Chapter 8

# China-ASEAN FTA: An Investment Perspective

*CHEN Huiping*

Investment negotiations between China and ASEAN[1] are currently undergoing as a part of the China-ASEAN Free Trade Area (CAFTA) negotiations. Despite good intentions on the part of all parties in the establishment of CAFTA, actual negotiations are proving to be more complex than expected. The source of the said complexity with regard to the investment negotiations is both procedural and substantive. This paper will focus mainly on the procedural aspects. For the purpose of completion, a discussion on the substantive aspects will be reserved for the future. We will probe into the origins, the status quo, and the difficulties and obstacles of the investment negotiations; the principles and strategies for the success of the investment negotiations; and finally, the significance and implications of the investment negotiations.

## Origins of the Investment Negotiations

The investment negotiations originated directly from the CAFTA negotiations between China and ASEAN and constitute an inherent part of the CAFTA negotiations.

### Initiation of CAFTA

China and ASEAN countries are adjoining countries with a history of close relations with respect to culture, economy and people. But regrettably, this kind of friendly relationship deteriorated during the end of World War II and 1970s, as a result of the Cold War and the founding of a socialist China. Since the 1980s, however, economic and political relations between the parties improved gradually with

---

[1] The Association of South-East Asian Nations (ASEAN) was formed in 1967 by Indonesia, Malaysia, Philippines, Thailand and Singapore, joined by Brunei in 1984 (known as ASEAN-6), and by Cambodia, Laos, Myanmar and Vietnam in the 1990s (known as CLMV or new ASEAN members).

the establishment of diplomatic relations between China and ASEAN countries, resulting in a particularly rapid development of the economic and trade relations in the post-Cold War era. In recent years, ASEAN has become the fifth biggest trade partner of China; and China is the sixth biggest trade partner of ASEAN. Still, great margin exists for further and closer economic cooperation between the two sides.

The "September 11" event dramatically tipped the balance, which existed in the previous international political regime to reconstruct a new international pattern in which geopolitics is an important element. As for China, it consistently maintains the position of "Peaceful Rise" even though it has made great success in her economic development with her "Open Door" policy. However, the quick rise of China in both the economy and politics brings about jealousy and isolation from some countries; and the "China Threat" thought is prevalent even in some Asian countries. Therefore, to keep a rapid economic growth and to stand firmly in international politics, it is necessary and urgent for China to assist in creating a peaceful, steady and prosperous Asia, which is a crucial region for China from the geopolitics perspective.

Following the current international economic trend of globalization of trade and regional economic integration, China is actively participating at all levels in the international economic community by entering the WTO in 2001 and initiating to establish free trade areas with some countries and regions.

Prompted by the combination of such bright prospects in the field of China-ASEAN trade and investment cooperation, the constant change of international affairs, and the need for a steady political situation in South-East Asia, the leaders of China and ASEAN have stepped onto the definition platform with goals aimed at strengthening economic and political relations between the two sides.

In December 1997, officials from both China and ASEAN had their first informal China-ASEAN summit to discuss and explore the development of China-ASEAN friendship and cooperation that was oriented towards a good-neighborly partnership of mutual trust in the 21st century. They intended to promote and strengthen bilateral and multilateral cooperation between them through mutual cooperation in trade and investment. Two years later, China signed the framework agreements on bilateral cooperation with the ten ASEAN countries with the purpose of widening cooperation in the fields of trade and investment. In November 2001, the leaders of China and ASEAN had a meeting in Brunei to set up for the first time a goal to establish CAFTA within ten years. For the purpose of establishing CAFTA, six rounds of negotiations had been held since. In November 2002, the Framework Agreement on Comprehensive Economic Cooperation between the Association of South East Asian Nations and the People's Republic of China (China-ASEAN Framework Agreement or Framework Agreement) was signed in

Phnom Penh, Cambodia, with the specific goal of establishing CAFTA by 2010 with the ASEAN-6 and by 2015 with the four new ASEAN members.

Formal negotiations on CAFTA began immediately after the signing of the China-ASEAN Framework Agreement. Practical considerations during the negotiations required a slight change in some points of the Framework Agreement and the addition of some new points as well. Approximately one year later, in October 2003, a Protocol to Amend the Framework Agreement (China-ASEAN Protocol) was signed by Chinese and ASEAN leaders. Further in scope and depth than CAFTA, the two parties decided to establish a strategic partnership for peace and prosperity while expressing the will to facilitate the negotiations on CAFTA and secure the successful establishment of CAFTA by 2010.

### Necessity to Incorporate Investment Issue into CAFTA

As asserted, regional integration cannot be fully achieved if investment arrangement is not included.[2] With globalization and increased multilateral and regional cooperation, competition for FDI is no longer confined to between countries, but increasingly between regions. Therefore, more countries are likely to submit to a regional approach in attracting FDI.[3] Being a free trade area, CAFTA covers not only trade but also investment. According to the China-ASEAN Framework Agreement, the objectives of CAFTA include strengthening and enhancing economic, trade and investment cooperation between the parties, and creating a transparent, liberal and facilitative investment regime.[4] One specific article for investment is provided by the Framework Agreement, which requires parties to enter into investment negotiations so as to achieve the above-mentioned CAFTA objectives.[5] It is not CAFTA that was the first to incorporate the investment issue into a free trade area, which as the term suggests, seems to be a specific area only for free trade. Rather, it is a universal tendency for trade regimes and trade areas to contain an investment element for the following reasons.

---

[2] Kee Hwee Wee and Hafiz Mirza, "ASEAN Investment Cooperation: Retrospect, Developments and Prospects", p. 20, http://www.gapresearch.org/finance/ASEAN%20Investment%20Cooperation.pdf (accessed 19 February 2005).

[3] Kee Hwee Wee and Hafiz Mirza, "ASEAN Investment Cooperation: Retrospect, Developments and Prospects", p. 2, http://www.gapresearch.org/finance/ASEAN%20Investment%20Cooperation.pdf (accessed 19 February 2005).

[4] Article 1(a) and (b) of the Framework Agreement on Comprehensive Economic Cooperation between the Association of South East Asian Nations and the People's Republic of China (ASEAN-China Framework Agreement).

[5] Article 5 of the ASEAN-China Framework Agreement.

The international economy has two pillars, i.e., transnational trade and transnational investment, which are complementary and mutually supportive. The concept of comprehensive economic cooperation incorporates free trade along with free movement of investment. At the same time, the international economic community is "walking on two legs" — one is the global trade liberalization based on the WTO multilateral trade regime, and the other is the regional economic integration based on regional free trade arrangements. The increasing complementarity and mutual promotion between trade and investment leads to the tendency for the "so-called" trade regime or trade area to extend from pure trade in goods in the past to all levels of economic activities covering trade in services and free investment. For example, GATT, the predecessor of WTO, did not involve an investment issue. But the investment issue was raised in the Uruguay Round, which was negotiated from 1986 to 1993. The resulted WTO regime therefore covers investment issues in its TRIMs and GATS agreements. Moreover, a working group on the Relationship between Trade and Investment was established in the WTO to advocate negotiations on multilateral investment rules in the WTO regime. In the same way, the newly concluded regional free trade agreements also cover the issue of investment liberalization, such as the North American Free Trade Agreement (NAFTA).

It is natural for CAFTA to follow this new practice. The anticipated free trade area between China and ASEAN countries dealing with both trade and investment issues reveals China and ASEAN's expectation that CAFTA is not only a pure trade regime, but a comprehensive regional economic integration as well. There is great potential and prospect for them to have investment cooperation. Although generally speaking, China and ASEAN are both net-capital-import countries, ASEAN has more direct investment in China at the current stage. However, with the rapid development of China's economy and its new policy and strategy of "Going Abroad", China's direct investment in ASEAN countries will quickly grow. Consequently, the establishment of the CAFTA's integration of the two sides into an economic bloc will not only promote mutual investment between China and ASEAN, but also attract foreign investments from the US, European countries, Japan, and the like.

### *Investment Negotiations an Integral Part of CAFTA Negotiations*

Given that investment is incorporated into CAFTA, creating a China-ASEAN investment agreement as a part of CAFTA agreements is an inherent goal of the CAFTA negotiations. Since trade and investment are technically separate issues, separate negotiations for each of them are necessary and practical. The pattern designed for the construction of the legal framework for CAFTA clearly reveals that investment negotiations are an integral part of CAFTA negotiations.

The pattern for CAFTA agreements is the internationally prevailing pattern of "framework agreement + protocol + annexes", which is a strategy for hard issues such as environmental protection. This pattern calls for a progressive signature of agreements in situations when clear common objectives among the contracting parties have been established but no specific agreements could be reached within a short period as to the details of the parties' rights and obligations. Hence the parties can set up a framework agreement on the forward-looking common objectives. Thereafter, they can then proceed to the negotiations on the specific matters until they come to an agreement. The resulting agreements will then be included in the existing framework agreement as protocols or annexes to be executed. This pattern was adopted by ASEAN countries when they were constructing their economic cooperation area. Naturally, this pattern was also followed in the establishment of CAFTA.

For CAFTA, the framework agreement is the China-ASEAN Framework Agreement signed in November 2002, which expresses the common wishes of the leaders of both sides to establish CAFTA and sets up the general objectives for CAFTA. This Framework Agreement contains a total of 16 articles stating the objectives, scope, measures and timeframes of CAFTA. As to its objectives, it particularly addresses the objectives of promoting investment cooperation and creating an investment regime. The Framework Agreement sets up the basic structure for CAFTA and is the legal basis for CAFTA. It is not an exaggeration to say that it serves as a roadmap for the Free Trade Area.[6]

The protocol of CAFTA is the China-ASEAN Protocol signed in December 2003, which seeks to revise some contents of the Framework Agreement and insert new contents concerning the Early Harvest Program and the bilateral or plurilateral agreements outside the CAFTA agreements.

The annexes of CAFTA include two types. One type comprises the lists given by specific countries or concluded by countries at or before the signing of the Framework Agreement with regard to Early Harvest Program and the priority activities of economic cooperation between the two sides. The Framework Agreement has already contained four such annexes. The other type consists of the agreements to be concluded through negotiations after the Framework Agreement on matters which are not clearly stated in it. These matters are left for future negotiations and the resulting agreements will be included as annexes to be enforced, according to the Framework Agreement.[7] For example, the Agreement on Trading in Goods

---

[6] China-ASEAN FTA Factsheet, http://www.mti.gov.sg/public/FTA/frm_FTA_Default.asp?sid=179&cid=1902 (accessed 23 July 2004).
[7] Article 13(1) of the China-ASEAN Framework Agreement.

of the Framework Agreement on Comprehensive Economic Cooperation between ASEAN and China, signed on 29 November 2004 as a result of the negotiations on trade in goods, is attached as an annex. The agreements to be signed during the negotiations on services and investments will become annexes to the Framework Agreement too.

The Framework Agreement has only one article, i.e., Article 5, which deals directly with investment issues. Article 5 specifies that "to promote investment and to create a liberal, facilitative, transparent and competitive investment regime, the parties agree to: (1) enter into negotiations in order to progressively liberalize the investment regime; (2) strengthen cooperation in investment, facilitate investment and improve transparency of investment rules and regulations; and (3) provide for the protection of investments".[8] This article is too simple and general for an investment agreement. This is why investment negotiations are needed to negotiate on an overall and comprehensive investment legal regime and to formulate an acceptable investment agreement that is reflective of prospective trends.

CAFTA covers such matters as trade in goods, services, investment liberalization, investment facilities and economic operation. Currently, initial CAFTA negotiations between China and ASEAN are primarily in the fields of trade in goods and services, and investment. The investment negotiations are an integral part; and the investment agreement to be reached will be regarded as an annex to the China-ASEAN Framework Agreement.

## Status Quo of Investment Negotiations

The investment negotiations were initiated for a comprehensive investment agreement between China and ASEAN to be a part of the CAFTA agreements. To this end, a special working group was set up to carry out detailed discussions on the framework of the agreement. Several meetings have been held by the working group and a consensus has been reached as to the general elements and principles of the investment agreement. However, crucial controversies with regard to some investment issues, such as market admission and the application of pre-establishment national treatment, bring the negotiations to a deadlock.

### Parties to Negotiations

In both CAFTA negotiations and its associate investment negotiations, China is one party, while ASEAN and its members constitute the other party. ASEAN acts

---

[8] Article 5 of the China-ASEAN Framework Agreement.

as an entity consisting of representatives from all its members. On ASEAN's part, both ASEAN member countries and the ASEAN Secretariat are responsible for sending representatives for the negotiations. For the Chinese constituents, the Ministry of Commerce of China (MOC, whose predecessor before 2003 is the Ministry of Foreign Trade and Economic Cooperation of China) is the main body acting on behalf of China. The Chinese delegation to the negotiations comprises representatives from the MOC, Ministry of Foreign affairs, State Development and Reform Commission, and the like.

## Organizations Responsible for Negotiations

The organization responsible for the general CAFTA negotiations is the China-ASEAN Trade Negotiation Committee (China-ASEAN TNC or TNC).[9] It was set up by the third China-ASEAN Economic Senior Officials' meeting held in Beijing in May 2002, to hold indepth discussions on the framework of CAFTA, even before the Framework Agreement was officially signed then.[10] The CAFTA negotiations have started in trade in goods, services and investments.

The organization responsible for the investment negotiations is the China-ASEAN Trade Negotiation Committee Working Group on Investment (TNC-WGI), which was set up pursuant to the Framework Agreement, which provides that other bodies may be established if necessary to coordinate and implement any economic cooperation activities.[11] Both China-ASEAN TNC and TNC-WGI should report regularly to the ASEAN Economic Ministers and the Minister of the MOC on the progress and outcome of their negotiations.[12] By the same token, the ASEAN Secretariat and MOC shall jointly provide the necessary secretariat support to the China-ASEAN TNC whenever and wherever negotiations are held.[13]

## Anticipated Timeframes for Negotiations

The China-ASEAN TNC for CAFTA negotiations had its first meeting in May 2002 and 17 rounds of China-ASEAN TNC meetings have been held so far. The negotiations on trade in goods produced on 29 November 2004 the Agreement on Trading in Goods of the Framework Agreement on Comprehensive Economic Cooperation between ASEAN and China.

---

[9] Article 12(1) of the China-ASEAN Framework Agreement.
[10] *China Daily*, 18 May 2002.
[11] Article 12(2) of the China-ASEAN Framework Agreement.
[12] Article 12(3) of the China-ASEAN Framework Agreement.
[13] Article 12(4) of the China-ASEAN Framework Agreement.

As to the investment negotiations, the Framework Agreement stipulates that they shall commence in 2003.[14] Actually, the TNC-WGI started negotiations on an investment agreement in June 2003 and seven meetings have been held so far. The first meeting held in June 2003 discussed the objectives of the investment negotiations. The second meeting held in August 2003, with a general introduction to each other's investment laws and policies. The third meeting held in November 2003 had a preliminary discussion of the main elements and principles of the anticipated investment agreement. The fourth meeting was held in January 2004, with detailed discussions on some agreed elements and principles of the investment agreement. The fifth meeting was held in March 2004, and differences between the two sides with regard to the opening-up of industries and national treatment emerged. The sixth meeting held in June 2004 discussed investment liberalization. The seventh meeting in July 2004 resulted in no agreement on the formality and substance of the framework of the investment agreement. The frequently-held TNC-WGI meetings show that both parties have sincerity and strong desire to reach an investment agreement.

No specific deadline for investment negotiations is stipulated by any legal document between China and ASEAN. The Framework Agreement only states that the negotiations shall be concluded as expeditiously as possible for implementation in accordance with the timeframes to be mutually agreed.[15] The investment negotiations were originally set for completion by end of 2004,[16] but did not. ASEAN had insisted that a deadline should be set for the completion of the entire negotiations and expected to conclude negotiations by June 2005 for the final signing of the China-ASEAN Investment Agreement. However, China did not agree to having a deadline.

*Template for Proposed Investment Agreement*

The Framework Agreement has set up basic contents and objectives for the investment agreement, which should in turn serve to realize the general objectives of CAFTA.

During the TNC-WGI meetings, it has been agreed that the Framework Agreement on ASEAN Investment Area (AIA Framework Agreement) would be used as a template for the negotiated investment agreement.[17] The AIA Framework

---

[14] Article 8(2) of the China-ASEAN Framework Agreement.
[15] Article 8(3) of the China-ASEAN Framework Agreement.
[16] China-ASEAN FTA Factsheet, http://www.mti.gov.sg/public/FTA/frm_FTA_Default.asp?sid=179&cid=1902 (accessed 23 July 2004).
[17] China-ASEAN FTA Factsheet, http://www.mti.gov.sg/public/FTA/frm_FTA_Default.asp?sid=179&cid=1902 (accessed 23 July 2004).

Agreement in 1998 and its Protocol in 2001 constitute investment instruments among ASEAN countries with the objective of establishing a competitive ASEAN Investment Area containing a more liberal and transparent investment environment, while ensuring the realization of free flow of investments by 2020.[18] This timeframe was later accelerated to 2010 by ASEAN members. The AIA Framework Agreement is a comprehensive agreement with comparatively high standards of investment treatment and investment protection. China accepted it as the template for lack of a better alternative. This was China's first negotiation over a multilateral investment agreement, as what it has already completed are bilateral investment treaties (BITs) with other countries.

## Current Status of Negotiations

The past seven TNC-WGI meetings touched upon a variety of issues over the investment agreement, such as the objectives, main elements and principles of the agreement; the opening-up of industries, national treatment and investment liberalization.

At the very beginning of the negotiations, China and ASEAN did not agree on what the intended investment agreement would be like. After several meetings, China and ASEAN came to an agreement as to the contents of the final investment agreement. Confidential sources reveal that the contents should include, *inter alia*, the following principles and elements: (1) the definitions of terms such as "investment" and "investor", as well as the scope of coverage and objectives; (2) the liberalization of investments including the opening-up of industries, national treatment, MFN treatment and prohibition of performance requirements; (3) protection, facilitation and promotion of investments, including adequate protection, expropriation, compensation for losses, transfer and repatriation of profits, subrogation and transparency; (4) dispute settlement mechanisms, including state-to-state and investor-to-state mechanisms; and (5) general provisions including safeguard measures, general exceptions, institutional arrangements, consultations, relations with other agreements, modification of schedules, annexes and action plans, amendments, entry into force, duration and termination, and programs and action plans. They also agreed that the proposed agreement will focus on investment protection and investment facilitation.

Although China and ASEAN have already had a common framework for the contents of the investment agreement, no further agreements on the details of the above-mentioned elements and principles could be reached subsequently. Present

---

[18] Article 3 of the the Framework Agreement on ASEAN Investment Area (AIA Framework Agreement).

negotiations are not proceeding as smoothly as expected because of key differences over some investment issues.

*Key Differences in Negotiations*

When the negotiations touch upon the substantive issues of the investment agreement, i.e., the specific rules for the above-mentioned elements, differences are becoming clear. For example, the two parties could not agree on the definitions of investment and investor. The liberalization of investments is a core controversy since China contends that time is not ripe for it to open up its industries and to grant national treatment to foreign investors. As to the concrete rules and standards for investment protection, the two parties are far from reaching a consensus.

With these essential controversies, investment negotiations are currently in a deadlock. However, the recent conclusion of the negotiations on trade and the signing of the trade agreement will stimulate a quicker resumption of the negotiations on investment.

## Existing Difficulties and Obstacles in Investment Negotiations

There are two underlying reasons which give rise to the negotiations deadlock: differences in opinion over the substantive rules in the investment agreement and the procedural difficulties between — and within — China, ASEAN and its member countries.

*Difficulties and Obstacles within China*

Firstly, the decision-making process of the Chinese Central Government is resistant to a quick and smooth negotiation process as well as an expedited agreement conclusion with ASEAN. China is a central state whose government makes all overall and final decisions, as well as formulates uniform foreign political and economic policies. The establishment of CAFTA is a part of a political goal resulting from an idealized enthusiasm espoused by political leaders. When they decided to establish free trade area with ASEAN countries, they lacked a complete analysis and consideration as to the potential problems and difficulties posed by the proposed investment negotiations. The result is that such underestimated-yet-actual difficulties prolonged the investment negotiations. Furthermore, with limited authority, Chinese delegation to the negotiations had to refer to the central government for all final considerations and decisions, and this in turn further slows down the negotiations.

Secondly, the complex composition of Chinese delegation to the negotiations adds to the low efficiency of decision-making on the part of the Chinese. The Chinese delegation comprises representatives from several different state agencies, such as the Ministry of Commerce, Ministry of Foreign affairs, State Development and Reform Commission, Ministry of Finance, and the like. Representatives from MOC come from different departments such as the Department of International Trade and Economic Affairs, Department of Treaty and Law, Department of Foreign Investment Administration, and Department of Foreign Economic Cooperation. The complexity of representatives with various agency interests possessing different levels of authority slows the agreement process for every specific investment issue.

Thirdly, the complicated allocation of policy-making authority over foreign investment issues among the central government and local governments results in inconsistent foreign investment policies. Thus, it is very difficult for China to promise to confer a uniform national treatment to foreign investors at various levels of administrative districts.

Fourthly, China's current ideology and legal practice towards the free flow of foreign investment makes it premature to conclude a liberalized investment agreement with ASEAN countries. With the painful history of being colonized or semi-colonized by some developed countries, the Chinese people highly cherish their political and economic state sovereignty. By the same token, they are very wary of any hint of possible invasion by economic powers. This explains why China never promises to give national treatment to foreign investors or investments in her national laws, regulations, or bilateral investment agreements with other countries. In short, China's allowance of immediate and full investment liberalization is still premature at the present stage.

Fifthly, the two-fold reality of relatively small and imbalanced flows of investment between China and ASEAN prevents China from investing in a greater interest in creating a liberalized investment regime. China is the largest recipient of FDI among developing countries and the second largest in the world.[19] The main FDI sources for China are the United States, Japan and the EU. Compared to these countries and regions, the inflows from ASEAN to China are relatively small. Furthermore, Chinese investments into ASEAN are even less by comparison. This reality lessens China's interest in creating a liberalized investment agreement with ASEAN.

## Difficulties and Obstacles within ASEAN and its Member Countries

Firstly, the complexity of composition of representatives from ASEAN countries also adds to the complicated decision-making process within ASEAN. Although ASEAN acts as an entity in its negotiations with China, it actually consists of

---

[19] Chia Siow Yue, "ASEAN-China Free Trade Area", paper for presentation at the ACP Conference in Hong Kong, 12–13 April 2004, p. 2.

investment officials and representatives from ten ASEAN member countries, as well as representatives from the Investment and Enterprise Unit of the ASEAN Secretariat. Moreover, representatives from some ASEAN member countries like Lao PDR, Myanmar and the Philippines have at times been absent from the TNC-WGI meetings. Thus, it is a challenge for these countries to reach a uniform consensus as a voice of ASEAN.

Secondly, discrepancies in economic and political situations and development among the ten ASEAN countries make it hard to coordinate their different interests and formulate a uniform position towards China. The ASEAN-6 countries are comparatively more developed than the four new members, both economically and politically. Seven of the ten ASEAN member states are WTO members (Brunei, Indonesia, Malaysia, Myanmar, Philippines, Singapore and Thailand) but three of them are not. Nine of them (except for Brunei) have signed bilateral investment treaties with China. Thus, complicated relations among them bar substantial progress in the investment negotiations.

Thirdly, the mutually competitive — rather than complementary — investment relations between China and ASEAN make ASEAN countries cast doubt on the realization of a free movement of investments within the free trade area. China and ASEAN are both major recipients for foreign direct investments, rather than significant mutual investors. Furthermore, ASEAN is a net investor in China at the present stage.[20] The mutually competitive rather than complementary investment relations between China and ASEAN prevent the movements of investments between the two sides. The Open-up Policy and the investment incentives carried out by China has redirected the flow of American, European and Japanese investments from ASEAN countries into China. Consequently, some ASEAN countries have incurred losses, which in turn breed dissatisfaction or even hostility against China. In this context, some ASEAN countries are doubtful as to whether the future China-ASEAN investment agreement will promote intra-area investments and increase outside investments into the area; and whether they will actually benefit from it.

However, it must be noted here, that China's "Going Abroad" policy has actually encouraged Chinese companies such as Sinopec, Haier and TCL to invest abroad, especially into countries such as Indonesia, Vietnam and the Philippines in Southeast Asia. It could be reasonably expected that more Chinese investments will be directed to ASEAN countries if the investment agreement is signed and the CAFTA is established.

---

[20] "Forging Closer ASEAN-China Economic Relations in the Twenty-first Century", a report submitted by the ASEAN-China Expert Group on Economic Cooperation, October 2001, p. 12, available at the ASEAN Secretariat website (accessed 15 October 2004).

## Difficulties and Obstacles between China and ASEAN

Firstly, China and ASEAN countries have different expectations from the investment agreement due to significant differences in their domestic investment regimes. China only expects the agreement to cover investment protection and investment facilitation with broad principles rather than specific elements. ASEAN expects to make the agreement a comprehensive one, including concrete and detailed elements covering not only investment protection, investment facilitation but also investment liberalization. Some ASEAN countries such as Singapore, Malaysia and Brunei are comparatively developed countries equipped with rather advanced and comprehensive domestic investment regimes, which enable greater breadth and depth in their openness to foreign investments. Therefore, they have high expectation on the investment agreement requiring both investment protection and investment liberalization, with the purpose of opening up the Chinese market. For the ASEAN countries as a whole, free flow of investment within them is the objective of the AIA Framework Agreement; and some have already realized such objectives. Furthermore, the main reason for ASEAN to have investment cooperation with China is to protect the interests of ASEAN investors in China and to open up China's investment regime for ASEAN investors.[21] Hence, ASEAN expects to have a greater emphasis on increased liberalization rather than a mere investment protection. Certainly, ASEAN is unwilling to produce a China-ASEAN agreement which only replicates the individual BIT that China has already signed with most ASEAN countries.

By contrast, China has no such objectives. As to the opening-up of industries, China's existing foreign investment system is far from being complete and perfect. Conditions also abound for foreign investments in China. In addition, the number of fields and sectors available towards foreign investment is also quite limited. As for granting national treatment, China's unique ownership system (public-owned enterprises, private-owned enterprises and mixed-ownership enterprises) allows for different treatment even among domestic investors. As it is, discrepancies in investment policies and practices already exist between and among various governments of different levels and provinces. As such, it is quite a challenge to confer national treatment to foreign investors. Under such circumstances, discussions over national treatment and investment liberalization with ASEAN countries remain at bay for China.

---

[21]"Forging Closer ASEAN-China Economic Relations in the Twenty-first Century", a report submitted by the ASEAN-China Expert Group on Economic Cooperation, October 2001, p. 12, available at the ASEAN Secretariat website (accessed 15 October 2004).

Secondly, the different negotiating experiences and bases of China and ASEAN countries contribute to their different attitudes toward the investment negotiations. For ASEAN countries, they have already experienced multilateral negotiations on investment and the investment liberalization negotiations in the process of establishing the ASEAN Investment Area. In addition, they also learn a lot from the five-year operation of the Investment Area. Before the investment negotiations with China, ASEAN required that the AIA Framework Agreement — with a rather high standard of investment liberalization — should be the basis and the template for the negotiations. By contrast, China lacks both experience and textual basis with respect to the negotiations on multilateral investment agreement to serve as reference. What it does have, however, is experience in the negotiations on bilateral investment agreement and the bilateral investment treaties (BITs). Hence, China's stance with ASEAN is timid, overcautious and indecisive.

However, the above difficulties and obstacles do not mean that China and ASEAN have nothing in common and that the negotiations are destined to fail. On the contrary, both are developing countries, both share a common willingness to establish an investment regime and a free trade area. The bilateral investment treaties (BITs) between China and ASEAN countries (except for Brunei) constitute a common basis for investment protection rules. There is every reason to have success in the negotiations.

## Principles for Success of Investment Negotiations

It would seem that to make the investment negotiations successful, the following principles should receive special attention during the investment negotiations.

### Principle of "The Big Picture Perspective"

The "Big Picture" perspective means that the overall objectives of both China and the CAFTA should be taken into consideration during the investment negotiations.

The overall political objective of China is to advocate a multi-polar world and multilateralism to dilute US unilateralism in world and regional affairs.[22] One of the applications of the objective is the signing of a series of political and economic agreements or declarations with ASEAN such as the framework agreement of friendship and cooperation towards the 21st century, the ASEAN Treaty of Amity,

---

[22] Sheng Lijun, "China-ASEAN Free Trade Area: Origins, Developments and Strategic Motivations", *ISEAS Working Paper: International Politics & Security Issues* Series No. 1, p. 7 (2003), http://bookshop.iseas.edu.sg (accessed 15 November 2004).

the Declaration on the Conduct of Parties in the South China Sea, and the efforts on the establishment of CAFTA.

With the boom in the Chinese economy, two completely different thoughts prevail among the ASEAN countries, i.e., the "China Threat" and the "China opportunity". In order to keep up with the pace of its economic development, China needs to maintain a peaceful, steady and prosperous peripheral environment in an ever-changing international situation. "Peaceful rise" is China's aim; so China wishes to share the benefits from its economic development with its ASEAN neighbors through the establishment of a free trade area. This is a "win-win" approach (to be discussed later), since it will not only eliminate the "China Threat", but will create a steady and safe peripheral neighborhood for China. Therefore, the investment negotiations with ASEAN should first serve this overall strategic objective. To this end, China should keep its grasp on the key issues instead of debating over trivial matters in the negotiation process.

The general objective of CAFTA is to strengthen and enhance the economic, trade and investment cooperation between China and ASEAN, and thus promote peace, prosperity and progress among China and ASEAN countries. Investment negotiations are part of the CAFTA negotiations; and therefore they should serve this objective. Once CAFTA is established, it will be the biggest free trade area in the world with a population of around 1.8 billion (China's 1.3 billion and ASEAN's 500 million people). Such a big economic bloc has economic significance for China, ASEAN, as well as the rest of the world. From this perspective, it is necessary for China to consider the general interests of China-ASEAN as a whole.

*Principle of "A Win-Win Objective"*

The principle of "win-win" should be the key principle for CAFTA investment negotiations. The main goal of China seeking the establishment of CAFTA is to strengthen its economic and political relations with ASEAN through benefits sharing. The direct purpose of CAFTA is to promote the mutual development of both sides. In this sense, "win-win" is the common target of both China and ASEAN. The negotiations should take into account the mutual benefits that the anticipated investment agreement would bring. If each side of the negotiations keeps the "win-win" objective in mind and think more for the other side, she would be willing to compromise or give concession in controversial issues, thus making the negotiations easier to conclude. Therefore, the principle of "win-win" should be the dominant principle for CAFTA investment negotiations.

*Principle of "Flexibility"*

Flexibility is the specific application of the principle of "win-win". Although both China and ASEAN countries are developing countries, various and obvious differences in economy, politics, religion and culture exist between them, and even among ASEAN countries themselves. If every minute detail should be agreed upon during the investment negotiations, the target of "win-win" will not be reached since it is simply impossible. Instead, the differences among the countries should be dealt with flexibly. For example, if some countries strongly consider some limited sectors as their sensitive sectors, it is better for other countries not to force them to open them up for foreign investment. Tactical and flexible measures may be used to bypass them so as to prevent them from becoming obstacles in the investment negotiations. The China-ASEAN Framework Agreement also supports the view of flexibility.[23] The principle of flexibility should also be taken into consideration when China is dealing with differences among the ASEAN countries. The new ASEAN members, including Cambodia, Laos, Myanmar and Vietnam, have less developed economy compared with the old ASEAN members, and thus they enjoy special and favorable treatment inside ASEAN. For those countries, it is better for China to give them special and differential treatment with the principle of flexibility instead of requiring superficial, nominal equity or reciprocity. In addition, the China-ASEAN Framework Agreement calls for the provision of special and differential treatment and flexibility to the new ASEAN members in several articles.[24]

The principle of flexibility also applies to the strategies during the negotiations. In consideration of China's difficulties and actual situations in liberalizing its investment regime, it is not appropriate for ASEAN to require a radical change of China's laws and regulations with regard to investment liberalization. Flexibility will gain ASEAN eventual success.

*Principle of "Keeping Your Eyes on the Prize"*

To be successful, the negotiators should know the purposes and the direction of the negotiations. Hence the primary goal should be to determine and unify their expectations and objectives for the negotiations. China originally intended to conclude an agreement focusing on investment protection without touching upon the issue of investment liberalization. After consultations, the two sides agreed that the agreement should be comprehensive in covering both investment protection

---

[23] Article 2(d) and (e) and Article 8(3) of the China-ASEAN Framework Agreement.
[24] Preamble, Article 2(d) and Article 8(3) of the China-ASEAN Framework Agreement.

and liberalization. The agreement should be a detailed, transparent, steady, free, facilitative, open, predictable and comparative investment legal regime. Keeping this in mind will facilitate the investment negotiations through a deep, thorough, practical and effective process.

China should accept the liberalization objective in the agreement. It should be forward-looking in the negotiations because the China-ASEAN Framework Agreement itself is a forward-looking pact, with the objective of closer economic relations between China and ASEAN oriented towards the 21st century.[25] Furthermore, the CAFTA between China and the old ASEAN members will finally be established by 2010 and that between China and the new ASEAN members will be established by 2015. This means that the earliest time for the realization of investment liberalization in CAFTA will not be until 2010. Moreover, both China and ASEAN promised to APEC that investment liberalization within the Asia-Pacific area would be realized no later than 2020.[26] Therefore, China would be committed to free investment in the following ten or more years. Moreover, there remain several more years for China to adjust its domestic situations. It is believed that corresponding reforms would have been carried out and the difficulties and obstacles of investment liberalization will be overcome by 2010. Keeping such considerations in mind, China will feel more comfortable on make promises on investment liberalization in the investment convention.

As investment liberalization cannot be realized overnight, China will need to approach it on a step-by-step and phase-by-phase basis. For example, we can have different timeframes for a gradual and progressive conferment of national treatment and opening-up of its industries. The China-ASEAN Framework Agreement itself admits that investment liberalization is a progressive process.[27] The establishment of CAFTA is also a progressive process. Furthermore, the so-called investment liberalization is never complete without some restraints. All investment agreements with promises of investment liberalization allow for some forms of exceptions or reservations. This approach can be borrowed for the CAFTA investment agreement.

*Principle of "Balance"*

The balance between the rights and obligations of both the host states and foreign investors should receive attention during the investment negotiations. Normally, international investment agreements do not impose obligations on the investors;

---

[25] This objective is clearly stated in the Preamble of the China-ASEAN Framework Agreement.
[26] Para. 6, APEC Economic Leaders' Declaration of Common Resolve, Bogor, Indonesia, 1994.
[27] Article 5 of the Framework Agreement on Comprehensive Economic Cooperation between the Association of South East Asian Nations and the People's Republic of China.

rather, they confer upon them rights, in accordance with the Vienna Convention on the Law of Treaties. However, they do impose heavy obligations on host states towards foreign investors. Hence, it is important to balance the rights and obligations in order to avoid failure in the negotiations. The failure of OECD's Multilateral Agreement on Investment (MAI) during 1996–1998 is an example.

## Strategies for Success of Investment Negotiations

In view of the above-mentioned difficulties, obstacles and principles, the following strategies may be adopted by both China and ASEAN during the investment negotiations so as to ensure the success of the negotiations.

### China's Strategies

Firstly, China should embrace a long-term perspective towards investment agreement with ASEAN. China has to remember that regional investment agreement is different from BITs and thus the contents and experience with regard to BITs should not serve as the only guidelines in the negotiations with ASEAN. A comprehensive high-standard investment agreement with gradual investment liberalization could be acceptable. With this new vision in mind, the Chinese delegation could be authorized with broader — as well as uniform — authority among representatives from various state agencies.

Secondly, China should study the AIA Framework Agreement, since the AIA Framework Agreement is the template for the negotiations. It is necessary for China to strengthen its understanding, study the AIA Agreement, and communicate in a timely fashion with ASEAN about the implications of some articles and the legal consequences thereof. ASEAN Investment Area was initiated in 1995 to promote intra-ASEAN investment flows and enhance ASEAN as an attractive and competitive site for foreign direct investments.[28] The AIA will be established as a competitive area by 2010, thus coinciding with the establishment of CAFTA.

Thirdly, China should study the legal systems for foreign investment of each ASEAN country. Each ASEAN country has its own unique situations and requirements towards foreign investment, which China should deal with flexibly so as to ensure an acceptable agreement by all ASEAN countries.

Fourthly, China should study the BITs between China and ASEAN countries, as well as the BITs concluded by Malaysia and Singapore with other countries. The BITs between China and ASEAN countries (except for Brunei) provide the

---

[28] Preamble of the Framework Agreement on the ASEAN Investment Area.

common basis for concluding investment agreements. The past investment negotiations experience shows that Malaysia and Singapore are the leading countries in investment negotiations. China would benefit from discovering their requirements, expectations and positions with respect to the agreement, by studying their BITs. In this way, China can "hit the bull's eye" in the investment negotiations.

Fifthly, China should initiate its own draft for investment in order to reflect its desired positions. The initiative of ASEAN to take AIA Framework Agreement as a template does not necessarily mean an automatic incorporation of its contents by the future CAFTA investment regime. Though China does not have an existing multilateral investment agreement to be presented to the negotiations, a draft by think-tank experts can express our overall positions in the negotiations.

Last but not least, the approach of bilateral or pluralateral negotiations added by the China-ASEAN Protocol may be employed.[29] For those matters which are most difficult to reach with ASEAN as a whole, China may negotiate and settle them with individual ASEAN members by way of bilateral or pluralateral negotiations. Later on, the concluded agreement may be extended to ASEAN as a whole.

### Strategies for ASEAN and its Member Countries

Firstly, ASEAN and its member countries should coordinate their interests and formulate a common position towards China, as well as future investment agreement.

Secondly, ASEAN should treat China as a unique country and confer upon it a more flexible treatment. Seemingly, China is a country with rapid economic development. However, it is in essence experiencing much political and economic transition. As such, it should be accorded special attention and unique treatment. In the long run, such changes are sure to bring about greater stabilization, reconciling both appearance and infrastructure. With this understanding, ASEAN will likely gain more confidence in its negotiations with China.

Thirdly, ASEAN should expect to exercise patience in negotiating with China, rather than rushing directly to a liberalized investment agreement. A Chinese idiom agrees well with this universal saying, "Haste makes waste". It *will* take time to explain investment liberalization proposed by ASEAN to Chinese leaders, industries and people. China *will* require a large amount of time to understand and evaluate the impact of investment liberalization upon the entire country; coordinate its relevant government agencies and industries; and make corresponding amendments to its laws and regulations. Once a promise is made, China will deliver. The amendment of its relevant laws and regulations even before its entry into WTO

---

[29] Article 4 of the China-ASEAN Protocol.

serves as a good example. Furthermore, China's accession to the WTO has resulted in the quick liberalization of foreign investment by the legal regime. It is highly reasonable to expect China to further liberalize its legal regime in the near future.

Fourthly, ASEAN should sign an investment agreement with China containing commitments of gradual investment liberalization. This is a tactful way to solve the controversial issues in terms of investment liberalization. The key obstacle to China's unwillingness to sign the investment agreement is its inability to promise a full investment liberalization at present, i.e., it cannot, at the moment, promise to open up most of its industries to foreign investment and to confer national treatment, especially pre-establishment national treatment to foreign investment. It is impossible and impractical to force China to promise full investment liberalization; but it is quite acceptable for China to undertake commitments of gradual investment liberalization. Therefore, ASEAN can adopt this strategy as a precursor to avoiding unnecessary frustration. A good illustration would be the bottom-up approach for national treatment used by GATS, which is a good way for gradual investment liberalization. Other approaches such as standstill, rollback, no backtracking, sensitive exclusion list and Temporary Exception list are options as well.

### Strategies for Both Parties

For both parties concerned, mutual understanding and communications are essential. Flexible measures could be taken for the same objectives and goals. Furthermore, compromise is necessary during negotiations. The successful experiences of multilateral agreements related to investment such as MIGA and ICSID show that compromise among different parties is an essential element.

## Significance and Implications of Investment Negotiations

The investment negotiations and their final success have great significance and implications for China and ASEAN — individually and corporately — as well as the developing countries.

### Significance and Implications for China

This China-ASEAN investment negotiation marks China's first attempt at regional or multilateral investment agreement. As such, the final product will be the first multilateral investment agreement that China has entered into. The significance and implications of the negotiations and agreement for China are self-evident.

The investment negotiations will, in themselves, aid China in accumulating experience in this regard. China's history is void of multilateral investment negotiations. By contrast, ASEAN has gleaned many benefits from its past experiences in negotiations. For example, the pattern of "framework agreement + protocol + annexes" of CAFTA agreements is borrowed from an ASEAN practice. Additionally, the AIA Framework Agreement serves as the template for China-ASEAN investment agreement. Similarly, China will accumulate experiences from the investment negotiations, which will in turn draw future benefits towards other negotiations.

The ongoing China-ASEAN investment negotiations will give Chinese practitioners and scholars the impetus to study the existing problems and corresponding strategies, and create practical and complete CAFTA agreements. In turn, it will promote the signing of other regional or even global investment agreements in future.

China's investment environment will be further improved and thus will attract more FDI inflows from both ASEAN and other countries. Prior to and after China's entry into the WTO, fundamental changes have already taken place with regard to China's foreign investment laws and regulations, according to China's obligations and commitments under its protocol for entry into the WTO. The conclusion of China-ASEAN investment agreement will further bring about improvement of China's domestic laws and regulations. The improvement of China's investment environment and the great potential of China's domestic market will absorb more and more foreign investments.

Chinese overseas investments in ASEAN countries would receive greater legal protection once the investment agreement is signed. With China's "Going Abroad" strategy, increasing investments from China will flow to the neighboring ASEAN countries and would be protected by the agreement.

### *Significance and Implications for ASEAN and its Member Countries*

The end product of the investment negotiations with China would become a benchmark for ASEAN, since this will also be ASEAN's first investment agreement with another country outside ASEAN that it has concluded as an entity. In addition to China, ASEAN is undergoing free trade area negotiations with some other countries, such as India, Japan, Australia and New Zealand. The ASEAN-India FTA negotiations commenced in January 2004, and is to conclude by 30 June 2005.[30] The ASEAN-Japan negotiation on the Comprehensive Economic Partnership (CEP) Agreement was expected to commence at the beginning of

---

[30] Article 8 of the Framework Agreement on Comprehensive Economic Cooperation between the Association of Southeast Asian Nations and the Republic of India.

2005.³¹ ASEAN, Australia and New Zealand had planned to commence negotiations towards a free trade agreement (FTA) in early 2005.³² It is clear that ASEAN's experience in the investment negotiations with China will enrich its FTA negotiations with other countries.

The final conclusion of the investment agreement will provide ASEAN investors with a larger space to perform and secure them favorable position and sufficient protection. Even the integrated ASEAN is still too small in terms of its total GNP, and it will benefit ASEAN if it enters into economic partnerships with leading economies and economic blocs.³³ China has a huge and dynamic economy and its growing demand for ASEAN goods and services could serve as a new engine for growth. With the creation of CAFTA, companies operating in ASEAN would have 1.7 billion consumers, with a combined gross domestic product of US$ 1.5 trillion to US$ 2 trillion as their potential market.³⁴ Such a huge market and China's investment regime would be opened up for ASEAN investors. Furthermore, ASEAN investors can take advantage of China's first regional investment agreement to be in a more favorable position compared with other countries or regions. Thus it is much easier for ASEAN investors to access the Chinese market while gaining added legal protection for their investments and interests in China.

The future investment agreement would direct more investments from China or from other countries into ASEAN countries. Existing barriers to investments of both ASEAN and China would be reduced and finally eliminated by the proposed investment agreement, thereby promoting FDI inflows to ASEAN from both China and other countries. Moreover, in consideration of China's rapid economic development, China is now carrying out a new strategy of "Going Abroad". With the encouragement of this new policy, China's investment abroad will soon increase dramatically. If a closer economic relationship between China and ASEAN is established, that is, if CAFTA is established and the investment agreement between the two sides is concluded, ASEAN countries will become the most desirable market for China's overseas investment in future. Currently, most of the Chinese investments in ASEAN flow mainly into four ASEAN new members, i.e., Vietnam, Lao PDR,

---

³¹ Article 10 of the Framework for Comprehensive Economic Partnership between the Association of Southeast Asian Nations and Japan.
³² Julie Glasgow, "ASEAN-Australia-New Zealand Summit and launch of negotiations towards Free Trade Agreement", http://www.asialine.dfat.gov.au/asialine/Asialine.nsf/WebSpecialFeatures/1?OpenDocument (accessed 20 February 2005).
³³ Chia Siow Yue, "ASEAN-Chia Free Trade Area", paper for presentation at the ACP Conference in Hong Kong, 12–13 April 2004, p. 5.
³⁴ ASEAN Secretariat, "ASEAN-China Free Trade Area Negotiations to Start Next Year", News Release, 30 October 2002, http://www.aseansec.org/13125.htm (accessed 30 August 2004).

Cambodia and Myanmar.[35] Therefore, the economic integration between the two sides would certainly stimulate trade and investment in ASEAN countries.

The confidence of foreign investors in ASEAN will be partly restored if ASEAN signs the investment agreement with China. The 1997 Asian financial crisis significantly eroded the confidence of foreign investors in ASEAN countries and thus greatly decreased the FDI inflows to ASEAN countries. In contrast, China is a driving engine in absorbing FDI, whose inflows had grown more than tenfold from 1990 to 2000[36] and is now the largest recipient of FDI in developing countries. ASEAN countries will share and benefit from China's quick development if the free flow of investment between the two parties is realized and the economic integration between them is established. With this vision realized, foreign investors' interests as well as opportunities to invest would increase.

Apart from trade and investment, this investment agreement would benefit ASEAN's economic development as a whole. The economic integration between China and ASEAN will stimulate growth and thus attract increased foreign capital. Consequently, employment will be on the rise and the economic welfare of each country will be enhanced.

### Significance and Implications for Both Sides

Generally speaking, the proposed CAFTA and the investment agreement would have obvious political and economic significance. CAFTA will contribute to improving political and social relations between ASEAN and China, building on existing geographic proximity, as well as historical and cultural ties. ASEAN-China collaboration will contribute to a balance of power in East Asia and provide for a larger and more effective voice in international fora.[37] The investment agreement will bring China and ASEAN towards a more intimate economic relation. CAFTA will create the world's biggest economic bloc with a population of around 1.8 billion, including China's 1.3 billion and ASEAN's 500 million people. All these will bring about regional peace and prosperity in China and Southeast Asia. The win-win deal will help to eliminate the "China Threat" perception in ASEAN countries, which is of grave concern to China.

---

[35] "Forging Closer ASEAN-China Economic Relations in the Twenty-first Century", a report submitted by the ASEAN-China Expert Group on Economic Cooperation, October 2001, p. 23, available on the ASEAN Secretariat website.

[36] "Forging Closer ASEAN-China Economic Relations in the Twenty-first Century", a report submitted by the ASEAN-China Expert Group on Economic Cooperation, October 2001, p. 4, available at the ASEAN Secretariat website (accessed 15 October 2004).

[37] Chia Siow Yue, "ASEAN-Chia Free Trade Area", paper for presentation at the ACP Conference in Hong Kong, 12–13 April 2004, p. 15.

The investment agreement and the creation of CAFTA will facilitate foreign investments among the regional countries and into the region. Both theories and reality show that geo-economic changes in various parts of the world have helped to bring about greater and different styles of competition for FDI,[38] and that regional integration and economic cooperation would help to improve the competitiveness of regions and of the constituent countries in attracting FDI.[39] This is the precise purpose for the creation of CAFTA. Therefore, the construction of China-ASEAN investment regime itself is not an end in itself but a means to an end. The real purpose of the investment regime is, taking the China-ASEAN Free Trade Area as a whole, to facilitate intra-area investments and foreign investments to the area, so as to promote economic development of both ASEAN and China through a steady, transparent and open investment regime. Therefore, CAFTA is expected to increase intra-regional trade and investment, improve efficiency and economy of scale and develop a larger market for capital and talent, thereby creating more opportunities for businesses.[40] With lower market risk and uncertainty, not only will more ASEAN and Chinese companies be willing to invest within the integrated market, but US, European and Japanese companies — which are also interested in making inroads into the Asian market — will also be attracted to invest in the integrated market. The integration of ASEAN with China can entice more foreign corporations since, standing on their own grounds, each lacks sufficient pulling power. A good example in this regard is the AIA. After the AIA was established in 1999, foreign investments to the AIA during 2001–2002 increased by 13.4%.[41] Lastly, enterprises will invest more into research and development and hence promote technological innovation.[42]

From the long-term perspective, the signing of the China-ASEAN investment agreement will help to form a unified position and perspective in the investment

---

[38] Kee Hwee Wee and Hafiz Mirza, "ASEAN Investment Cooperation: Retrospect, Developments and Prospects", p. 2, http://www.gapresearch.org/finance/ASEAN%20Investment% 20Cooperation.pdf (accessed 19 February 2005).

[39] Kee Hwee Wee and Hafiz Mirza, "ASEAN Investment Cooperation: Retrospect, Developments and Prospects", p. 3, http://www.gapresearch.org/finance/ASEAN%20Investment% 20Cooperation.pdf (accessed 19 February 2005).

[40] ASEAN Secretariat, "ASEAN-China Free Trade Area Negotiations to Start Next Year", News Release, 30 October 2002, http://www.aseansec.org/13125.htm (accessed 30 August 2004).

[41] ASEAN Secretariat, Joint Press Statement of the Fifth ASEAN Investment Area Ministerial Council, 11 September 2002, Bandar Seri Begawan, Brunei Darussalam, http://www.aseansec.org/12614.htm (accessed 19 February 2005).

[42] "Forging Closer ASEAN-China Economic Relations in the Twenty-first Century", a report submitted by the ASEAN-China Expert Group on Economic Cooperation, October 2001, p. 31, available on the ASEAN Secretariat website.

legal systems for developing countries and benefit them in future international investment agreements. Both China and ASEAN countries are developing countries and the success of the establishment of CAFTA and the conclusion of an investment agreement will allow them to have a larger voice in international trade and investment affairs on issues of common interest.[43] It will also help to accumulate experience in text drafting and negotiations for future WTO investment negotiations, and even for future global international investment agreements. Moreover, it will help to integrate ASEAN and China as an entity for developing countries, so as to enable a unified position in the international arena of investment negotiations, thereby strengthening the negotiating power of developing countries and creating a fairer global investment agreement from the perspective of developing countries.

---

[43]"Forging Closer ASEAN-China Economic Relations in the Twenty-first Century", a report submitted by the ASEAN-China Expert Group on Economic Cooperation, October 2001, p. 2, available on the ASEAN Secretariat website.

# Part V

## Issues in China-ASEAN Regional Cooperation

## Chapter 9

# China's Border Trade with Newer ASEAN Members: Problems and Prospects

*LIAO Shaolian*

Border trade is an important part of foreign trade in many countries. Due to the residents' needs for commodity exchange and the convenient geographical proximity of the countries, border trade[1] between China and neighboring newer ASEAN members (NNAMs, i.e., Vietnam, Laos and Myanmar) has had a rather long history. It has also significantly contributed to the economic and social development in the border regions of the contiguous countries. This chapter first reviews the development process of border trade since the 1980s. After discussing the positive effects of border trade on the local economy and the trends over the past decade, it analyzes the problems which caused fluctuations since the late 1990s. Finally, it explores the prospects for the future and points out that border trade between China and NNAMs is experiencing a dramatic change in the direction of development, operation mechanisms and future role.

---

[1] In China, there are mainly two types of border trade: exchange market trade and small-volume trade. Exchange market trade refers to the selling and buying of goods/commodities in kind in the border markets of contiguous countries. For example, a Chinese village located near the border (within 20 kilometers) opens a market for the inhabitants of a neighboring Vietnamese village to sell or buy products and vice versa (this includes barter trade). In China, though there is no ceiling for the value of small-volume trade, only approved small-volume trade companies can engaged in this type of border trade. In Vietnam, there are three types of border trade, namely formal trade, small-volume trade and exchange market trade. The Ministry of Trade-approved imports and exports of commodities/goods through border ports at both the international and national levels are regarded as formal trade. Imports and exports approved by provincial governments and traded through border ports are regarded as small-volume trade. The ceiling is half a million dong per deal. Exchange market trade includes barter trade and the commodities are mainly for daily consumption and production. In practice, it is sometimes hard to distinguish between formal trade and small-volume trade because the commodities of formal trade often go through local ports and in the form of small-volume trade. Due to the difference in the definition of border trade and lack of data on exchange market trade, the statistics on border trade provided by the countries concerned vary all the time.

## Overview of Border Trade Between China and NNAMs

*Favorable Factors*

China's two southwestern provinces (Yunnan Province and Guangxi Autonomous Region) share borders with Myanmar, Laos and Vietnam. The inhabitants on both sides of the border share the same cultures[2] and have a long history of close economic contact. Along the border, there are official ports as well as numerous local passageways.[3] The cross-border distances between villages/counties are often more or less the same with or even shorter than the distances between domestic villages/counties. As border regions are mostly mountainous, cross-border trade in many cases offers great savings in terms of transportation costs and convenience. For example, people from Pingxiang in Guangxi, China, and those from Nan Sam in Lang Son province, Vietnam, would choose to buy commodities in either the domestic market or cross-border market, depending on which market had the lower prices. Therefore, border trade is based mainly on price differences. A small deal may yield high profits because of price differences and low transportation costs. When demand exceeds supply in the cross-border market, a local businessman would amass the commodities in short supply through wholesale and retail channels and export them across the border to satisfy the market needs of the other side. Due to its flexibility and convenience, border trade has been popular in the border regions.

The development of border trade between China and its neighboring countries is not only attributed to geographical and cultural factors, and the economic need for commodity exchange. The improvement of bilateral relations between China and the neighboring countries, and the encouragement and support of local governments of the countries concerned have also played an important part. In 1991, China and Vietnam signed a treaty on border trade between the two countries. In accordance with the treaty, 21 pairs of border ports were opened. Aside from these ports, there were 26 exchange markets on the Yunnan side and 23 on the Vietnamese side.[4] Since then, the two countries have signed a series of documents,[5] which have

---

[2] For example, minority ethnic groups account for 66% of the population in the six northern provinces in Vietnam. The same ethnic groups live in the nearby areas on the Chinese side. They share the same languages and cultures.

[3] In Guangxi Autonomous Region, which is adjacent to Vietnam, there are 261 passageways through which inhabitants in the border regions can pass freely. (Fan Honggui, "Anti-drug Policies and Measures in Vietnam", *Around Southeast Asia*, Supplementary Issue, 2000.)

[4] Wang Shilu, ed., *Contemporary Vietnam* (Sichuan People's Press, 1992), p. 394.

[5] For example, Temporary Agreement on Solving Border Issues in November 1991, MOU on Inspection of Import and Export Commodities in February 1992.

created favorable conditions for the development of border trade. In September 1996, the Vietnamese government decided to open Mong Cai in Quang Ninh province, and adopted preferential policies there. One year later, Dong Dang, Huu Nghi Quan (Friendship) and Thanh (in van Lang County) in Lang Son province were opened, and these were followed by the opening of some border ports in Cao Bang province in May 1998.[6] The signing of a land-borderline treaty in December 1999 indicated the final resolution of land-border disputes between the two countries and greatly improved the business environment in the border regions. In June 2001, Vietnam further adjusted its policy on trade with China and lowered duties and fees in border trade. At the same time, the Vietnamese government encouraged businessmen from more prosperous areas to set up enterprises in border ports of its northern provinces. The Lao government established the Muoding border trade zone and implemented a series of preferential policies. Myanmar reduced its service charge rate in border trade from 10% to 8% in 1988.[7]

Due to the above-mentioned factors, border trade between China and NNAMs experienced rapid progress since the early 1990s.

### Development Process of Border Trade

Border trade between China and Vietnam began rather early. In the beginning of the 1950s, the two countries signed a treaty on small-volume trade in border regions.[8] However, due to deteriorating bilateral relations between the two countries, trade relations were suspended in 1978. Border trade resumed in the second half of 1982. Though the Vietnamese government prohibited the Vietnamese inhabitants in the border regions from doing business with the neighboring Chinese in the early 1980s, regular markets were opened at the foot of mountains on the Chinese side and barter trade was carried out. Since the Vietnamese government prohibited border trade with China and there were still many landmines scattered in the border regions, the number of people engaged in border trade was very limited. The Vietnamese exchanged leather and agricultural products for Chinese commodities of daily use (e.g., cloth, electric torches, pharmaceuticals, rubber shoes and wine). In September 1983, the government of Guangxi, China, opened some border trade markets in safe and convenient locations. Due to large price differentials,

---

[6] According to Vietnamese regulations, Chinese inhabitants with border certificates or border travel passes issued by the Chinese local governments can enter freely the Vietnamese economic zones in the border regions.
[7] Guo Kuan, *Entering Myanmar* (Yunnan Fine Arts Press, 2004), p. 71.
[8] Nong Lifu, "A Review of China-Vietnam Border Trade and Policy Recommendation", *Around Southeast Asia*, Supplementary Issue (2000).

residents of the border regions engaged in border trade rapidly became prosperous, encouraging more and more Vietnamese to enter the business. In 1988, bilateral relationships in the border regions improved, and the Vietnamese government formally allowed the residents in border villages and towns to cross over to China to visit their relatives and exchange commodities of daily use. In 1989, small-volume border trade resumed, and a series of regulations on border crossing and border trade were formulated. At the same time, the Chinese government implemented preferential and flexible border trade policies and improved the infrastructure. From then on, more and more Vietnamese and Chinese inhabitants engaged in border trade, which resulted in a substantial increase in the volume of trade. At that time, border trade was the main form of bilateral trade between Vietnam and China and about 70–80% of border trade were carried out between Vietnam and Guangxi Autonomous Region. In 1989, the border trade volume between Guangxi and Vietnam reached RMB 450 million yuan.[9]

The bilateral relationship between Vietnam and China was normalized in November 1991, resulting in an agreement on border affairs and the opening of 21 pairs of ports along the border. Agreements on cross-border transportation of commodities and border trade were signed in 1994 and 1998 respectively. By 1999, 25 pairs of ports had been opened.[10] Besides these, there were 59 pairs of border passageways and 13 border trade markets. All these provided favorable conditions for the development of border trade and other forms of economic exchange. In order to further promote border trade, the Chinese government implemented policies on the exemption and reduction of customs duties and established tax-free zones. With the improvement of bilateral relations and the implementation of preferential policies, border trade volumes between Vietnam and China increased steadily from US$247 million in 1991 to US$998 million in 1995. The volume of border trade between Vietnam and Guangxi rose from RMB 1.9 billion yuan in 1992 to RMB 3.1 billion yuan (Table 9.1). The total volume of border trade between Vietnam and Yunnan also increased from RMB 1 billion yuan in 1989 to RMB 3 billion yuan in 1995.[11] In the Jiegao Border Trade Zone (Yunnan Province, China) near Mujie of Myanmar, border trade volume increased from RMB 1 million yuan in 1988 to RMB 1.2 billion yuan in 1992, and further to RMB 2.5 billion yuan in 1995.[12]

---

[9] Cao Daming and Qi Huan, *Entering Vietnam* (Yunnan Fine Arts Press, 2004), p. 115.
[10] Amongst them, four pairs of international ports, seven pairs of ports at the national level, and 14 pairs of ports for small-volume trade were opened. (*Around Southeast Asian*, Supplementary Issue, p. 82, 2000)
[11] *Social Sciences in Yunnan*, Vol. 5, p. 20 (2003).
[12] *Southeast Asia*, Vol. 1, p. 16 (2003).

Table 9.1  Border Trade Volumes between Vietnam and Guangxi (1988–1999)

(RMB 100 million yuan)

| Year | 1989 | 1990 | 1991 | 1992 | 1993 | 1994 | 1995 | 1996 | 1997 | 1998 | 1999 |
|---|---|---|---|---|---|---|---|---|---|---|---|
| Volume | 4.5 | 7.6 | 18.9 | 26.0 | 26.0 | 26.4 | 26.5 | 29.0 | 30.9 | 27.6 | 33.9 |

Source: Bureau of Border Trade, Guangxi Autonomous Region; Nong Lifu, "Review of China-Vietnam Border Trade and Policy Recommendation", *Around Southeast Asia*, supplementary issue, p. 47 (2000).

Table 9.2  Border Trade between Vietnam and Yunnan (1996–2004)

(US$ million)

| Year | Total Border Trade | Import from Yunnan | Export to Yunnan |
|---|---|---|---|
| 1996 | 14.60 | 8.58 | 6.02 |
| 1997 | 12.54 | 10.17 | 2.28 |
| 1998 | 27.29 | 22.22 | 5.07 |
| 1999 | 45.94 | 38.08 | 7.86 |
| 2000 | 59.74 | 53.20 | 6.54 |
| 2001 | 82.86 | 64.80 | 18.06 |
| 2002 | 83.18 | 56.49 | 26.69 |
| 2003 | 86.62 | 58.92 | 27.46 |
| 2004 | 107.41 | 60.10 | 47.31 |

Source: Bureau of Foreign Trade and Economic Cooperation, Yunnan Province; Cao Daming & Qi Huan, *Entering Vietnam* (Yunnan Fine Arts Press, 2004) p. 74; *Report On Situation of Border Trade*, Border Trade Office, Bureau of Foreign Trade and Economic Cooperation, Yunnan Province, 11 January 2005.

Since the late 1990s, however, border trade between China and NNAMs fluctuated noticeably (Table 9.2). The fluctuations can be attributed mainly to two factors. On the one hand, the financial crisis in Southeast Asia beginning in 1997 brought about a decline in market demand in border trade. On the other hand, a series of problems in the administration of border trade and fierce competition from other countries hindered the development of border trade. (These will be discussed in more detail below.) Border trade volume between Vietnam and China dropped from around US$1 billion in 1995 to US$0.5 billion in 2002.[13] The decline was more evident in imports from China. Small-volume trade between Guangxi and Vietnam declined to RMB 1.29 billion yuan in 2000, a decline of 16.8% over the previous year, while imports from China dropped to RMB 0.74 billion yuan in

---

[13] Cao Daming and Qi Huan, *Entering Vietnam* (Yunnan Fine Arts Press, 2004), p. 33.

2000, a decline of 30.3% over the previous year.[14] In 2001, exports from Yunnan province to Myanmar, Laos and Vietnam was only US$0.23 billion, a drop of 17.2% over the previous year.[15] In 2001, border trade volume between Vietnam and Guangxi further declined by 2.5%, while imports from China decreased by 9.9%.[16] Similarly, imports from Yunnan declined from US$64.8 million in 2001 to US$56.49 million in 2002.[17] Border trade between Yunnan and Laos showed a similar trend of development (Table 9.3). However, due to market expansion in all countries involved, border trade between China and NNAMs increased again in the past two years.[18] In 2004 (Table 9.4), border trade between Yunnan and NNAMs reached US$524.07 million, an increase of 25% over the previous year, with an increase of 22.1% for export and 29.3% for import.[19]

Table 9.3  Border Trade Volumes between Yunnan and Laos (1995–2002)

(US$ million)

|        | 1995  | 1996  | 1997 | 1998 | 1999 | 2000 | 2001 | 2002 | 2003  | 2004  |
|--------|-------|-------|------|------|------|------|------|------|-------|-------|
| Total  | 24.78 | 13.56 | 9.28 | 9.80 | 9.88 | 9.64 | 7.90 | 8.44 | 13.08 | 16.02 |
| Export | 19.78 | 9.12  | 5.90 | 4.18 | 4.54 | 3.74 | 3.71 | 2.48 | 6.59  | 8.90  |
| Import | 5.00  | 4.44  | 3.38 | 5.62 | 5.34 | 5.90 | 4.19 | 5.96 | 6.49  | 7.12  |

Source: Statistics Department, Bureau of Foreign Trade and Economic Cooperation, Yunnan Province, Li Ping, *Entering Laos* (Yunnan Fine Arts Press, 2004) p. 36; *Report On Situation of Border Trade*, Border Trade Office, Bureau of Foreign Trade and Economic Cooperation, Yunnan Province, 11 January 2005.

Table 9.4  Border Trade between Yunnan and NNAMs in 2003 and 2004

(US$ million)

| Year | Total Trade | Export to NNAMs | Import from NNAMs |
|------|-------------|-----------------|-------------------|
| 2003 | 419.27      | 252.78          | 166.49            |
| 2004 | 524.07      | 308.76          | 215.31            |

Source: *Report on Situation of Border Trade*, Bureau of Foreign Trade and Economic Cooperation, Yunnan Province, 20 January 2005.

---

[14] Nong Lifu, "Foreign Trade between Guangxi and Southeast Asia: Present and the Future", *Around Southeast Asia*, Vol. 3–4, p. 98 (2002).
[15] *Southeast Asia*, Vol. 4, p. 58 (2002).
[16] Gu Xiaosong, ed., *China-ASEAN FTA and Guangxi* (Guangxi People's Press, 2002), p. 142.
[17] Cao Daming and Qi Huan, *Entering Vietnam* (Yunnan Fine Arts Press, 2004), p. 74.
[18] *Report On Situation of Border Trade*, Bureau of Foreign Trade and Economic Cooperation, Yunnan Province, July 2004.
[19] *Report On Situation of Border Trade*, Bureau of Foreign Trade and Economic Cooperation, Yunnan Province, July 2004.

The development of border trade between China and NNAMs is by no means balanced. Due to relatively more favorable economic and natural conditions, Vietnam shows a greater demand for Chinese products, especially in the border area between Vietnam and Guangxi, where the volume of border trade accounts for about 80% of the total border trade between the two countries. Out of the total border trade volume of Yunnan province between 1991 and 1998, trade with Myanmar accounted for 82.8% (US$1.4 billion), 12.0% for Vietnam (US$0.2 billion) and 5.2% (US$0.9 billion) for Laos. In the past few years, Myanmar accounted for 75–76% of Yunnan's total border trade with NNAMs.[20]

## Characteristics of Border Trade

In the past decade, border trade has displayed the following features.

First, both the forms of border trade and commodity structure have become more diversified. In the initial stages, barter trade was the main form of border trade. Later, exchange market trade and small-volume trade dominated, and entrepôt trade developed gradually over the last few years. Some commodities are produced in the border counties and villages, rather than in more central parts of the countries, and transported to the border regions. For example, Vietnamese businessmen would go to see commodity samples in factories in the Chinese border counties to place orders, while Chinese trading companies also sent people to Vietnam to order goods. Before the 1990s, Vietnam mainly imported commodities for daily use, processed food, construction materials, farm tools and medicines. Starting from the mid-1990s, the dominant commodities changed to means of production, e.g., machinery, equipment, electrical products, electronic products and construction materials. In the past few years, diesel engines, walking tractors, and industrial chemicals occupied a rather large proportion of border trade. However, primary products such as food products and industrial raw materials are still major exports from Vietnam.

Second, different kinds of participants are now involved in border trade. At the initial stage, border trade was carried out mainly by individual inhabitants in the border regions. Since the 1980s, in China and Vietnam, state-owned enterprises (SOEs) and collectively-owned enterprises have become the major actors. As for SOEs, some are enterprises owned by the central government but most are owned

---

[20] In 2002, Myanmar accounted for 75.3% (US$0.3 billion) of Yunnan's total border trade with NNAMs. (Guo Kuan, *Entering Vietnam*, Yunnan Fine Arts Press, pp. 29–30, 2004) In 2004, Myanmar occupied 76.4% of Yunnan's total border trade with NNAMs. (*Report On Situation of Border Trade*, Border Trade Office, Bureau of Foreign Trade and Economic Cooperation, Yunnan Province, 11 January 2005.)

by local governments of border provinces. Recently, with the reform of foreign trade, private companies are playing a more active role. In 2004, border trade volume of private companies in Yunnan province reached US$250 million, an increase of 87% over the previous year, and accounted for 47% of total border trade of the province.[21] Enterprises from other parts of the country and even businessmen from Hong Kong and Taiwan have also come to engage in border trade.

Third, Chinese Renminbi (RMB) is the currency used most often in border trade because of its stable value and low exchange rate risk. The use of Renminbi can help make up for the lack of hard currency such as US dollars in NNAMs.

Fourth, the form of account settlement is changing from cash payment to settlement in banks. Since payment in cash is inconvenient (e.g., time consumed in counting cash, risk in carrying large amounts of cash and danger of receiving fake cash), more and more trading companies and individuals prefer to settle their accounts through the banks. In 1996, China and Vietnam signed an agreement on account settlement cooperation in border trade. Accounts can be opened in the banks of the other country.[22] In practice, a small proportion of the accounts in border trade are settled in official banks and the rest in underground private banks. The private banks are mainly engaged in exchange of money and account settlement. In Vietnam, the private banks have legal status[23] and keep close business contacts with their Chinese counterparts. Money is deposited in local private banks on both sides and they settle the cross-border accounts through telephone calls. Thus, the recipients can easily draw the money remitted from the other side in their local private banks. This kind of business transaction has become more and more popular in the border regions due to quicker settlements, simpler procedures and lower costs.

*Positive Impacts of Border Trade*

Border trade has played an important role in the economic and social development of the border regions.

Firstly, it contributes to higher economic growth, change in economic structure and increase of government revenue. In the past, agriculture in the northern

---

[21] *Report On Situation of Border Trade*, Border Trade Office, Bureau of Foreign Trade and Economic Cooperation, Yunnan Province, 11 January 2005.
[22] When paying money to their Chinese counterparts, the Vietnamese companies have to deposit Vietnamese dong in their own banks. The amount should be equivalent to value in RMB which is converted from Vietnamese dong to US dollars and then to Chinese RMB. That means the money has to be converted twice: from Vietnamese dong to US dollars and from US dollars to Chinese RMB.
[23] In China, strictly speaking, the so-called underground private banks are illegal. However, there are more than 300 such banks in Pingxiang and Dongxing bordering Vietnam.

provinces of Vietnam occupied a very large proportion of the economy, while the shares of industry and service sector remained consistently small. Since the border area was opened and border trade developed, the northern provinces experienced rapid economic growth (Table 9.5). The economic structure also underwent dramatic changes. Tourism, commerce, banking, transportation, telecommunications, etc. developed rapidly. Originally, these border provinces relied heavily on financial support from the central government for their economic development. Since border trade became the major economic impetus in these regions, many new infrastructure projects were generated, e.g., highways, bridges, an electricity supply system, telecommunications networks, schools and clinics. Taking Lao Cai in Vietnam as an example, since the border port can retain the duty and value-added tax collected as part of the local government revenue, it can afford to quicken its steps in developing local infrastructure, which, as a result, greatly promotes its economic development. Mujie in Myanmar, which shares its border with Ruili of China, was originally a small village but has become an important border city due to rapid development of border trade. The same changes also happen in China. In Pingxiang, Guangxi, there are more than 130 hotels now and the output value of the service sector accounts for two thirds of local GDP. The government revenue rose from RMB 0.8 million yuan in 1980 to RMB 47.5 million yuan in 1995, and 70% of the revenue came from tax and fees of border trade.[24]

Secondly, border trade has greatly promoted imports and exports in the border regions, which has attracted more foreign investments. Though the quality of the commodities in border trade is relatively low, they generally suit the market needs, and border trade has been the major form of foreign trade in the border regions.[25] Furthermore, the development of border trade and the local economy are also important factors in attracting foreign capital. The preferential treatments offered

Table 9.5 Annual Average GDP Growth Rates of Northern Provinces in Vietnam (1991–1999) (%)

| Province | Lang Son | Cao Bang | Guang Ninh | Lao Cai | Ha Giang* |
|---|---|---|---|---|---|
| Annual Growth Rate | 7.51 | 6.5 | 8.23 | 6.3 | 10.3 |

*Annual average growth rate between 1996 and 2000.
Source: *People's Daily* (Vietnam), 1 September 2000.

[24] Nong Lifu, "A Review of China-Vietnam Border Trade and Policy Recommendation", *Around Southeast Asia*, Supplementary Issue (2000).
[25] Border trade between Yunnan and NNAMs accounted for two thirds of total foreign trade of Yunnan Province in 2001.

in these regions have contributed a great deal to the increase in foreign investments. Many investment projects are closely linked to border trade.

Lastly, border trade helps to increase the income of local inhabitants and raise their living standards. Since the local economy in the border regions has diversified and more job opportunities have been created, the local people are now employed in different economic sectors and many even run their own businesses. The incidence of poverty has declined, while the proportion of rich households has increased. A large proportion of the people now live a better life. According to a Vietnamese survey in 1996, among the 720 households in Dong Dang, Lang Son province, 80% had motorbikes and there were dozens of automobiles. Among the 50,000 households in Mong Cai, Quang Ninh province, there were more than 4,000 motorbikes, 160 automobiles and 664 boats.[26] In the border town of Aidian in Guangxi, 80% of the local people built new houses in the past decade.

## Existing Problems in Border Trade

Though border trade has played an important role in promoting economic development and increasing the income of inhabitants in the border regions, there have been fluctuations since the late 1990s. A series of difficulties and problems have prohibited development at the pace of the 1980s and early 1990s.

### Weak Administration of Border Trade

In China's border regions, import quotas set by the central government are usually quite restrictive and are transmitted to the grassroot level very late. There is a big gap between the quotas and market demand. The delayed arrivals of quotas and licenses have caused local trading companies to miss many import opportunities, resulting in a constant trade imbalance between the countries involved. For example, in 2001, the Honghe District's export value in its "small-amount trade" reached US$45.4 million but its import value was only US$7.2 million. The import-export ratio in Hekou Port has long been 1:12.[27] The persistent trade imbalance has become one of the major factors that induced the Vietnamese government to adopt its import-substitution policy and restrict the imports of many Chinese products.

Although China implements joint inspection (commodity inspection, health quarantine, and animal and plant quarantine) in the border ports and has simplified the inspection procedure, there are no regulations specifically for border

---

[26] Guangxi Institute of Southeast Asian Studies, *Southeast Asian Economic and Trade Information*, 1998, No. 24.
[27] Cao Daming and Qi Huan, *Entering Vietnam* (Yunnan Fine Arts Press, 2004), p. 147.

regions. Therefore, trading companies engaged in small-volume trade in the border regions have to go through the same procedures as other trading companies in the A-class ports (at the national level). Aidian Township in Guangxi, though a B-class port, follows the management system of an A-class port. Higher requirements and complicated administrative regulations tend to hinder the development of border trade. Moreover, a lack of effective administration in border trade also brings about a series of problems such as smuggling, illegal trade, and evasion of taxes, which results in a serious loss of government revenue. The same problems also exist on the Vietnamese side. According to their estimation, 40–50% of their tax revenue from border trade is lost each year. Since the smuggled products are sold at cheaper prices, legitimate Vietnamese products cannot compete with them, seriously hindering the development of these industries.

Furthermore, there is a considerable lack of dispute settlement institutions in dealing with business issues in the border regions. Border trade relies heavily on credit. In practice, the buyers usually pay a deposit before the goods change hands. This is actually very risky in the border regions, where cheating and refusal to repay debts are rather common. The absence of dispute settlement mechanisms between the neighboring countries in fact encourages these illegal actions.

## Difference of Policies in Contiguous Countries

The difference in trade policies of the countries involved also hinders the smooth-functioning of border trade. The Vietnamese government implements a series of more preferential and less stringent policies in the development of their border ports and economic development zones, and these include duty drawback, financial subsidies and reduction or exemption of land taxes. Compared with these more flexible policies, many Chinese policies seem more rigid and show a lack of decentralization. The local governments have not been given enough decision-making power in formulating their policies based on their actual local conditions. Some Chinese companies, which originally planned to settle down in ports such as Hekou and Tianbao, changed their minds and decided to set up enterprises in Vietnam due to more favorable government policies on the Vietnamese side.[28]

In addition to a lack of transparency in government policies, Myanmar's trade policies change all the time and are implemented arbitrarily.[29] In 1997, Myanmar

---

[28] For example, investors are exempt from land usage tax in Qingshui Port in Vietnam.
[29] The government decided that from 1 March 2002, it would not extend mature import and export licenses to foreign trading companies in Myanmar and would not issue any more licenses to foreign companies. When importing some specific products in the border region, the importers have to apply in person to the Bureau of Border Trade in Rangoon. On 4 August 2004, the transactions tax rate for all import commodities was raised to 25%, ten times the rate in 2001.

began to adopt the policy of encouraging exports and restricting the imports of many commodities of daily use by enforcing complicated application procedures and raising customs duties and charges. Due to strict import volume control and constant devaluation of its currency,[30] the exports of Chinese commodities to Myanmar in border trade were adversely affected.

Laos suffers consistently from an unfavorable balance of trade with China, so it has to resort to both tariff and non-tariff barriers in restricting the inflow of Chinese products. This is also regarded as an effective measure for increasing government revenue. Originally, Laos practised a 50% tariff reduction on many Chinese commodities, but beginning in 2003, no tariff reduction was offered for the import of Chinese motorcycles and electrical household appliances. Moreover, the existence of a large number of checkpoints, repeated inspection and the collection of charges are also obstacles businessmen have to confront.

It should be noted that the impact of border trade on China and NNAMs varies due to a difference in economic scale. Since China is much bigger than any of the NNAMs, the increase of cheaper imports from the latter does not create much rivalry in the related Chinese industries. However, the influx of cheaper Chinese products into the NNAMs will pose a direct threat to them. It is understandable that the governments of NNAMs are rather concerned, and have taken a series of measures to protect their own domestic industries and markets.

### Account Settlement Irregularity in Border Trade

In border trade, account settlement has not been standardized yet and it has not been integrated into the settlement system of banks.[31] Therefore, it is hard for financial authorities to practise effective supervision over border trade and prevent illegal foreign exchange outflow, arbitrage of exchange, etc. Since the value of Vietnamese dong has been unstable,[32] both Chinese and Vietnamese businessmen and companies prefer using Chinese Renminbi for quoting prices and settling accounts. However, in some areas (e.g., Pingxiang in Guangxi), Vietnamese businessmen prefer to trade in Vietnamese dong in settling accounts and the Chinese counterparts who are usually new hands in border trade have to agree, due to fierce

---

[30] The exchange rate of 100 Myanmar Kyat to Renminbi in the market was 25 yuan in 1989, 7 yuan in 1995, 0.7 yuan at the end of 2000, 1.3 yuan at the end of 2001 and 0.7 yuan at the end of 2002.

[31] According to the estimation of Vietnam, only 4.8% of the accounts in border trade with China are settled in banks (Vietnam, *Commercial News*, 2 June 2000.)

[32] The exchange rate of Vietnamese dong to Renminbi was 1580–1630:1 in 1999 and dropped to 1840:1 in 2002.

competition in the market where supply exceeds demand. The Chinese businessmen usually change the Vietnamese dong they earn to Renminbi in "underground private banks", since it is risky to hold unstable Vietnamese dong. As these private banks handle a large proportion of account settlements, not only do their exchange rates become the prevailing exchange rates in border trade, but a rather large share of trading companies' profits also flow into them. Furthermore, the private banks on both sides do not have to rely on direct cash transactions; instead, they settle their accounts according to the schedule agreed upon by both parties. Therefore, it is hard, if not impossible, to estimate the actual amount of accounts settled. This way of settling accounts evades the supervision of financial and taxation authorities and customs, which unquestionably jeopardizes the operation of formal financial institutions. This makes illegal activities such as money laundering, smuggling, and tax evasion more convenient.

*Burdens in Paying Taxes and Fees*

It often happens that fees are levied not according to regulations, but to the opinions of local officials. It has been found that ports of the same country adopt different policies. The repeated payment of taxes and fees increases the burden of trading companies and inhabitants engaging in border trade. In small-volume trade via designated ports of China, 50% of customs duties and value-added taxes (VAT) were levied on imported commodities, with the exception of wine, cigarettes, cosmetics and some other commodities. However, in practice, when the imported commodities were sold in domestic markets, the tax administration would still levy VAT. Originally, the import of 162 commodities was exempted from duty in border ports in Yunnan Province. The levy of 50% of customs duty and VAT already increased import costs substantially, so the actual payment of full duties and VAT made the import of many commodities unprofitable, which exerted a direct impact on border trade volume. In Yunnan Province, jade used to be an important import commodity in border trade. Due to high duties and fees, its import dropped sharply in the past few years.

It is very difficult for Chinese trading companies, if not impossible, to get a refund of duties under official regulations since they are unable to present the various kinds of receipts required in the refund application procedure. A large proportion of export commodities are produced by township and village enterprises, which cannot issue VAT receipts and the overwhelming majority of transactions in border trade is settled in cash without any settlement records in either Chinese or foreign banks. Therefore, claiming tax refunds is extremely problematic for Chinese companies engaging in border export business.

## Weaknesses of Border Trade Companies

In addition to unfavorable external factors such as poor infrastructure (transportation, communication, etc.), small-volume trade companies have to face many other difficulties. As they are usually small in size, they suffer from shortage of capital and it is hard for them to obtain loans from banks. They cannot afford to set up their own factories, so they have to export products produced in other areas. At the same time, the lack of coordination amongst these companies compounded by fierce competition leads to the raising of purchasing prices and lowering of selling prices. They usually have little access to information on foreign markets and policies, or the credit rating of foreign companies. As they often sell products on credit and the buyers often fail to pay them on time, most of the debts turns bad. In the case of Guangxi, it is estimated that total unpaid foreign debt is as high as over RMB 1 billion yuan.[33]

The low quality of products in border trade also affects its healthy development. Although some commodities do pass through the inspection processes of the appropriate government departments and satisfy the quality requirements set either by the central or local governments, many others fail to receive any form of inspection. For example, the service life of Chinese diesel engines sold to Vietnam is only half of that of Japanese engines (though the former's price is also half that of the latter's accordingly). In particular, the products produced by township and village enterprises in border regions (e.g., garments and other commodities for daily use) are of low quality, so consumers in the neighboring countries frequently complain.

## Prospects for the Future

Despite the existence of the above-mentioned problems in border trade, there is still great potential for its further development. In addition to geographical proximity and the tradition of close economic contacts in the border regions, the marketing environment and transportation system have seen vast improvement over the past two decades. More importantly, both the inhabitants and local governments in those regions show great enthusiasm in strengthening economic linkages[34] and pin hopes on border trade for further economic development in the border regions. Though the central government of Vietnam does not strongly encourage border

---

[33] Nong Lifu, "A Review of China-Vietnam Border Trade and Policy Recommendation", *Around Southeast Asia*, Supplementary Issue, p. 50 (2000).

[34] For example, since 2001 the major border ports of China, Vietnam, Laos and Myanmar take turns holding joint border trade fairs in the fourth quarter of each year.

trade,[35] the local governments in the northern provinces attach great importance to border trade and other forms of economic contacts with Yunnan and Guangxi.[36] The fact that China has promised to offer preferential treatment to the export products of Laos and Myanmar (including duty exemption for many products of these countries), will contribute to the resurgence of border trade.[37] Also, regular exchange visits of local officials and some coordination mechanisms, which have already begun, will also help the smooth development of economic relations in the border regions.

However, it has to be noted that border trade between China and NNAMs is experiencing a dramatic change in the direction of development, operation mechanisms and the role it is to play in future.

As time goes by, countries in the region have to quicken their steps in trade liberalization and facilitation. China, as a member of the WTO, and Vietnam, which will become a WTO member in the near future, particularly have to unify trade policies implemented in different regions. As border trade is becoming part of formal trade, the share of border trade in total foreign trade will gradually decline and it can no longer rely heavily on preferential policies in its development. Faced with new challenges, the border regions will have to pay more attention to its methods of advancement in the future, e.g., specialization of border markets and increase in the quality of commodities.

It can be expected that border regions will be more closely linked to the world market and face more intense competition. Therefore, they have to undergo further reforms and opening-up. Over the last few years, trade relations between China and NNAMs have been developing very quickly,[38] and China has provided a huge market for their products.[39] It has created a favorable environment for the further

---

[35] The central government worries about the worsening of unfavorable balance of trade, outflow of valuable foreign exchange, illegal cross-border trade and smuggling.

[36] For example, preferential policies have been adopted in the border ports of Mong Cai in Quang Ninh province, Dong Dang in Lang Son province, Thanh Thuy in Cao Bang province, and Lao Cai in Lao Cai province. (*People's Daily*, Vietnam, 8 September 2000.)

[37] According to bilateral memorandums of understanding signed between China and Laos, Myanmar and Cambodia respectively in November 2002. China promised to begin providing special treatments in preferential tariffs to the three countries in January 2004. Many products of these countries can be exported to China free of tariffs.

[38] According to the statistics of Vietnam, bilateral trade between Vietnam and China reached US$4.83 billion in 2003, an increase of 30% over the previous year. It further rose by 37% for the first half of 2004 over the same period of last year. The target of US$5 billion in bilateral trade for 2005 is to be achieved ahead of time.

[39] For the first half of 2004, border trade between Yunnan Province and Vietnam reached US$44.6 million, an increase of 37% over the same period of the previous year. The growth rate of import from Vietnam was as high as 119.9% over the same period of the previous year.

development of border trade. However, since the level of consumption is rising and market demand is becoming more diversified in all countries involved on both sides, the quality of products and ability to supply sufficient export commodities to neighboring countries must increase.

From a long-term point of view, the economic development of border regions should not rely too much on loosely regulated border trade. The border regions concerned should strive for a more diversified and comprehensive development. At present, a large proportion of the products in border trade are actually produced in provinces other than the border regions themselves and most of the products imported are sold to other provinces.[40] This shows clearly that the relationship between border trade and local economic development is still not extensive enough. When market demand is relatively stable, the establishment of local processing industry would be the natural outcome. It will increase value-added, reduce transportation costs and create more job opportunities for local residents. At the same time, it will also bring about developments in infrastructure, finance, telecommunications, tourism, etc. In short, a more diversified economic structure will produce a tremendous positive, direct impact on the social and economic development of the border regions.

In order to further develop economic exchanges between the contiguous areas, trans-national coordination and cooperation are essential. Means and ways of dealing with policy differences and other obstacles mentioned above have to be thoroughly explored and discussed amongst the government officials concerned. For example, Mengla County in Yunnan province and the adjacent county of Muoding in Laos adopt different travel regulations. In 2003, Muoding Port of Laos began to practise the system of visa application on arrival, while Muohan on the Chinese side has not yet adopted this policy. This difference brings about problems and inconvenience to businessmen and tourists who intend to enter China. Jinshuihe of China is a port at the national level, while the contiguous Matlutang on the Vietnamese side is a port at the provincial level, where businessmen from a third country cannot travel freely. This will prohibit the expansion of border trade and tourism.

In conclusion, although border trade between China and NNAMs has benefited the development of border regions, it is also undergoing changes in direction of development and operation mechanisms, and this will affect its future role it is

---

[40] For example, 80% of the products imported via Jiegao Port in Yunnan Province are sold to other provinces while 85% of the exports are from other areas.

to play in the countries' economies. It is expanding from exchange market trade and small-volume trade to various other forms of economic relations, including entrepôt trade, transnational economic and technological cooperation. If properly handled, it will continue to contribute to comprehensive development and advancement in the border regions.

# Chapter 10

# Transboundary Environmental Issues of the Mekong River: Cooperation or Conflicts among the Riparian Countries

*LU Xixi*

## Introduction

While developed countries are progressing towards dam removal and river rehabilitation, the developing regions of Asia continue to construct dams and reservoirs to meet their increasing demand on energy. This is particularly true for China. The number of large dams constructed nationwide is estimated at over 80,000 (WCD, 2000), which places China on the top of the list in terms of number of dams built. Dams have been constructed for flood control, hydropower, irrigation and storage of water for domestic purposes.

In the Upper Mekong river (or Lancang Jiang), China has proposed a cascade of eight dams, and two of them have been completed to meet its need for electricity. China has also signed an agreement with Myanmar, Laos and Thailand to improve the navigation of a section of the river from Jinghong to Luang Prabang in Laos for boosting business trade among China and the riparian countries. Navigation improvement works will primarily consist of blasting and removal of reefs and sand bars in the river channel. The trans-boundary environmental issues arising from these hydraulic engineering projects have become a main concern for communities located downstream of the dams. The Lower Mekong countries are concerned that the construction of the cascade dams in China will reduce downstream water flow and water quality, while environmental groups and local communities are worried about the effects of these structures on fish species, water fluctuation and river bank collapse (MRC, 2003).

Often, reports about these issues are not consistent. In fact, most of them are contradictory (Table 10.1). Hence, a holistic review on these environmental issues is necessary. This chapter aims to examine the current developments in some of the main environmental issues arising from the hydro-power development in China,

Table 10.1 List of the Cascade Dams in the Upper Mekong Basin, China (after Mekong River Commission, 2003)

| Name of Project | Installed Capacity (MW) | Annual Generation (GWh) | Total Storage Capacity (million m$^3$) | Catchment Area (km$^3$) | Average Flow (m$^3$s$^{-1}$) | Commissioning |
|---|---|---|---|---|---|---|
| Manwan | 1,500 | 7,870 | 920 | 114,500 | 1,230 | 1986–1992 |
| Dachaoshan | 1,350 | 7,090 | 880 | 121,000 | 1,230 | 1996–2003 |
| Xiaowan | 4,200 | 18,540 | 15,130 | 113,300 | 1,220 | 2001–2012 |
| Gongguoqiao | 750 | 4,670 | 510 | 97,300 | 985 | 2010–2012 |
| Nuozhadu | 5,500 | 22,670 | 24,670 | 144,700 | 1,750 | 2012–2013 |
| Jinghong | 1,500 | 8,470 | 1,040 | 149,100 | 1,840 | |
| Ganlanba | 150 | 1,010 | — | 151,800 | 1,880 | |
| Mengsong | 600 | 3,740 | — | 160,000 | 2,020 | |
| Total | 15,500 | 74,060 | | | | |

and potential conflicts and areas for cooperation between China and the riparian ASEAN countries.

## The Mekong River

The Mekong River, which originates in the Tibet-Qinghai plateau, flows through a distance of approximately 4,500 km, before it enters the South China Sea at the Vietnam Delta (Figure 10.1). The first 2,000 km, the upper basin, flows through the Chinese territory while the lower basin covers an area of 600,000 km$^2$ in Myanmar, Thailand, Laos, Cambodia and Vietnam. Though abundant in biodiversity, the Mekong River is one of the most undeveloped rivers in the world, hence providing these countries a great asset for harnessing the river for hydropower development, irrigation project, flood control and domestic uses, which is crucial and beneficial for economic development, regionally and locally.

Administratively, the Mekong river basin is divided into two sub-basins: the Upper Mekong Basin (24% of total drainage area) and the Lower Mekong Basin (76% of total drainage area). The lower basin currently supports a population of about 60 million people, and this is expected to increase to 90 million people in 2025. Correspondingly, electric power demand in the whole Mekong region is estimated to increase by 7% annually until 2022, requiring a fourfold increase in current electric generating capacity (MRC, 2003). In view of the future demand and economic viability of hydropower for the Mekong region, numerous projects to tap the hydroelectric potential of the Mekong River have been planned by individual countries; in tandem, research examining the potential environmental and social ramifications of these hydropower projects is also growing steadily.

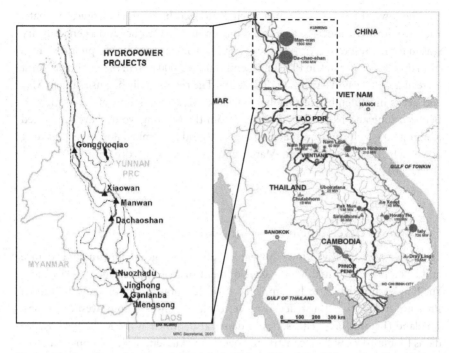

**Figure 10.1** Locations of the Cascade Dams in the Mekong River (after Mekong River Commission, 2003 and He and Zhao, 2001) (inset)

The northern part of the lower Mekong river basin is mountainous with steep-sided slopes, with relatively few upland plains. In Luang Prabang, Lao PDR, the Mekong River is marked by relatively sharp bends and flows through a rock-cut channel partially filed with flood alluvium. Reaching the Korat Plateau, the river cuts deeply into the rim of the plateau, forming sheer cliffs above the river and deep underwater canyons, before turning eastwards to flow past Vientiane and subsequently along the Lao PDR-Thailand border. This stretch of the Mekong River is characterized by rapids, interspersed with alluvial reaches. Anastomosing of the river occurs at the section where the river reaches the border of Cambodia, with large permanent bars dividing the channels (Gupta *et al*., 2002). In Cambodia, the Mekong is connected to the Tonlé Sap (Great Lake) via the Tonlé Sap River. During the dry season, the lake drains into the Mekong via Tonlé Sap River. As the flood season progresses, the Mekong River rises to above the lake level, and the flow in Tonlé Sap River reverses and fills the lake instead. At Phnom Penh, the Mekong River separates into two main channels and smaller distributaries at Phnom Penh, flowing out to sea through an extensive delta south of Vietnam.

The lower Mekong study area is characterized by a largely tropical climate, with two distinct seasons — a wet season from June to October and a generally dry season for the rest of the year. In the lower basin, mean annual precipitation varies from over 3,000 mm in Lao PDR and Cambodia to 1,000 mm in the semi-arid Korat Plateau in northeast Thailand (MRC, 2003). The river usually begins rising in May and peaks in September or October, with the average peak flow at 45,000 $m^3 s^{-1}$. Between June and November, discharge from the Mekong would have amounted to about 80% of its total annual discharge. Around November, flows start receding and reach the lowest levels in March and April, at approximately 1,500 $m^3 s^{-1}$ (Kite, 2001).

## Hydropower Development in Lancang-Mekong Basin

The Upper Mekong River flows through deeply-incised gorges in Yunnan Province and has a huge potential for the development of hydropower energy (Dore and Yu, 2004). China has already completed two hydropower dam projects in Yunnan Province and plans to construct six more, to supply power to Southwest China and Thailand (Figure 10.1). This series of dams, also known as the Mekong Cascades dams (Table 10.2), will be constructed over a 750 km length in the upper basin of the Mekong River (Lancang Jiang) in Yunnan, with a total gradient change of 800 m (Plinston and He, 1999). Manwan Dam was the first to be constructed in the cascades project and started to store water in 1992. The second dam, Dachaoshan, was completed and began operation in 2003. Construction of the third dam, Xiaowan, which will be one of the highest dams in the world at 292 m, commenced in December 2001 and is slated for completion in 2012 (IRN, 2002). The reservoirs of both completed and projected dams in the Yunnan Province are expected to have a total storage capacity of over 40 $km^3$, impounding up to more than half of the mean annual runoff of the entire basin, and the entire cascade will have a combined installation capacity of 15.55 million kW (MRC, 2003) (Table 10.2).

It should also be noted that China is not the only party who wishes to develop hydropower for economic purposes. Other riparian countries are doing the same, such as the controversial dams of Nam Theun 2 and Theun Hinboun in Laos. Laos is aiming to become a major regional exporter of hydropower, and is planning several hydropower plants to export electricity to Thailand (ADB, 2004). Today there are concrete plans for 11 dams on the Lower Mekong proper, and several times that number on its tributaries. Thailand has a fast-expanding economy, and an increasing demand for energy. Several of the proposed dams in neighboring countries are meant for export of power to Thailand.

Table 10.2  Summary of Players/Stakeholders in the Mekong River Region

| Players/Stakeholders | Main Functions | Agenda |
|---|---|---|
| **Multilateral Organizations/Forums** | | |
| Mekong River Commission (MRC) | Promotes cooperation among Mekong countries in all fields of sustainable development, utilization, management and conservation of the water and related resources of the Mekong River Basin (MRC Mandate, 1995). | Aims to serve as a regulatory authority overseeing the development of the Mekong River basin, and the management of transboundary environmental issues. However, its powers have been limited by China and Burma's refusal to join the Commission. |
| Asian Development Bank (ADB) | Provides monetary financing to Mekong countries as part of the Greater Mekong Subregion (GMS) programme. | Emphasizes economic development in the Greater Mekong Subregion through development and sharing of resource base among countries and the promotion of freer flow of goods and people. |
| ASEAN Mekong Basin Development Co-operation Initiative | ASEAN and ASEAN+3 (China, Japan, Korea) initiative which seeks to integrate farming, mining, forestry, industry, transport, telecommunications, energy, tourism, trade and investment activities in the Mekong basin. | Committed to regional economic development of the Mekong region, with attention currently on developing an integrated electricity grid connecting hydropower dams in China, Buram and Lao PDR to markets in Thailand and Vietnam. |
| Golden Quadrangle Forum (China, Burma, Thailand, Lao PDR) | Supports the improvement of channel navigability to facilitate shipping and trading flows between China and Mekong countries. | Transportation and trade initiative. |
| National and Local Governments | A range of national and local agencies are involved in the resource management and development of the Mekong basin, representing state, sectoral or inter-sectoral interests. | Focused on rapid resource-based export-oriented production. Although resource management and exploitation approaches are similar for most of the Mekong riparian countries, the significance of the Mekong basin within every national territory is unique, and each maintains its own perspective with regard to the use of resources by other countries, which will have cross-border impacts. |

Table 10.2 (Continued)

| Players/Stakeholders | Main Functions | Agenda |
|---|---|---|
| **Corporate/Foreign Investors** | | |
| Hydropower Developers/Consultants E.g., Huaneng Group (China), Norpower (Norway), Electricité de France, Transfield Holdings (Australia), etc. Mining companies | Design and construction of hydropower projects mainly on the tributaries of the Mekong River, e.g., the Lancang Jiang, usually in joint-venture projects with the national government. Mineral ore extraction and sand mining from riverbeds. | To secure large-scale, high-profitability dam development projects (usually under Build-Own-Transfer schemes) aimed at tapping the hydropower potential of the Mekong River. To meet the large demand for raw materials needed for construction projects in developing cities, especially in China. |
| Others: Agroforestry, Transportation and Tourism infrastructure development companies | Agroforestry: Extraction of timber, cultivation of crops and rearing of livestock within the Mekong basin. Transportation and tourism: setting up of infrastructure to increase efficiency of transportation networks and develop tourist areas. | Agroforestry: Mainly for export purposes. Transportation and tourism: to increase capital inflows to the Mekong region. |
| **Non-governmental Organizations** | | |
| E.g., International Rivers Network (IRN), Southeast Asian River Network (SEARIN), Rivers Watch (RWESA), River Protection Network, Rak Chiang Khong (RCK), etc. | To raise awareness on environmental and socio-economic issues in the Mekong basin, and to serve as a consolidatory opposition to "unsustainable" developments within the basin. | Determined by each NGO and the group/groups which they represent. Examples include protection of fisheries, conservation of riparian forests, protection of local trades, etc. |
| **Local Communities** | | |
| E.g., fishermen, farmers, riparian inhabitants. | Small-scale fishing, crop cultivation, extraction of drinking water, and utilization of other non-food resources provided by the rivers, such as electricity and transport. | To sustain their livelihoods by using resources provided by the rivers. They are often not empowered enough to influence the large-scale developments that will impact them, and there is hardly any trickle-down effect from these developments. |

## Navigation Improvement Project

Besides the hydropower development in the Mekong basin, a navigation improvement project is being carried out in the upper basin too. The Upper Mekong River Navigation Improvement Project is an agreement signed by China, Myanmar, Thailand and Laos in April 2000. It is funded by the Chinese government as part of a grand scheme to improve navigability between Simao, China to Luang Prabang, Laos. More commercial ships will be able to navigate along the Mekong River, hence improving trade linkage between China and these ASEAN countries. It is anticipated that the volume of cargo and passenger traffic along the river will reach 1.5 million tons and 400,000 people by 2010 respectively (The Joint Experts Group, 2001).

In order to improve navigability of the river, natural obstacles such as rapids will be blasted. In the first stage of the project, 11 rapids and ten scattered reefs along the 331-km section of the Mekong from the China-Burma border to Ban Houayxai in Laos were removed. The second and third phases of the project involved channelisation of the river (IRN, 2002a). However, this project has invoked strong protests and anger from many parties (local communities and NGOs), both for political and environmental reasons, despite the agreement signed among the governments of four riparian countries. The project has since been suspended for re-evaluation after the first phase of blasting of the rapids and reefs.

## Environmental Issues Arising from Developments in the Basin

Although proponents of the hydropower projects argue that the projects may bring about many benefits, such development in the upper basin may have adverse impacts on a series of environmental variables (see Table 10.1) (WRR, 2001; Neave, 2002). This chapter aims to look at water discharge, water level fluctuation, sediment concentration and sediment flux, channel and river bank stability, and aquatic ecology, as indicators of the extent of environmental disturbance caused by the anthropogenic developments so far.

### Water Discharge and Water Level Fluctuations

Apart from providing renewable energy, the Mekong cascades are supposed to provide better flood control during the wet season and increased water supply in downstream areas during the dry season. Chapman and He (1996) estimated that the impact of Manwan and Dachaoshan dams on the water discharge are insignificant due to their small capacity, and significant changes will only be noted after the operation of Xiaowan dam. When Xiaowan dam is completed, dry season flows can be increased by up to 70%, as far as 1,000 km downstream in Vientiane, Laos due to the impoundment of discharge during the wet season (He and Chen, 2002; IRN, 2002). This would be beneficial downstream in terms of irrigation and navigation development, hydropower transmission and possible flood control through flow regulation by the cascade reservoirs.

Yet the reality is that the Mekong River has recently been hit by lower water levels. This has been perceived by many NGOs and local communities as a result of the two upper stream dams in China. However, a report released by MRC suggests that the recent droughts of the Mekong River were mainly due to the dry weather in combination with forest clearing rather than the Chinese dams (MRC, 2004; Phouthonesy, 2003).

Multi-year records indicate that the annual minimum water discharge in Chiang Sean, Thailand (nearest hydrological station to China) has been decreasing rather than increasing, especially after the completion of Manwan Dam in 1992. When the reservoir of Manwan Dam was filled in the 1992–1993 dry season, Thai authorities reported unusually low water levels in the province of Chiang Rai (IRN, 2002). Ironically, the historical data also show that the annual minimum water level at same place has been increasing. As the water discharge was calculated based on the water level, this inconsistency casts doubt about the quality of the data.

The annual mean and annual maximum water discharges did not experience significant changes in the lower Mekong river within the eight-year period (1992–2000) during the operation of Manwan Dam (Lu and Siew, submitted), apart from the sharp decrease in the discharge across all stations in 1992 when the dam was closed for infilling.

Changes in water level fluctuations were observed only during the dry seasons. Wet season water level fluctuations did not reveal any discernable differences between pre- and post-dam years, and appeared to be unaffected by the operation of the dam. Analysis of day-to-day changes in the water level yielded similar results, with dry season water level changes displaying greater sensitivity to dam operation (Lu and Siew, submitted). The magnitude of dry-season fluctuations in the post-dam period was augmented relative to pre-dam fluctuations. It was also noted that the water level fluctuations have been exacerbated since the blasting of reefs in 2001 as part of the navigation channel improvement project (SEARIN, 2003).

*Sediment Concentration and Sediment Flux*

Similar to the impact on water discharge, the impact of the dams on sediment downstream is also not consistent among the authors (Table 10.1). Sediment will be trapped behind the dams and can only be released downstream when the gates are opened. Kummu et al. estimated that trapped efficiency is as high as 90% for Manwan Dam. Lu and Siew noted that there has been a declining trend in suspended sediment concentration along the entire length of the lower Mekong Basin since the 1960s. The rate of decline was more obvious after the closure of Manwan Dam for infilling in 1992. However, it is important to note, especially for tropical and arid regions with prevalence to soil erosion, that even if sediment load is greatly reduced by dams, this does not necessarily lead to clearer water in downstream reaches (Brandt, 2000). Other factors may play a part in concealing the effects of upstream dams.

Though sediment concentration has been declining, estimation of sediment load for the lower Mekong River is challenging because continuous sediment measurements data are not available. Kummu et al. found that sediment load has declined almost 50% in Chiang Saen and in lesser proportion in other areas, whereas Lu and Siew found similar declines for the same areas (Table 10.4). However, at least two areas have actually experienced increases in sediment load as demonstrated by Lu and Siew (2004). Tyson (2001) argued that sediment load may increase after construction of the dams due to potential in-channel erosion and slumping.

### Channel and River Bank Stability

Any changes in the water and sediment input may result in changes in a river's morphology. The lower Mekong Basin, especially in Cambodia and Vietnam, is characterized by a large area of alluvial floodplains. The profile and course of the river, and river stability may be influenced by several factors such as water discharge, water level fluctuation and sediment load. Local community and NGOs complained that the river bank erosion (or collapse) was due to the water level fluctuation caused by the dams in the upper stream and rock blasting within the river. However, more evidence is required to support these claims as the water level changes mainly occurred in the dry season rather than in the wet season.

A hypothetical analysis reveals that frequent collapses of the river bank, if truly occurring as claimed, could be the consequence of the reduction of the sediment concentration. The decrease in sediment load in the lower course may increase erosional power or transport capacity (Brandt, 2000), and may eventually change the river's morphology. As sediment load decreases due to the impoundment of sediment in the upper course, the river in the lower basin may increase its transport capacity (Friedman et al., 1998). This causes the soil foundation of the river bank to be loose, hence hindering the development of infrastructure such as bridges and buildings for better accessibility (Goh, 2004).

The release of relatively sediment-starved, high-energy water from Manwan and Dachaoshan dams is likely to cause channel scouring and possibly, coarsening of the bed material until equilibrium is reached and material cannot be moved by the flows (Batalla et al., 2004). Dam-related sediment starvation effects are already obvious for distances more than 600 km downstream. Apart from dams, reef blasting for navigation improvement along the river will also create impacts on bank erosion. River bank erosion might be worsened with the entry of large navigation vessels, which can undercut the river banks with large waves generated from their wakes. With shorelines in estuarine areas at Tan Chau and My Thuan reportedly experiencing considerable erosion (Pham et al., n.d.), the exacerbation of erosion activity is expected when reef blasting is resumed, and more dams begin operations on the upper basin of the Mekong River.

*Aquatic Ecology*

Development of dams in the upper stream may be detrimental to fish migration and breeding, replenishment of groundwater, and seasonal flooding of farmlands over vast areas. Tonle Sap Lake in Cambodia depends greatly on the annual flooding cycle of the Mekong River. Developments may have caused the water level in the lake to be lowered. The fishes in the lake provide as much as 70% of protein intake for the Cambodians and forms 10% of the gross domestic products (ADB, 2004). A disturbance in the flooding cycle will affect the ecological system of the lake.

The reduction in sediment load and the loss of rich nutrients may result in poorer soil on the floodplain, which will affect the agriculture industry. With low soil fertility for planting, farmers may have to resort to use of chemicals to enhance crop quality and in the long run, crop yield may decline due to the intensive use of artificial fertilizers.

Environmentalists have pointed out that the Navigation Improvement Project will affect the livelihood of people who are highly dependent on the Mekong River and the river's biodiversity. As rapids are natural breeding grounds for many migratory species, the destruction of reefs would destroy such breeding and nursery grounds, and eventually leads to possible extinction of some species. For example, for rare species like the Mekong Catfish, destruction of the rapids would directly disturb the catfish population in the Mekong River.

Also, the destruction of reefs in the river would imply that oxygen source in the river will be reduced. This will greatly affect the survival of the fishes in the river, which will in turn affect the populations depending on the river for their livelihood. Cambodia and Vietnam did not participate in this navigation improvement project. However, the impacts are felt by these countries in the lower basin.

Furthermore, the frequent water fluctuations in the dry season will impact the growth of *kai*, a vegetable found in the Mekong Basin. The Thai and Lao people are highly dependent on this crop for their source of protein. With the disturbance in the Mekong Basin, such sources of protein may be reduced, taking away another means of livelihood for the people.

## Conflict and Cooperation Between China and ASEAN-Mekong Countries

### Multi-Stakeholders of the Mekong River

The environmental issues highlighted above are clearly trans-boundary in nature, involving riparian countries sharing a single resource, and represent the contestation of interests among multiple stakeholders (Figure 10.2). The utility and value of the Mekong River and its resources held by some of the multi-level stakeholders in the production of material and non-material benefits are varied. Although there is general consensus among different stakeholders of the importance of economic development in the Mekong region, each stakeholder has its own interpretation of the cost-benefit relationship with regard to economic development and

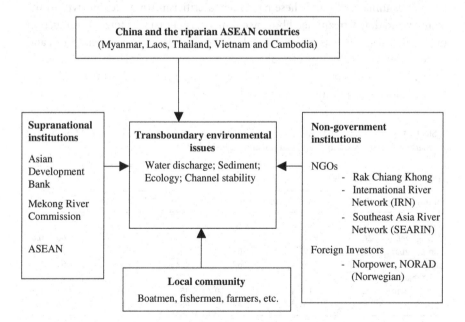

Figure 10.2 Multi-stakeholders of the Mekong River Basin

environmental effects. These differences in perspectives are not limited to country and resource sector interests, but are based on wider social, cultural and economic differentiation, within or beyond country level (Hirsch and Cheong, 1996). In many past and present development projects on the Mekong, countries have mainly adopted unilateral approaches, over-emphasizing the stratified interests of certain powerful groups, ignoring the interests of poorer countries, and local community stakeholders comprising fishermen, traders, boat operators, bean sprout farmers, stone sellers, etc. (Lee *et al.*, 2004).

A case in point would be China's hydropower development project on the Lancang River, the upper section of the Mekong mainstream. The Chinese went ahead with the dam construction project without consulting the ASEAN countries which the Mekong River flows through. Planning for the project started in the early 1980s when China opened its doors for economic developments. ASEAN countries came to know about the project only in the early 1990s, when construction for the first dam was well underway. Countries and populations downstream of the proposed dams have begun to voice their concerns over the construction, especially after changes in the river regime were noted with the operation of two completed dams (Quang, 2003; Kummu, submitted; Lu and Siew, submitted). Apart from China, other ASEAN countries have also been constructing dams on the Mekong's tributaries unilaterally, and these have incited criticism from factions within the country and downstream neighboring countries. As a first step towards mitigation of these often conflicting interests, a stakeholder approach offers a socially just and ecologically sustainable way of analyzing the various interests (Hirsch and Cheong, 1996). This involves the identification of various stakeholders, and the analysis of their agenda, perceptions or concerns (Table 10.3).

Table 10.3 Pre- (1962–1992) and Post-Manwan Dam (1993–2000) Comparison of Mean Annual Discharge and Estimated Sediment Load (Lu and Siew, submitted)

| Time Period | 1962–1992 | | | | 1993–2000 | | | |
|---|---|---|---|---|---|---|---|---|
| Parameter | Q [$m^3 s^{-1}$] | SL [MT/yr] | n | $r^2$ | Q [$m^3 s^{-1}$] | SL [MT/yr] | n | $r^2$ |
| **Location** | | | | | | | | |
| Chiang Saen | 2676 | 74.1 | 292 | 0.77 | 2653 | 34.5 | 60 | 0.65 |
| Luang Prabang | 3965 | 97.3 | 69 | 0.63 | 3924 | 47.2 | 35 | 0.61 |
| Nongkhai | 4440 | 74.4 | 567 | 0.72 | 4732 | 76.1 | 151 | 0.70 |
| Mukdahan | 7508 | 97.5 | 1018 | 0.65 | 7949 | 131.1 | 193 | 0.75 |
| Khong Chiam | 9298 | 166.4 | 310 | 0.78 | 8562 | 104.4 | 34 | 0.88 |
| Pakse | 9598 | 151.3 | 34 | 0.77 | 9862 | 113.5 | 58 | 0.83 |

Q: mean annual discharge; SL: estimated annual sediment load; n: no. of samples; $r^2$: correlation.

The summary table shows that various institutions hold different perceptions on the Mekong Basin and it will be very difficult to decide who has the power and authority to charter future development, and who is to be responsible for the various impacts affecting the region. In the following section, some potential areas of conflict between China and ASEAN, arising from the rapid development taking place in the Mekong Basin, will be examined.

*Potential Areas of Conflict Between China and ASEAN*

China's hydropower projects on the Lancang Jiang in the upper basin of the Mekong River are likely to create not only extensive environmental effects, but also political and social consequences. Once the series of dams are in operation, China will be able to control the amount of water flowing downstream in the Mekong River through regulation of discharge. The regulation of flow at the discretion of the Chinese is a critical issue for the rest of the riparian countries, especially those furthest downstream, but little has been done to highlight their concerns. However, on the other hand, ASEAN countries have also been actively proposing and constructing large-scale dams within their national boundaries — this will definitely add on to and exacerbate the effects from the Chinese dams upstream.

Water-based conflicts arising from these developments can be overt as well as subtle. Overt forms of conflict tend to be associated with the decisions and actions by water management authorities, which often involve violence, social dislocation and destruction of resources and assets (Miller, 2003). Other forms of conflict usually stem from effects that are not immediately obvious during the course of dam construction. These effects are usually those inflicted on water quality, water level fluctuations, availability of water and fisheries.

A decline in water quality holds severe implications for stream ecology and riparian communities, as it would not only affect the health of aquatic food resources in the river, but would also pose a health risk to villagers reliant on the river for fresh water. The decrease in the availability of food sources for a rapidly expanding population will inevitably cause conflicts over land and fishing rights. Frequent water level fluctuations are another cause of conflict, as a change in level (as little as 15 cm) may cause the loss of a harvest if it is suddenly inundated or deprived of water. River bank structure also weakens and becomes more susceptible to erosion with increased water level fluctuations, resulting in a further loss of arable land for downstream communities. For instance, in China's Lancang development, daily fluctuations below the Manwan Dam are between 3–4 m, and more than 100 cave-ins and slides have occurred below the dam as a result of infilling and discharge (Oxfam Hong Kong, 2002). Water availability is also crucial in the Mekong Basin, as a decline in the availability of water, particularly in the dry season, places further stress on a scarce resource needed for farming and domestic consumption purposes. Tourism along the Mekong River was also affected by the low water levels. It was reported that Laotian tour operators had to cancel ten river excursions in March 2004 alone (Wain, 2004). As noted earlier, contestation over the impacts of the Chinese dams are becoming more frequent and serious, especially with the recent serious decrease in water levels in the lower Mekong, which MRC attributed to climate change but was heavily contested by SEARIN and local communities (Sukin, 2004). Downstream flooding during the wet season is also a potential cause of conflict, as the release of water from reservoirs upstream that have reached their maximum holding capacity may exacerbate flooding downstream, causing the loss of property and lives.

*Potential Areas for Cooperation*

The inherent contradiction in viewing the Mekong River as a transboundary shared resource while maintaining sovereignty over national development projects on the river is perhaps the main obstacle to effective cooperation. For instance, China's recent efforts at reforestation in Yunnan (Lancang Jiang basin) after a period of extensive degradation since the 1950s are at odds with their channel navigation improvement project on the Mekong, which may render large tracts of undisturbed forest in Laos and Burma susceptible to large-scale logging, by making this remote section of the river navigable. As Goh (2004) rightly points out, the main problem with the Chinese view is that it does not take a basin-wide perspective, and focuses only on impacts within Chinese territory. Apart from this, there is also a lack of formal agreements between China and ASEAN to safeguard other riparian countries from unilateral developments that have basin-wide implications.

Other politico-economic obstacles also hinder cooperation between China and the Mekong countries. China has been providing more financial aid to poorer countries like Laos and Cambodia to improve links with and to assert her presence in Southeast Asia. Such "in-debtedness" gratuity perpetuates the unequal economic and political relationship between China and these ASEAN countries, and places China in an upper position with regard to decisions on development within the Mekong region. By being the economic powerhouse in the Mekong region, possessing the technical expertise (as foreign hydropower developers) to develop the cascade dams and not to mention its location on the headwaters of the Mekong River, China has the potential ability to determine the distribution of benefits or costs among the riparian countries. If China's development schemes on the Mekong River were allowed to proceed unchecked under their rhetoric of regionally shared economic benefits, there would be significant environmental, geopolitical, economic and social consequences on smaller players in the region, bringing into question the real benefits that hydropower development can bring for countries in the lower basin. Therefore, a common platform for potential cooperation on the basis of equal partnership would be through the Asian Development Bank's (ADB) Greater Mekong Subregion (GMS) program, which is working towards the establishment of a regional power grid to replace coal-fired plants with hydropower plants. ADB (2004) believes that as some countries have less power resources while others lack the technology and capital, it is essential and beneficial for them to cooperate to optimise the shared resources. Twelve hydropower projects have been proposed to connect to the grid, including two dams on the Upper Mekong in China. Critics of the regional power grid counter that hydropower projects would disrupt the Mekong River ecosystem, forcibly displacing populations and undermine the basic rights to water of the people who can least afford it (IRN, 2004). Previous lessons from dam developments have shown that unless all levels of stakeholders, especially local representatives, are included in the consultative process prior to establishment of the infrastructure, the project will meet with strong resistance. Comprehensive and accurate impact assessments and proper cost-benefit analyses need to be conducted by independent agencies to evaluate the feasibility of these large-scale projects, following which, mitigation and just compensation frameworks must be drawn up and implemented to minimize damage to the livelihoods of the riparian population. If these large-scale developments are deemed unsustainable, alternatives at various levels other than large-scale projects should be considered. For instance, a community-based method of generating electricity through micro hydro turbines has proven to be highly successful in Mae Kham Pong village in Chiang Mai Province, Thailand. In addition to providing the entire village with electricity, they are also able to supply electricity to nearly 100 households in neighboring villages (Watershed, 2002). This is a potential area for the GMS program to explore, as it is more sustainable, much lower in cost, and allows for local management and ownership, which is in line with ADB's rights-based policy.

Apart from ADB's GMS Program, the ASEAN-Mekong River Development Cooperation Program also serves as a forum for potential cooperation in sustainable development in the Mekong Basin. In the fifth Ministerial meeting in 2003, China pledged its willingness to promote cooperation with all countries in the region, and also indicated that Yunnan Province would play an increasingly important role in this cooperation (Phouthonesy, 2003). This presents an opportunity for increased communication and information sharing between China and ASEAN in terms of development in the Lancang River and the impacts on downstream areas in the Mekong Basin.

With emerging tensions over deteriorating natural resources, and increasing conflicts over the spillover effects created by national, unilateral activities on the Mekong River, it is clear that the way forward to a more sustainable utilization of the river's resources is through informed cooperation and integrated management, not a mere collaboration formed solely to exploit the resources of the Mekong River. The ADB's GMS program and ASEAN's Mekong Basin Development Project, both in which China is actively participating, currently emphasize mainly the economic aspects of cooperation, and downplay environmental issues associated with resource exploitation. It is important that these high-profile programs, which tend to garner greater governmental support, adhere to the holistic policies which they are supposed to promote, and ensure that environmental and social areas are given equal emphasis. Information sharing and cooperative efforts among ADB, ASEAN and consultative bodies like MRC would also need to be strengthened.

## Conclusion

Water discharge and sediment fluxes in the lower Mekong River serve as important indicators of the response of the river to the construction of dams upstream, and the extent to which the river is able to subsume these effects to maintain its natural equilibrium.

From the water discharge trends observed in the lower Mekong River, it is clear that the hydrologic effects of the Manwan Dam were largely restricted to the time when the reservoir was filled in 1992, and after that the impact has been very minimal. However, dry season flows showed a declining trend, and water level fluctuations in the dry season increased considerably in the post-dam period.

Areas along the upper-middle and lowermost reaches of the Mekong River have experienced a decline in suspended sediment concentration. In contrast, sediment load in areas located mid-length of the river have remained stable or even experienced an increase, possibly due to the remobilization of extensive alluvial storage, suggesting that the presence of such sediment sinks in mid-stream areas aid in buffering downstream areas from changes in sediment flux upstream.

The frequent river bank collapse or erosion was due to the increase in sediment transport capacity as a result of sediment decline. The release of relatively sediment-starved, high-energy water from the Manwan and Dachaoshan dams is likely to cause channel scouring and possibly, coarsening of the bed material until equilibrium is reached and the material cannot be moved by the flows.

Communities downstream of the cascade dams have reported drastic changes in flow velocity, sedimentation, and most acutely, water level fluctuations in recent years. This paper found that some issues are supported by evidences, while certain issues require more careful studies, given the complex nature of the transboundary issues.

However, it would seem that a propagation of dam-induced effects downstream is likely after the completion of all the dams, given the influence of the Lancang contribution in the middle-lower reaches of the Mekong. This has been demonstrated in the other large Chinese rivers such as the Yangtze and Yellow rivers (Lu, 2004; Lu *et al.*, 2003). These effects could be further augmented by water diversion projects and numerous tributary dams occurring along the length of the lower Mekong River.

Therefore, dialogue and cooperation among the countries in the Mekong River basin are very necessary to avoid potential conflicts from the transboundary environmental issues arsing from such economic activities. Though economic development is vital for both China and the riparian ASEAN countries, it has to be compromised with environmental impact and hence social issues that arise from unplanned developments, which disregard the welfare of people and their environment.

## Acknowledgements

This study was funded by the National University of Singapore (NUS) research grant R-109-000-044-112, and China 973 Program (Project No. 2003CB415105-6). The author would like to extend his appreciation and thanks to Dr. Zou Keyuan for inviting him to the symposium and for making it a great success, and to Dr. Carl Grundy-Warr, Mr. Uffer Poulsen, Ms. Siew Ruyan and Ms. Teo Yen Peng for their kind assistance and support.

## References

Asian Development Bank, *The GSM beyond Borders, Regional Strategy and Program 2004–2008* (2004).

Batalla, R. J., Gomez, C. M., Kondolf, G. M., "Reservoir-induced Hydrological Changes in the Ebro River Basin (NE Spain)", *Journal of Hydrology*, Vol. 290, pp. 117–136 (2004).

Brandt, S. A., "Classification of Geomorphological Effects Downstream of Dams", *Catena*, Vol. 40, pp. 375–401 (2000).

Chapman, E. C., He, D., *Downstream Implications of China's Dams on the Lancang Jiang (Upper Mekong) and their Potential Significance for Greater Regional Cooperation, Basin-Wide* (Sydney: Australian National Mekong Resource Center, 1996).

Dore, J., Yu, X., *Yunnan Hydropower Expansion: Update on China's Energy Industry Reforms and the Nu, Lancang and Jinsha Hydropower Dams* (Kunming, PRC:

Chiang Mai University's Unit for Social and Environmental Research & Green Watershed, 2004).
Wain, B., "Mekong River at Risk", *Far Eastern Economic Review* (26 August 2004).
Friedman, J. M., Osterkamp, W. R., Scott, M. L., Auble, G. T., "Downstream Effects of Dams on Channel Geometry and Bottomland Vegetation: Regional Patterns in the Great Plains", *Wetlands*, Vol. 18, pp. 619-633 (1998).
Goh, E., "China in the Mekong Basin: The Regional Security Implications of Resource Development on the Lancang Jiang", *Institute of Defence and Strategic Studies (IDSS) Working Paper No. 69*, http://www.ntu.edu.sg/idss/WorkingPapers/wp69.pdf (2004).
Gupta, A., Lim, H., Huang, X., Chen, P., "Evaluation of Part of the Mekong River Using Satellite Imagery", *Geomorphology*, Vol. 44, pp. 221–239 (2002).
He, D. M., Chen, L. H., "The Impact of Hydropower Cascade Development in the Lancang-Mekong Basin, Yunnan", *Mekong Update & Dialogue*, Vol. 5(3), pp. 2–4 (Australian Mekong Resource Centre, 2002).
Hirsch, P., Cheong, G., "Natural Resource Management in the Mekong River Basin — Perspectives for Australian Development Cooperation", Final Overview Report to AusAID, http://www.usyd.edu.au/su/geography/hirsch/ (1996).
International Rivers Network (IRN), "China's Upper Mekong Dams Endanger Millions Downstream", Briefing Paper 3, http://www.irn.org (October 2002).
International Rivers Network (IRN), "Navigation Project Threatens Livelihoods, Ecosystem", Briefing Paper 2, http://www.irn.org (October 2002a).
International Rivers Network (IRN), "Sizing Up the Grid: How the Mekong Power Grid Compares Against the Policies of the Asian Development Bank", http://www.irn.org (January 2004).
Lee, W. L., Lo, D. N., Tan, A. L., Ye, J. E., Zheng, L. Z., "Thammachat or Money: A Study of the Mekong River to the Local Communities of Chiang Khong", unpublished GE3230 field studies report (2004).
Kite, G., "Modelling the Mekong: Hydrological Simulation for Environmental Impact Studies", *Journal of Hydrology*, Vol. 253, pp. 1–13 (2001).
Kummu, M., Sarkkula, J., Varis, O., "Sedimentation and Mekong Upstream Development: Impacts on the Lower Mekong Basin", *Geomorphology* (submitted).
Lu, X. X., Siew, R. Y., "Water Discharge and Sediment Flux Changes in the Lower Mekong River", *Geomorphology* (submitted).
Lu, X. X., "Vulnerability of Water Discharge in Large Chinese Rivers", *Regional Environmental Change* (in press, 2004).
Lu, X. X., Ashmore, P., Wang, J., "Seasonal Water Discharge and Sediment Load Changes in the Upper Yangzte, China", *Mountain Research and Development*, Vol. 23, pp. 56–64 (2003).
MRC, "The Present Low Flows in the Lower Mekong Basin" (Mekong River Commision, 2004).
MRC, *State of the Basin Report* (Mekong River Commission, 2003).
MRC, *Lower Mekong Hydrologic Yearbook* (Mekong River Commission, 2000).
Miller, F., "Water Conflict and Governance in the Mekong Basin: Diverse Interests, Power and Participation", *Mekong Update & Dialogue*, Vol. 6(4), pp. 4–6 (Australian Mekong Resource Centre, 2003).

Neave, M., "Responses to Feature on 'Potential Physical and Environmental Impacts of Dam Building in the Lancang-Mekong Basin, Yunnan'", *Mekong Update & Dialogue*, Vol. 5(3), pp. 4–5 (Australian Mekong Resource Centre, 2002).

Oxfam Hongkong, "Master Report on 'Findings of the Study on the Social, Economic and Environmental Impacts of the Lancang River Manwan Power Plant'", http://airc.ynu.edu.cn/English_site/Eng_publicat/eng_publicatmain.asp (December 2002).

Quang, M. N. P. E., "Hydrologic Impacts of China's Upper Mekong Dams on the Lower Mekong River", http://www.mekonginfo.org/mrc_en/activity.nsf/0/cf3b1a24510ba45085256de4006d6c29?OpenDocument (2003).

Pham, B. V., Lam, D. N., Ho, D. D., "Using Remotely Sensed Data to Detect Changes of Riverbank in Mekong River, Vietnam", http://www.gisdevelopment.net/application/naturalhazards/floods/nhcy0009.pf.htm (accessed 28 February 2005).

Phouthonesy, E., "China to Continue Promoting ASEAN-Mekong River Basin Cooperation — Mekong Drought Not Caused by Chinese Dams", *People's Daily Online*, http://www.vientianetimes.org.la/Contents/2004-65/Mekong.htm (accessed 20 August 2003).

Plinston, D., He, D., *Water Resources and Hydropower: Policies and Strategies for Sustainable Development of the Lancang River Basin* (Manila: Asian Development Bank, 1999).

Sukin, K., "Mekong River Basin on Drought Alert", *The Nation*, http://www.nationmultimedia.com/search/page.arcview.php?clid=4&id=94756&usrsess (accessed 5 March 2004).

The Joint Experts Group, "The Navigation Channel Improvement Project of the Lancang Mekong River from China–Myanmar Boundary Marker 243 to Ban Houei Sai of Laos", Report on Environmental Impact Assessment (2001).

Tyson, R., "Downstream Ecological Implications of China's Lancang Hydropower and Mekong Navigation Project", International Rivers Network (IRN), http://www.irn.org (2001).

Southeast Asia Rivers Network (SEARIN), "Lancang Development in China: Downstream Perspectives from Thailand", http://www.searin.org (2003).

*Watershed*, Vol. 7 No. 3, "Towards Ecological Recovery and Regional Alliance" (March–June, 2002).

World Commission on Dams, "A New Framework for Decision-Making", Report on Dams and Development (2000).

*World Rivers Review*, "Proposed Mekong Dam Scheme in China Threatens Millions in Downstream Countries", p. 5 (June 2001).

# Chapter 11

# Are China and ASEAN Competing for FDI?

*CHEN Wen*

## Introduction

The economic relationship between China and ASEAN is deep-rooted on account of history, geography, and past migration in the region.[1] But only after the normalization of bilateral diplomatic relations and the adoption of the opening up and reform policies of Chinese government, did their economic relations start to grow steadily. With the high economic growth of China and its increasing role in the world economy, ASEAN countries have become more and more cautious about China's presence, not only in the export markets, but also competition in attracting foreign investment.

As early as in 1987, Chia (1987) expressed her concern that China would compete with ASEAN to attract foreign investors from Japan, the United States, Europe and the NIEs to engage in its manufacturing. Yang *et al.* (1989) and Luo *et al.* (1989) also paid attention to China's increasing competitiveness with ASEAN countries in the realm of foreign capital enticement due to the deepening of China's economic restructuring. While Tan (1992) argued that the rather similar stage of development of China and the ASEAN countries would lead to competition in attracting FDI from the NIEs and OECD countries.

China became the first FDI recipient among the developing countries in 1992. Regarding China's absorption of the bulk of FDI flowing to the developing countries and the entry into the WTO, the concerns have been building up in recent years. Fan and Dickie (2000) noted that the increased appeal of FDI in China drew potential investment away from ASEAN-5[2] and some other scholars from Southeast Asia mentioned that China was competing for foreign direct investment

---

[1] Please find the details at Wong (1984 and 1987).
[2] ASEAN-5 comprises Indonesia, Malaysia, the Philippines, Singapore and Thailand.

with ASEAN as much as for export in the other markets due to China's deeper integration into the world economy caused by its accession to WTO (Wang, 1999; Chirathivat, 2001 and 2002a; Panitchpakdi and Clifford, 2002). "China is grabbing much of the new foreign investment in Asia, leaving its once-glittering neighbors — Thailand, South Korea and Singapore — with crumbs",[3] as the media pointed out, "China's gain (in FDI inflow) would be its neighbors' pain".[4] Even ASEAN officials have also expressed their concerns over the decline of FDI in their countries and the substantial shift of investment to China over the past few years. For example, Mahathir Mohamad pointed out: "There's not much capital going around. Whatever there is gets sucked in by China".[5] He also complained that "China is an economic threat for Southeast Asia. It is already a threat in terms of attracting foreign direct investment".[6] What's more, ASEAN leaders have decided to build a common market by 2020 for the sake of not losing out to regional giants — China and India — in attracting foreign direct investment.

Is China really squeezing ASEAN FDI inflow? If so, to what extent is China competing with ASEAN? If not, why did ASEAN FDI inflow decrease during the last few years? And what about the outlook of the Chinese outward direct investment in ASEAN? This paper seeks to answer these questions. The next section gives a brief description about the situation of FDI in China. Analyses of the source and the extent of China's FDI inflow follow in the third section. China's "round-tripping" phenomenon of investment and the factors for the decrease of ASEAN inward FDI will also be emphasized in this section. The fourth section outlines the prospects of Chinese outward direct investment in ASEAN. The conclusion and policy implications are presented at the end.

## FDI in China

### The Growth of FDI Inflow

FDI has played an important role in China's economic development and the strategy of foreign investment absorption is considered a major component of China's fundamental principle of opening up to the outside world. When the Chinese-Foreign Equity Joint Venture Law was promulgated in 1979, the doors of China

---

[3] "Economic Juggernaut: China is Passing US as Asian Power", *New York Times*, 29 June 2002.
[4] Fairlamb, D. and J. Mehring, "Why China is so alluring", *Business Week*, Issue 3874, 15 March 2004.
[5] Chandler, C., "Coping with China: As China Becomes the Workshop of the World, Where does that Leave the Rest of Asia?", *Fortune*, 16 January 2003.
[6] Frost, S., "Chinese Outward Direct Investment in Southeast Asia: How Much and What are the Regional Implications?", *SEARC Working Paper Series No. 67* (2004).

were opened to foreign investors. During the past years, we can see that the value of Chinese realized FDI inflow has kept increasing, except in 1999.

From the value of FDI inflow and the growth trend, three phases of high FDI inflow can be elicited for the past 26 years (see Table 11.1). The first five years of the implementation of the Joint Venture Law experienced moderate FDI inflow due to foreign investors' uncertainty over China's business climate and investment policy. Then came the first "biggish" FDI inflow in 1984. After a series of Joint Venture Regulations to transparentize the investment environment issued by the Chinese government in 1983 and 1984, the realized FDI inflow in 1984 registered US$1.42 billion compared to US$916 million in 1983, over 50% more than that in the previous year. But the growth rate of absorption of FDI slackened in the next few years. As shown in Table 11.1, the total realized FDI value for 1979 to 1991 is only US$25.06 billion. After 1991, China entered a period of high growth in

Table 11.1  Foreign Direct Investment in China

| Year | Number of Projects | Contractual Amount | | Realized Amount | |
|---|---|---|---|---|---|
| | | Million US$ | Change % | Million US$ | Change % |
| 1979–82 | 920 | 4958 | — | 1769 | — |
| 1983 | 638 | 1917 | — | 916 | — |
| 1984 | 2166 | 2875 | 50.0 | 1419 | 54.9 |
| 1985 | 3073 | 6333 | 120.3 | 1956 | 37.8 |
| 1986 | 1498 | 3330 | −47.4 | 2244 | 14.7 |
| 1987 | 2233 | 3709 | 11.4 | 2314 | 3.1 |
| 1988 | 5945 | 5297 | 42.8 | 3194 | 38.0 |
| 1989 | 5779 | 5600 | 5.7 | 3393 | 6.2 |
| 1990 | 7273 | 6596 | 17.8 | 3487 | 2.8 |
| 1991 | 12978 | 11977 | 81.6 | 4366 | 25.2 |
| 1992 | 48764 | 58124 | 385.3 | 11008 | 152.1 |
| 1993 | 83437 | 111436 | 92.2 | 27515 | 150.0 |
| 1994 | 47549 | 82680 | −26.0 | 33767 | 22.7 |
| 1995 | 37011 | 91282 | 10.4 | 37521 | 11.1 |
| 1996 | 24556 | 73276 | −19.7 | 41726 | 11.2 |
| 1997 | 21001 | 51003 | −30.4 | 45257 | 8.5 |
| 1998 | 19799 | 52102 | 2.2 | 45463 | 0.5 |
| 1999 | 16918 | 41223 | −20.9 | 40319 | −11.3 |
| 2000 | 22347 | 62380 | 51.3 | 40715 | 1.0 |
| 2001 | 26140 | 69195 | 10.9 | 46878 | 15.1 |
| 2002 | 34171 | 82768 | 19.6 | 52743 | 12.5 |
| 2003 | 41081 | 115070 | 39.0 | 53505 | 1.4 |
| Total | 465277 | 943131 | — | 501471 | — |

Sources: Figures for 1979 to 2002 are from *2003 Yearbook of China's Foreign Economic Relations and Trade*, figures for 2003 are from MOFCOM FDI statistics.

attracting FDI. The south trip of Deng Xiaoping in early 1992 reaffirmed China's opening and reform policy, which brought about an immediate acceleration in FDI inflow. Realized FDI inflow recorded 152.1% growth to US$11.01 billion in 1992 and 150% growth to US$27.52 billion in 1993. China has become the largest FDI recipient in the developing world since 1992. The macro-adjustment measures to control the overheated economy led to slow growth in realized FDI inflow for the following years. In 1996, China's realized FDI inflow reached US$41.73 billion. The impending accession to the WTO brought the third resurging inflow of FDI to China in 2001, in spite of decreasing global FDI inflow. In 2002, China registered US$52.74 billion realized FDI inflow, ranking the largest FDI recipient in the world.

*Current Status of FDI Inflow in China*

According to UNCTAD, if Luxembourg[7] was excluded, China maintained its position as the largest FDI recipient in the world in 2003 with US$53.50 billion FDI inflow. Although China suffered from SARS, FDI inflow to China still increased by a small margin in 2003 — 1.44% over 2002. At the same time, 41,081 foreign-invested enterprises (FIEs) were newly set up, an increase of 20.22%. By the end of 2003, China has in total approved the establishment of 465,277 foreign-invested enterprises, with US$943.13 billion and US$501.47 billion for contractual FDI and realized FDI respectively (see Table 11.1). Among them, China-foreign joint ventures and cooperative enterprises were the main forms of FDI, accounting for 62.95% of the total foreign-invested enterprises, and investment in these enterprises accounted for 55.42% of the total contractual FDI and 58.36% of the total realized FDI.[8] And in the first eight months of 2004, there were US$43.56 billion realized FDI flowing into China, 18.77% up over the same period of last year.[9]

If we look at the sectoral distribution of the cumulative FDI, as shown in Table 11.2, manufacturing industry attracted most of the FDI by the end of 2003, holding 72.85% of approval investment projects and 63.66% of contractual FDI value. And with further relaxing of policies on ownership control, export sourcing as well as the expansion scope of business operations, the FDI inflow to China is likely to increase.

---

[7] Although Luxembourg received US$88 billion of FDI flows in 2003, most of the FDI inflow was trans-shipped to other destinations (UNCTAD, 2004).
[8] MOFCOM, *An Overview of Chinese Absorption of Foreign Direct Investment in 2003*, from MOFCOM website.
[9] According to the information by the Foreign Investment Department of the Ministry of Commerce.

Table 11.2  Sectoral Distribution of Cumulative FDI as of 2003

Value unit: US$100 million

| Sector | No. of Project | Share % | Contractual FDI Value | Share % |
| --- | --- | --- | --- | --- |
| Total | 465277 | 100 | 9431.31 | 100 |
| Agricultural, Forestry, Animal Husbandry and Fishery | 13333 | 2.87 | 180.36 | 1.91 |
| Mining | 524 | 0.11 | 16.81 | 0.18 |
| Manufacturing | 338952 | 72.85 | 6003.98 | 63.66 |
| Production and Supply of Power, Gas and Water | 654 | 0.14 | 56.82 | 0.60 |
| Construction | 10040 | 2.16 | 242.49 | 2.57 |
| Transport, Warehousing and Post | 5235 | 1.13 | 238.13 | 2.52 |
| Wholesale, Retailing | 23565 | 5.06 | 288.42 | 3.06 |
| Finance | 48 | 0.01 | 8.65 | 0.09 |
| Real Estate | 40941 | 8.80 | 1808.96 | 19.18 |
| Lease and Business Service | 15438 | 3.32 | 301.70 | 3.20 |
| Scientific Research, Technology Service and Geological Prospecting | 3528 | 0.76 | 41.59 | 0.44 |
| Resident Service and other Service | 10333 | 2.22 | 163.19 | 1.73 |
| Education | 1482 | 0.32 | 25.86 | 0.27 |
| Health Care, Social Security and Social Welfare | 1204 | 0.26 | 54.33 | 0.58 |

Source: MOFCOM FDI statistics.

## Has China Crowded Out ASEAN Countries' FDI?

As the value of FDI in China has kept rising, and China becomes the most attractive destination for FDI,[10] the competition for FDI would be intensified. What is the implication for ASEAN countries? Has China gained at the expense of ASEAN? Is China really squeezing ASEAN's FDI inflow? To answer the question, we will focus our analysis on only ASEAN-5 instead of ASEAN-10, since ASEAN-5 absorbed the lion's share of FDI into Southeast Asia (Figure 11.1).

### Share of FDI

From Figure 11.1, it seems that China has crowded out ASEAN-5 in FDI absorption during the period of 2000 to 2002, in terms of value. But if we compared their market shares, there are no indicators to confirm this view. During these years, the much-improved share of China in FDI inflow has not been accompanied by

---

[10] According to UNCTAD's 2004 survey, China was ranked as a top destination of global FDI flow, followed by India and Thailand (UNCTAD, 2004).

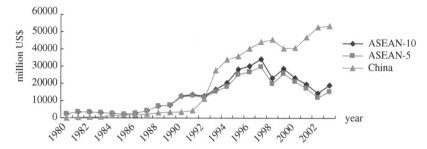

Source: Figures for the years 1998–2003 come from UNCTAD (2004), others come from UNCTAD FDI database.

**Figure 11.1** Global FDI Flow to China and ASEAN for 1980 to 2003

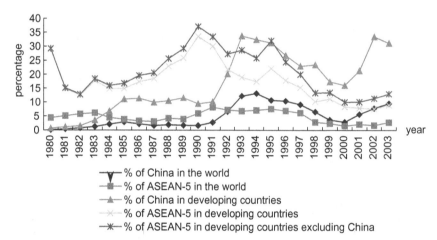

Source: The same as Figure 11.1.

**Figure 11.2** Shares of FDI Inflow to China and ASEAN-5 for 1980 to 2003

reduction in the shares of ASEAN-5 in the world and in all developing countries as well (Tan, 1999). For the period of 1980 to 2003, the trends of the shares of China and ASEAN-5 in the world were similar except for the years 1984, 1985, 1992 and 1993, according to Figure 11.2. Actually, the declined share of ASEAN-5 in 1992 did not seem to be caused by the increased share of China, but mainly due to its own share decline in the developing countries excluding China. That is to say, the share of ASEAN-5 in 1992 was lost to other developing countries instead of China. But the drop in share of ASEAN-5 in 1993 could be due to China, since the share of ASEAN-5 for FDI inflow in developing countries excluding China rose while that in developing countries decreased. The situations for 1984 and 1985 were the same as those in 1992 and 1993 respectively.

## Source of FDI Inflow

As shown in Table 11.3, by the end of 2003, Hong Kong ranked first in the world with US$222.58 billion and 44.38% of a cumulative realized FDI in Mainland China, followed by the US (8.79%), Japan (8.25%), EU (7.55%), Taiwan (China) (7.28%), ASEAN-5 (6.40%), the Virgin Islands (6.01%), Singapore (4.69%) and ROK (3.93%). During the period of 1995–2001, EU, the US, Japan, ASEAN, Hong Kong and Taiwan (China) accounted for 25.44%, 19.38%, 16.68%, 8.49%, 3.23% and 2.06% cumulative FDI respectively in ASEAN-5.[11] Therefore, there exists some

Table 11.3  Top 15 Investors in China as of 2003

Value unit: US$100 million

| Country/Region | Project | | Contractual FDI | | Realized FDI | |
| --- | --- | --- | --- | --- | --- | --- |
| | Number | % of Total | Value | % of Total | Value | % of Total |
| Hong Kong | 224509 | 48.25 | 4145.14 | 43.95 | 2225.75 | 44.38 |
| US | 41340 | 8.89 | 864.43 | 9.17 | 440.88 | 8.79 |
| Japan | 28401 | 6.10 | 574.87 | 6.10 | 413.94 | 8.25 |
| Taiwan (China) | 60186 | 12.94 | 700.29 | 7.43 | 364.88 | 7.28 |
| Virgin Islands | 8877 | 1.91 | 620.12 | 6.58 | 301.65 | 6.01 |
| Singapore | 11871 | 2.55 | 435.68 | 4.62 | 235.31 | 4.69 |
| ROK | 27128 | 5.83 | 366.53 | 3.89 | 196.88 | 3.93 |
| UK | 3856 | 0.83 | 208.42 | 2.21 | 114.38 | 2.28 |
| Germany | 3504 | 0.75 | 157.13 | 1.67 | 88.51 | 1.76 |
| France | 2302 | 0.49 | 79.15 | 0.84 | 61.48 | 1.23 |
| Macao | 8407 | 1.81 | 120.87 | 1.28 | 51.90 | 1.03 |
| The Netherlands | 1254 | 0.27 | 99.26 | 1.05 | 50.64 | 1.01 |
| Cayman Islands | 923 | 0.20 | 111.75 | 1.18 | 46.69 | 0.93 |
| Canada | 6941 | 1.49 | 119.87 | 1.27 | 39.21 | 0.78 |
| Australia | 6073 | 1.31 | 99.93 | 1.06 | 34.21 | 0.68 |
| Others | 29705 | 6.38 | 728.30 | 7.72 | 348.71 | 6.95 |
| ASEAN-5 | 21158 | 4.55 | 633.70 | 6.72 | 320.80 | 6.40 |
| EU | 16158 | 3.47 | 659.42 | 6.99 | 378.73 | 7.55 |
| Total | 465277 | 100 | 9431.31 | 100 | 5014.71 | 100 |

Source: MOFCOM FDI statistics.

---

[11] Calculated from ASEAN Secretariat, *Statistics of Foreign Direct Investment in ASEAN: Comprehensive Data Set 2002*, Table 3.1.1.

discrepancy in FDI sourcing between China and ASEAN-5. More than half of cumulative realized FDI in China comes from Hong Kong and other overseas Chinese diaspora, whereas over 60% of FDI in ASEAN-5 originates from the developed countries.

And, if we look at the main investors in ASEAN-5, US and Japan, we can see that the decline in shares of ASEAN-5 in the US and Japan was not due to China, except for Japan in the years 1993, 1995 and 2000 (Figures 11.3 and 11.4).

What's more, there is one important thing we need to pay attention to, that is the so-called "round-tripping".[12] It is generally acknowledged that the genuine

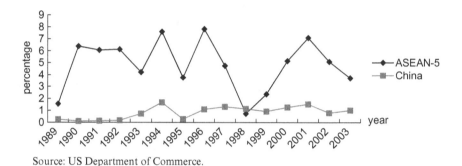

Source: US Department of Commerce.

**Figure 11.3** Shares of China and ASEAN-5 in the US FDI Outflow

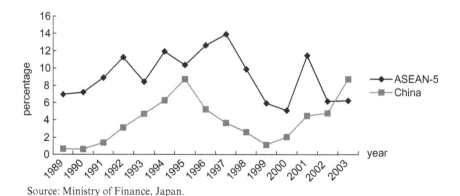

Source: Ministry of Finance, Japan.

**Figure 11.4** Shares of China and ASEAN-5 in Japan's FDI Outflow

---

[12] Round-tripping refers to the circular flow of capital out of China and the subsequent "re-investment" of the outgoing capital back to China again in the form of FDI seeking to benefit from the preferential treatment of foreign investors by Chinese government.

extent of FDI inflow to China is overstated due to "round-tripping". Though it is difficult to figure out precisely, there are some estimates. For example, according to the studies of Harrold and Lall (1993) and Lardy (1995), about 25% of China's FDI inflow in 1992 came from "round-tripping", while Huang (1998) figured that at least 23% of China's FDI inflow was accounted by "round-tripping" for the same year. Probably, the extent of "round-tripping" would be larger, as Hong Kong-based Political and Economic Risk Consultancy (PERC) pointed out, that about US$36 billion out of US$100 billion FDI inflow to the Mainland and Hong Kong in 2000 were genuine FDI, with most of these going to Mainland China (Wu et al., 2002). Actually, this kind of investment fund would definitely go to China instead of flowing to ASEAN. In addition, the importance of tax havens as sources of China's inward FDI has increased in recent years.[13] And some of the funds come from Taiwan, for the purpose of bypassing Taiwanese government restrictions on investment in the Mainland.

## FDI Performance

Though the amount of FDI that China has absorbed is very considerable, dominating 31.10% of FDI inflow to developing countries, China's FDI record is much less impressive relative to its economic size. China's inward FDI stock as a percentage of GDP in 2003 is 35.6%, compared to 57.2% and 161.3% for Malaysia and Singapore respectively. Furthermore, China is lower than some of the new ASEAN members, such as Cambodia and Vietnam (Table 11.4). According to UNCTAD analysis, China's inward FDI performance index[14] only ranked 37th for the period of 2001 to 2003, although it was a significant improvement over its previous rank of 50th (UNCTAD, 2004). And even though China's inward FDI performance index

Table 11.4  Inward FDI Stocks as a Percentage of GDP in 2003

Unit: %

| China | Indonesia | Malaysia | Philippines | Singapore | Thailand | Brunei | Cambodia | Lao | Vietnam | Developing World |
|---|---|---|---|---|---|---|---|---|---|---|
| 35.6 | 27.5 | 57.2 | 14.5 | 161.3 | 25.8 | 156.0 | 46.4 | 30.1 | 50.6 | 31.4 |

Source: UNCTAD, *World Investment Reports 2004: The Shift towards Service*, Annex Table B.6.

---

[13] In 2003, up to 10.8% of China's inward FDI (US$5.78 billion) came from British Virgin Islands, making it China's second largest FDI supplier.

[14] As defined by UNCTAD, the inward FDI performance index ranks countries by the amount of FDI they receive relative to their economic size. It is the ratio of a country's share in global FDI inflows to its share in global GDP. A value greater than one indicates that the country attracts more FDI in proportion to its economic size, a value below one shows that it receives less (UNCTAD, 2004).

Table 11.5  Inward FDI Performance Index in 2000–2002

| | China | Indonesia | Malaysia | Philippines | Singapore | Thailand | Brunei | Myanmar | Vietnam | Argentina | Brazil | Chile |
|---|---|---|---|---|---|---|---|---|---|---|---|---|
| Index | 1.331 | −0.528 | 0.923 | 0.818 | 4.755 | 0.753 | 5.531 | 0.707 | 1.327 | 0.692 | 1.538 | 1.736 |
| Rank | 50 | 139 | 70 | 90 | 6 | 80 | 4 | 84 | 39 | 85 | 37 | 34 |

Source: UNCTAD FDI database.

Table 11.6  FDI Inflows Per Capita to China and Selected Economies, 2000

Unit: US$

| Economy | FDI per capita |
|---|---|
| China | 30.1 |
| OECD average | 1320.9 |
| India | 2.2 |
| Indonesia | −21.6 |
| Malaysia | 162.8 |
| Myanmar | 5.3 |
| Philippines | 16.3 |
| Singapore | 1547.2 |
| Thailand | 54.0 |
| Vietnam | 16.7 |
| Argentina | 314.8 |
| Brazil | 195.4 |
| Chile | 241.6 |

Source: OECD, "Attracting Investment to China", *OECD Observer* (September 2003).

is above unity, it is much lower than Singapore's index (Table 11.5). Moreover, if population size is taken into account, China's inward FDI per capita is lower than some major developing economies, including some ASEAN countries (Table 11.6).

## *Main Factors for the Decrease of Inward FDI in ASEAN-5*

Inward FDI in ASEAN-5 has dropped tremendously both in volume and share of the world since the year 1998. But as analyzed above, ASEAN-5 did not lose out to China in FDI absorption. So, why has ASEAN-5 received less FDI inflow? As Wu *et al.* (2002) pointed out, "the post-1997 sharp drop in FDI to ASEAN-5 can be largely attributed to the Asian Financial Crisis, which has severely dampened investors' confidence in the region." The outbreak of the Asian Financial Crisis caused increased political risk and economic instability in the region, which turned foreign investors away (Cheng, 2001). Therefore, although the presence of natural resources, low-cost labor, adequate infrastructure and favorable policies to foreign investors are impetus to FDI inflow, sound macroeconomic and political stability are the baseline for attracting FDI, since "foreign investment is a coward because it only goes where it is safe".[15] The second factor, I think, is low economic growth or even minus economic growth, because the ultimate purpose of FDI is to obtain higher profit. We can see that the GDP growth rate in ASEAN-5 was −8.93%[16] in 1998, when the FDI inflow to ASEAN dropped tremendously. And the GDP growth rates in the following years were still low compared to those before the Asian Financial Crisis. Thirdly, the decrease of the world FDI flows is certainly another factor that caused less FDI inflow to ASEAN-5 in value. In addition, in the era of rapid development in regional cooperation, fragmented markets in the region caused by slow process of integration may be another contributor to the decrease of inward FDI. In contrast with the NAFTA and some other regional groups in Latin America, Southeast Asia countries lagged behind in the integration process in a way. In 2003, the ASEAN Investment Area (AIA) came into being. And in the same year, ASEAN leaders agreed to integrate their economies in a bid to create the so-called ASEAN Economic Community by 2020 to attract more FDI inflow, where the region of over 500 million people becomes a single market with a single production base.

## China's Direct Investment in ASEAN

The Chinese government began to encourage outward direct investment in 1990s. But compared to the tremendous inward FDI, Chinese outward FDI is minimal. With the prevailing high growth of Chinese economy and its deeper integration into the world, Chinese outward direct investment started to become a new source of foreign investment in the Asia and beyond (Wong and Chan, 2003; Vatikiotis, 2004; and Frost, 2004). As the "going out" strategy issued in 2000, more and more Chinese enterprises, especially those SOEs, are going abroad, investing in a wide range of businesses from trading to manufacturing and natural resource exploitation under the encouragement of the governments at all levels (Figure 11.5).

According to *the Statistical Communique of China on Overseas Direct Investment*, jointly issued by the Ministry of Commerce and the National Bureau of Statistics, by the end of 2003, the total amount of overseas direct investment by Chinese companies reached US$33.2 billion. According to the statistics of the Ministry of Commerce, by the end of 2003, some 857 projects had been set up by Chinese

---

[15] Speech of William R. Rhodes, chairman of Citigroup and Citibank, made at the Extraordinary Annual Meeting of World Economic Forum in June 2003.
[16] From ASEAN Finance and Macroeconomic Surveillance Unit (FMSU) database.

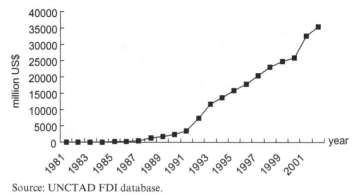

Source: UNCTAD FDI database.

**Figure 11.5** FDI Outflow Stock from China for 1981 to 2002

enterprises in ASEAN countries, with Chinese outward investment amounting to US$940 million. The value of Chinese investments in ASEAN has kept increasing at a high growth rate in recent years except 2002. From Table 11.7, we can see that seven ASEAN members are among the 30 economies ranked as China's top outward FDI destinations for the period from 1979 to 2002. As pointed out by Wong and Chan (2003), actual Chinese investment should be significantly higher than the official figures, since Chinese companies are known to circumvent official foreign currency controls by investing through offshore entities. It is believed that Chinese outward investment is likely to soar in the years ahead and the earliest beneficiaries will probably be China's neighbors — the ASEAN members who adjoin with China's geography and have deep cultural and historical roots with China. While China's inward FDI constituted 35.6% of its GDP and the world's outward FDI made up 23.0% of its GDP in 2003, its outward FDI as a percentage of GDP is only 2.6% (UNCTAD, 2004). With the high growth of the Chinese economy, there is a huge space for China's outward FDI to the world. Moreover, the setting up of the China and ASEAN Free Trade Area will no doubt bring more Chinese investments to ASEAN.

Nowadays, Malaysia, Singapore and Thailand have offices in China to attract Chinese investment (Dai, 2004), since Chinese enterprises are considered as important investors in ASEAN countries. Actually, some Chinese TNCs have set up foreign affiliates in Southeast Asia, such as TCL in Vietnam, Haier in Indonesia, Philippines and Malaysia, Konka and Changhong in Indonesia. These investments are mainly market-seeking, either for local markets or as a channel to other markets by circumventing trade barriers erected by foreign countries. Another reason for Chinese direct investment abroad is resource-seeking. With the high growth of

Table 11.7  China's Approval FDI Outflows, 1979–2002

Unit: million US$

| Rank[a] | Economy | 1999 Number of Projects[b] | 1999 Value | 2000 Number of Projects[b] | 2000 Value | 2001 Number of Projects[b] | 2001 Value | 2002 Number of Projects[b] | 2002 Value | 1979–2002 Cumulative Number of Projects[b] | 1979–2002 Cumulative Investment Value |
|---|---|---|---|---|---|---|---|---|---|---|---|
|  | Total | 220 | 590.6 | 243 | 551.0 | 232 | 707.5 | 350 | 982.7 | 6 960 | 9 340.0 |
| 1 | Hong Kong | 24 | 24.5 | 15 | 17.5 | 26 | 200.7 | 40 | 355.6 | 2 025 | 4 074.3 |
| 2 | US | 21 | 81.1 | 15 | 23.1 | 19 | 53.7 | 41 | 151.5 | 703 | 834.5 |
| 3 | Canada | 1 | 0.1 | 8 | 31.7 | 4 | 3.5 | 4 | 1.2 | 144 | 436.0 |
| 4 | Australia | 3 | 1.7 | 13 | 10.2 | 6 | 10.1 | 15 | 48.6 | 215 | 431.0 |
| 5 | Thailand | 3 | 2.0 | 6 | 3.3 | 9 | 121.3 | 5 | 3.9 | 234 | 214.7 |
| 6 | Russian Federation | 12 | 3.8 | 14 | 13.9 | 12 | 12.4 | 27 | 35.5 | 482 | 206.8 |
| 7 | Peru | 1 | 75.7 | 1 | 0.001 | 2 | 3.1 | — | — | 20 | 201.2 |
| 8 | Macao | 3 | 0.2 | 4 | 0.5 | 6 | 2.4 | 2 | 2.0 | 229 | 183.7 |
| 9 | Mexico | 2 | 97.0 | 1 | 19.8 | 1 | 0.2 | 1 | 2.0 | 45 | 167.4 |
| 10 | Zambia | 4 | 6.7 | 3 | 11.6 | 3 | 4.3 | 1 | 0.3 | 18 | 134.4 |
| 11 | Cambodia | 13 | 32.8 | 7 | 17.2 | 7 | 34.9 | 3 | 5.1 | 61 | 125.0 |
| 12 | Brazil | 1 | 0.5 | 3 | 21.1 | 4 | 31.8 | 8 | 9.3 | 67 | 119.7 |
| 13 | South Africa | 14 | 12.8 | 17 | 31.5 | 2 | 12.4 | 3 | 1.7 | 98 | 119.3 |
| 14 | Republic of Korea | 1 | 0.1 | 5 | 4.2 | 2 | 0.8 | 7 | 83.4 | 62 | 107.8 |
| 15 | Vietnam | 2 | 6.6 | 17 | 17.6 | 12 | 26.8 | 20 | 27.2 | 73 | 85.0 |
| 16 | Japan | 1 | 0.5 | 2 | 0.3 | 6 | 1.7 | 11 | 18.2 | 236 | 82.1 |
| 17 | Singapore | 6 | 2.9 | 6 | 1.0 | 3 | 0.4 | 6 | 2.1 | 172 | 71.7 |
| 18 | Myanmar | 1 | 6.6 | 7 | 32.9 | 3 | 1.8 | 5 | 15.8 | 38 | 66.1 |
| 19 | Indonesia | 0 | 18.9 | 1 | 8.0 | 2 | 0.6 | 6 | 3.7 | 59 | 65.0 |
| 20 | Mali | 1 | 1.2 | 1 | 28.7 | — | — | — | — | 5 | 58.1 |

Table 11.7 (Continued)

Unit: million US$

| Rank[a] | Economy | 1999 | | 2000 | | 2001 | | 2002 | | 1979–2002 | |
|---|---|---|---|---|---|---|---|---|---|---|---|
| | | Number of Projects[b] | Value | Number of Projects[b] | Value | Number of Projects[b] | Value | Number of Projects[b] | Value | Cumulative Number of Projects[b] | Cumulative Investment Value |
| 21 | Mongolia | 15 | 40.3 | 12 | 5.4 | 7 | 4.5 | 7 | 3.4 | 69 | 56.6 |
| 22 | Germany | 1 | 0.3 | 1 | 1.6 | 3 | 3.5 | 6 | 2.8 | 150 | 51.5 |
| 23 | New Zealand | — | — | — | — | 2 | 0.9 | 2 | 0.9 | 26 | 48.7 |
| 24 | Egypt | 5 | 3.8 | 3 | 9.7 | 2 | 1.4 | 3 | 16.3 | 27 | 48.5 |
| 25 | Oman | — | — | 1 | — | — | — | 1 | 17.5 | 70 | 47.2 |
| 26 | Papua New Guinea | — | — | 1 | 0.9 | — | — | — | — | 20 | 44.7 |
| 27 | Nigeria | 2 | 1.6 | 1 | 2.6 | 8 | 6.4 | 9 | 11.4 | 49 | 44.3 |
| 28 | Tanzania, United Rep. of | 3 | 16.3 | 1 | 1.0 | — | — | 2 | 0.4 | 19 | 41.3 |
| 29 | Kazakhstan | 7 | 17.2 | 5 | 7.7 | 1 | 0.3 | 3 | 26.9 | 51 | 39.6 |
| 30 | Lao PDR | 1 | 2.0 | 2 | 24.4 | 1 | 1.2 | 2 | 61 | 18 | 36.6 |

Note: [a] Ranked by cumulative investment value.
[b] The number of projects refers to approved investment projects involving Chinese enterprises.

Source: UNCTAD, *World Investment Report 2004: The Shift towards Service*, Annex Table A.1.12, p. 298 (2004).

the country's economy and the shortage of domestic resources, Chinese businesses begin to invest in ASEAN countries that are rich in natural resources. For example, In January 2001, China National Offshore Oil Corporation (CNOOC) invested US$585 million to acquire Indonesian oil and gas fields. At the same time, a joint venture with US$760 million investment was set in Sabah, Malaysia to produce pulp and paper.

## Conclusions and Implications

From the analyses above, we can see there is no obvious evidence to say that China has squeezed ASEAN FDI inflow. Though China has absorbed US$53.5 billion in 2003, almost twice than in 1993, the share of Chinese inward FDI in the world in 2003 is lower than that in the period of 1993 to 1996 (see Figures 11.1 and 11.2). And the improved shares of Chinese inward FDI have actually not been at the expense of ASEAN-5, but other developing countries and industrialized countries. What's more, we can foresee increasing China's direct investment in ASEAN with Chinese economic growth and the integration between China and ASEAN members.

Nevertheless, we cannot deny that the competition for FDI between China and ASEAN does exist. Though the deterioration of ASEAN macroeconomic environment is the baseline for the decrease of ASEAN absorption of FDI inflow, the presence of China opening up to foreign investors and improved mature business climate put ASEAN under pressure. If China had not received so much inward FDI, some portion of the FDI could have been channeled to ASEAN-5 instead (Wu et al., 2002). But if ASEAN does not improve its own investment climate, FDI would flow to other economies. Actually, as Mr. Severino (2001), ASEAN former secretary-general, pointed out: "Competition is what globalization is all about. It is what commerce is all about. It is what economics should be all about. Competition — free competition — is not bad or necessarily frightening, provided it is fair."

So, in order to be more attractive to FDI, ASEAN economies should not only adopt some policies to improve their own investment environments, but expedite the integration process of the ASEAN economic community, which will feature freer flows of goods and services, and freer flows of investment and movement of skilled labor. In addition, the integration of China and ASEAN would also bring more opportunities to ASEAN in attracting FDI inflows from China and other countries. Both China and ASEAN should concentrate their efforts on setting up regional production networks via deeper division of labor and cooperation of industrial activities across the region.

## References

ASEAN Secretariat, "Forging Closer ASEAN-China Economic Relations in the Twenty-first Century", a report submitted by the ASEAN-China Expert Group on Economic Cooperation (2001).

Bhaskaran, M., *China as Potential Superpower: Regional Responses* (Berlin: Deutsche Bank Research Report, 15 January 2003).

Cheng, J.Y.S., "Sino-ASEAN Relations in the Early Twenty-first Century", *Contemporary Southeast Asia*, Vol. 23, No. 3 (2001).

Chia, S.Y., "ASEAN-China Trade in Manufactured Goods", in S.Y. Chia and B.F. Cheng, eds., *ASEAN-China Economic Relations: Trends and Patterns* (Singapore: Institute of Southeast Asian Studies, 1987).

Chirathivat, S., "Interdependence between China and Southeast Asian Economies on the Eve of the Accession of China into the WTO", in I. Yamazawa and K. Imai, eds., *China Enters WTO: Pursuing the Symbiosis with the Global Economy* (Tokyo: Institute of Developing Economies, 2001).

———, "ASEAN–China Economic Partnership in an Integrating World Economy", *Chulalongkorn Review*, Vol. 14 (2002a).

———, "ASEAN–China Free Trade Area: Background, Implication and Future Development", *Journal of Asian Economics*, Vol. 13 (2002b).

Dai, Y., "China Eyeing Regional Investment", *China Daily*, 9 February 2004.

Fan, X.Q. and P.M. Dickie, "The Contribution of Foreign Direct Investment to Growth and Stability: A Post-Crisis ASEAN-5 Review", *ASEAN Economic Bulletin*, Vol. 17, No. 3 (2000).

Francis, S., *Regionalism in South East Asia: the Old and the New*, 2004.

Frost, S., "Chinese Outward Direct Investment in Southeast Asia: How Much and What are the Regional Implications?", *SEARC Working Papers Series No. 67* (2004).

Harrold, P. and R. Lall, "China Reform and Development in 1992–93", *World Bank Discussion Papers* (1993).

Huang Y.S., *Foreign Direct Investment in China: An Asian Perspective* (Singapore: Institute of Southeast Asian Studies, 1998).

Lardy, N., "The Role of Foreign Trade and Investment in China's Economic Transformation", *The China Quarterly*, Vol. 144 (1995).

Luo, S.H., Cheng, B.F. and Gao, T.S., "China's Changing Industrial Structure: Its Impacts on Economic Relations with ASEAN Countries", in S.Y. Chia and B.F. Cheng, eds., *ASEAN-China Economic Relations: Developments in ASEAN and China* (Singapore: Institute of Southeast Asian Studies, 1989).

Panitchpakdi, S. and M. Clifford, *China and the WTO* (Singapore: John Wiley & Sons (Asia) Pte Ltd, 2002).

Severino, R.C., "ASEAN and China — Partners in Competition", speech at the ASEAN Forum sponsored by the ASEAN Consulates at Guangzhou, 9 June 2001.

Tan, K.Y., "Industrial Restructuring in the East Asian Newly Industrializing Economies and the Implications for ASEAN and ASEAN–China Economic Relations", in S.Y. Chia and B.F. Cheng, eds., *ASEAN–China Economic Relations: In the Context of Pacific Economic Development and Co-operation* (Singapore: Institute of Southeast Asian Studies, 1992).

Tan, R., "Foreign Direct Investment Flows to and from China", *PASCN Discussion Paper No. 99-21* (1999).

UNCTAD, *World Investment Report 2004: The Shift towards Service* (New York and Geneva: UNCTAD, 2004).

Vatikiotis, M., "Outward Bound", *Far Eastern Economic Review*, 5 February 2004.

Wang, Z., "The Impact of China's WTO Entry on the World Labour-intensive Export Market: a Recursive Dynamic ECG Analysis", *The World Economy*, Vol. 22, No. 3 (1999).

Wong, J. and Chan S., "China's Outward Direct Investment: Expanding Worldwide", *China: An International Journal*, Issues 1 & 2 (2003).

Wu, F., Siaw, P.T. and Yeo, H.S., "Foreign Direct Investments to China and Southeast Asia: Has ASEAN been losing out?", *Economic Survey of Singapore* (Third Quarter 2002).

Wu, F. and Yeo, H.S., *China's Rising Investment in Southeast Asia: How ASEAN and Singapore can benefit* (Singapore Ministry of Trade and Industry, August 2002).

Yang, D.M., Yu, Y.D. and Shen, H.S., "Economic Reforms in China and their Impact on China-ASEAN Economic Relations", in S.Y. Chia and B.F. Cheng, eds., *ASEAN-China Economic Relations: Developments in ASEAN and China* (Singapore: Institute of Southeast Asian Studies, 1989).

Part VI

**The South China Sea and Maritime Issues**

# Chapter 12

# Regional Cooperation for Conservation and Management of Fishery Resources in the South China Sea

*Kuen-chen FU*

## Foreword

The South China Sea is a semi-enclosed sea, enclosed by China Mainland, China Taiwan, the Philippines, Indonesia, Singapore, Malaysia, Thailand, Cambodia and Vietnam. Its total area reaches 3.3 million square kilometers. It crosses 24 degrees (from about 3°30′ north latitude to 27°40′ north latitude) on the surface of the Earth. It consists of a "large marine ecosystem" (LME), with a large area of sub-ecological systems and habitats, including mangroves, seaweed beds, coral reefs, etc. Obviously, its fishery resources are very abundant. According to the records available, in the South China Sea area, there are 1,064 fish species, 135 shrimp species, 73 cephalopodous (头足类) species, and among them about 100 fish species are of primary commercial value.[1] Ribbon fish (trichiurus haumela — 带鱼类), ink fish (sepia — 乌贼类), butter fish (Psenopsis — 鲳类), golden thread (nemipterus virgatus — 金线鱼类) and other species of commercial value spread over the whole area. Bottom species with commercial value include: snail mullet (saurida — 蛇鲻), ribbon fish (带鱼), golden thread (金线鱼), grouper (epinephelus — 石斑鱼), etc. Mid- and upper-level fish would include: golden sardine (sardinella aurita — 金色小沙丁), wart (psenopsis anomala — 刺鲳), round scad (decapterus maruadsi — 蓝圆鲹), etc. In the middle and southern parts of the South China Sea, you will find tuna (thunnus — 金枪鱼), sail fish (Istiophorus — 旗鱼) and other oceanic fish species.[2] The original fish density in the South China Sea was very high. According to some scholars' estimation, before 1975 when large-scaled fishing gears were not used, the density of fish resources in the northern South China Sea reached 1,100 kg per square km. The number was 2,300 kg per square km

---

[1] http://www.gd-fishmarket.com/fishcondition.htm.
[2] The Chinese Fishery Resources Investigation and Classification Committee, *China Fishery Resources* (Zhejiang Province Science and Technology Publishers, 1990).

in the Beibu Bay area.³ And the number was 1,120 kg at the Paracel (Xisha) Islands.⁴

Since the 1970s, together with the improvement of powerful fishing gears, the fishery resources in the South China Sea decreased sharply. The environmental protection and fishery resources conservation have become an urgent topic for the littoral states around this semi-enclosed sea. All the known fish with commercial value in this region are straddling, existing in not only one state's EEZ, but also other states' EEZs or in the high seas. This indicates that an effective regional scheme for the conservation and management of the fishery resources in the South China Sea will have to be a scheme established multilaterally. The problem is that the sovereignty disputes in the region are some of the most difficult in the world. Without the ocean boundaries delimitated in the area, any cooperation scheme might not be feasible, or at least not enforceable. Fortunately, the political atmosphere in this region has recently improved with the awareness of the seriousness of the issue. Thus a feasible cooperation scheme for the above purposes is becoming more and more possible.

### Current Situation of Fishery Management in South China Sea

*Over-Fishing has Drastically Lowered the Resources Density and the Quality of Catches in the Region. Fishery Resources in the South China Sea are not as Abundant as in the Past*

The total catch in the South China Sea in 1950 was 80,000 tons. It was 425,000 tons in 1955. The figure varied between 400,000 and 800,000 tons from 1956 to 1979. Then, after 1980, due to the increasing number of fishing vessels operating in the region, the total catch jumped up. In 1999, the figure reached 3,340,000 tons,⁵ more than 40 times the number in 1950. Surprisingly, according to an investigation in 2000 by a Chinese study group on the Chinese continental shelves and EEZ, the total amount of fish caught by fishermen from China, Indonesia, Thailand, Philippines, Vietnam and Malaysia in the South China Sea region reached 12,200,000 tons.⁶ That was roughly 153 times the figure in 1950. And the potential total of fishery

---

³Wu Zhuang, "Suspension of Fisheries and Its Impacts in The South China Sea", paper presented in the Cross-Strait Cooperation and South China Sea Resources, Boao, Hainan Province, 12 January 2004.
⁴http://202.100.218.58/gov /thaiyang/hyb1-7.htm.
⁵http: //www.gd-fishmarket.com/fishcondition.htm.
⁶Wu Shi-chun and Guo Wen-lu, "A Preliminary Study on Regional Cooperation and Common Conservation in the South China Sea", presented at the South China Sea Resources and Cross-Strait Cooperation, Boao, Hainan, 12 January 2004.

resources in this region is 2,800,000 tons only.[7] Over-fishing has seriously threatened the future of fisheries in the South China Sea. Another investigation carried out by the Chinese Ministry of Agriculture indicated that the fish resources density in the northern part of the South China Sea, had dropped to 290 kg per square km in 1999, roughly 26% of its original density. The density in Beibuwan (Beibu Bay or Tonkin Bay), which used to be one of the best fishing grounds in the region, was 528 kg per square km, or about 23% of its original figure.[8] At the same time, the quality of catches in the South China Sea was also found to have worsened. For example, in the 1990s, the total catch of economic target species by bottom trawlers in the northern part of the South China Sea decreased about 20% compared with the 1970s. Most of the target species are now smaller, younger fish of less than one year old. Obviously, the catches have become less valuable.[9]

*Over-Lapping Maritime Zones Claimed by the Littoral States have Caused Various Disputes, and Hindered Mutual Trust Among the States Concerned*

Since the 1970s, the littoral states in the South China Sea have begun promulgating their domestic laws and regulations respectively for establishing their EEZs and continental shelves. Vietnam in 1977, Cambodia and the Philippines in 1978, Indonesia, Malaysia and Thailand in 1980, Brunei in 1982, all publicized their proclamations on establishing their EEZs.[10] China also proclaimed in 1998 its statutes on its EEZ and continental shelf.[11] Due to the fact that all these claims are seriously over-lapping, the territorial disputes in the region have become even more complicated. On the one hand, the number of illegal fishing activities has apparently increased. On the other hand, conflicts over jurisdiction among the states concerned have become more frequent. In October 2003 alone, more than 11 Vietnamese fishing vessels fished illegally in the Chinese South China Sea waters. The dynamites discovered from these vessels reached 101 kg. From February to October 2003, Chinese Taiwanese ocean patrol ships expelled 22 Vietnam fishing

---

[7] http://www.gd-fishmarket.com/fishcondition.htm.
[8] Wu Zhuang, "Impacts of the Fishery Suspension System in the South China Sea", presented at the South China Sea Resources and Cross-Strait Cooperation, Boao, Hainan, 12 January 2004.
[9] Wu Shi-chun and Guo Wen-lu, "A Preliminary Study on the Regional Cooperation and Common Conservation in the South China Sea," presented at the South China Sea Resources and Cross-Strait Cooperation, Boao, Hainan, 12 January 2004.
[10] Li Jin-min, "Impacts and Complication of the Sovereign Disputes Caused by the UNCLOS", proceedings of the Conference for the 20th Anniversary of the Signing of the UNCLOS, 15–16 August 2002, Xiamen University, Center for the Oceans Law Studies.
[11] The PRC Law of the Exclusive Economic Zone and Continental Shelves was adopted by the Third Meeting of the Standing Committee of the Ninth National Peoples Congress on 26 June 1998.

vessels from the South China Sea waters claimed by Chinese Taiwan.[12] According to incomplete statistics, there were 104 Taiwanese fishing vessels arrested and detained by the Philippines and Indonesian governments in the South China Sea area from 1992 to 1999.[13] When this author visited Haimen (Seagate) Town in Hainan Province in January 2003, local fishermen who used to fish in the waters of the Spratly Islands complained about the brutal treatments they received from the Philippines soldiers. The Chinese fishermen fishing in their forefathers' traditional fishing grounds are now facing danger of losing their livelihood in the same fishing grounds. The areas of occurrence of such disputes would cover almost the whole South China Sea area.

*Resource Assessment has not been Precise Enough. Conservation Techniques have been Outdated. The Whole Ecological System in the Area is at the Brink of Collapse*

Some of the littoral states in the South China Sea have been interested in exploiting the fishery resources, but not at all in assessment of the resources. Up to this moment, there has been no reliable, thorough and detailed estimation of the fish resources in the region. Different institutes might have publicized some different statistics. But the differences among these figures often astonished the observers. Take the year 2000 as an example. One institute estimated that the annual total catch in the South China Sea had reached 5,000,000 tons.[14] About two thirds of the fish species had been over-fished. However, another source indicated that the total catch had reached 12,200,000 tons.[15] That was more than two times the earlier cited figure. As to the original density of the fish resources in this region, the estimation in the northern waters of the South China Sea and of the Paracel Islands might be rough but still reliable. But the original density and current total catch in the middle and southern waters of the South China Sea are basically unknown.[16] Under this circumstance, it is simply not possible to decide the "Total Allowable Catch — TAC", the "Optimum Utilization — OU", the "Management Reference Point", the "Conservation Reference Point" and/or any other criteria required by the UNCLOS.

---

[12] http://www.news.sohu.com, 3 November 2003.
[13] Kuen-chen Fu, "Illegal Fishing and Potential Conflicts of National Jurisdictions — A Perspective of Taiwan," in *Ocean Management and The Law* (Taiwan: Winsoon Books, August 2003), p. 122.
[14] http://na.neaa.gov/lme/text/lme36.htm.
[15] Wu Shi-chun and Guo Wen-lu, "A Preliminary Study on the Regional Cooperation and Common Conservation in the South China Sea", presented at the South China Sea Resources and Cross-Strait Cooperation, Boao, Hainan, 12 January 2004.
[16] http://202.100.218.58/gov/thaiyang/hyb1-7.htm.

In terms of fishery conservation techniques, the littoral states in this area have adopted the so-called "input control methods". Namely, they control the number of fishing permits, the quantity of fishing vessels, and the types of fishing gears and fishing methods. China, in particular, has promulgated its regulations on "suspension of fishing" that prohibit fishing in specific areas and during specific period of time. China has also controlled its fishing capabilities and its total catch by decreasing its total number of fishing vessels. The goal is "zero growth".[17]

For development of the international fishery management, three stages may be observed. The first stage was emphasizing the "TAC". The main methods adopted in this stage were the "input controls". Fishing vessel numbers and fishing capabilities were controlled by various governments. The second stage was establishing thinking on the "Maximum Sustainable Yield — MSY". Still emphasized was a more reasonable quantity of fishing capabilities input into the area. The third stage is emphasizing the "Precautionary Principle" and "ecosystem" management. It may be accurate to say that the current international fishery management methodology is moving from the second to the third stage.[18] The "input controls" methods currently adopted by the states around the South China Sea are rather old and inadequate.

Another unfavorable factor influencing the fishery resources conservation in the South China Sea is the development of its non-living resources. In the 1970s, Vietnam, Philippines, Malaysia, Indonesia and Brunei initiated large area management of the oil and gas resources. The Wan-An Bank, the Reeds Bank and the Natuna Island vicinities are the most famous concession areas. These states carried out unilateral exploration and exploitation activities or signed bilateral concession contracts with Western oil companies, ignoring blatantly the sovereign claims made by the Chinese from both China Mainland and China Taiwan. Among these states, Malaysia alone has already drilled 18 oil wells and 40 gas wells in the historic waters claimed by the Chinese. Some of these wells locate deep, about 100 nautical miles, inside of the U-shaped lines demarcated by the Chinese government since 1947.[19] These unilateral moves have caused serious damages to the living resources in the South China Sea, worsening the already fragile environment in the region.

---

[17] Wu Shi-chun and Guo Wen-lu, "A Preliminary Study on the Regional Cooperation and Common Conservation in the South China Sea", presented at the South China Sea Resources and Cross-Strait Cooperation, Boao, Hainan, 12 January 2004.
[18] Stuart M. Kaye, *International Fisheries Management* (Kluwer Law International, 2001), pp. 44–88.
[19] Wu Shichun, "Chinese Energy Security and Oil Well Development in the South China Sea", presented at the South China Sea Resources and Cross-Strait Cooperation, Boao, Hainan, 12 January 2004.

In addition, Japanese tanks and ships carrying spent fuels from the nine nuclear power plants in Japan have also posed real threats to the South China Sea fishery resources. These high-level radioactive wastes are shipped from Japan to two European factories, one in the UK and another in France, for refinery. Yet, the Japanese government has always shipped these materials of extreme danger in secrecy. According to Article 22 of the UNCLOS, this practice of Japan is an obvious violation of the statutory international law. It endangers not only the people of the South China Sea, but also the living resources in the region.[20]

Apparently, conservation of the South China Sea fishery resources needs cooperation from all the littoral states in the region. It has to be established on the basis of relevant political, economic, legal and technical cooperation. These preconditions were not in existence, before the last ten years of the 20th century. Now it seems that, the time has arrived.

## Recent Development of Regional Politics

The international political arena changed substantially in the 1990s. Probably because of China's booming economy, littoral states of the South China Sea area became more interdependent with one another. Mutual trust has also been strengthened, together with escalated cooperation in more aspects. With this background, a cooperation scheme for conservation and management of fishery resources in the region has become feasible.

### *Willingness for Cooperation Grows in the International Community*

After the end of the Cold War, peace and development seem to have become the main concerns in international affairs. Various governments in the world have put economic development in the front burner. Economic integration and globalization have become common goals for every nation today. Members of the international community have sensed the reality that no one can strive and prosper without the participation of others. At the same time, new problems and difficulties have also occurred. Environmental crisis, international terrorism, and proliferation of mass destructive weapons and etc. demand cross-boundary efforts to deal with them. Thus the willingness to cooperate has grown substantially. The 1992 UN Environmental and Development Conference, the 1995 WTO, the 1997 Asian financial crisis, and the "9.11" terrorist attack in 2001, all marked the path of change.

---

[20] Kuen-chen Fu, *Ocean Management and The Law* (Taiwan: Winsoon Books, August 2003), p. 81.

## Mutual Trust Among the South China Sea States has also been Raised

Together with the change of the atmosphere, the littoral states around the South China Sea have also improved their relations. The fourth summit meeting of ASEAN leaders in 1992 achieved two agreements on strengthening economic cooperation and on more favorable custom treatments. After that meeting, cooperation among themselves for combating transnational crimes, anti-terrorism, and in other fields, was carried out more effectively. In the year 2003, during the ninth summit meeting of ASEAN leaders, a declaration of the second agreement was made in Bali, Indonesia. In that declaration, ASEAN leaders proclaimed that they would create a EU-styled "ASEAN Community", pushing the political, economical, social and cultural cooperation, and the region's integration, a big step forward.[21] In terms of China's relationship with ASEAN, the progress is also evident. Since 1975, China has recognized the status of ASEAN as a whole entity. Normalization of relations between China and ASEAN states began in the year 1980. In 1992, China became a "consultation partner" of ASEAN. This status was promoted to a higher status of "dialogue partner" in 1996.[22] Then, in the 1997 East Asian financial crisis, China played an important role of rescuer by insisting a "no depreciation policy" for its RMB. With this effort, China has successfully persuaded ASEAN states that China will not be a threat to regional security, but a reliable partner in the South China Sea region. Obviously, the so-called "10+1" and "10+3" frameworks are all results of such a growing confidence of the ASEAN states. On 4 November 2002, China and ASEAN signed the new agreement for a full-scale economic cooperation, intending for a China-ASEAN Free Trade Area in the year 2010. The new agreement has initiated the big project since 1 July 2003.[23]

## With Common Acknowledgement of the South China Sea Issues, Fishery Cooperation is Already on the Way

China declared her basic policy for resolving the South China Sea disputes in the 1980s. The main theme since then has always been that "the sovereignty has been mine; the disputes should be put aside; and the joint exploitation should be carried out." During a meeting with ASEAN foreign ministers in 1992, China consulted with the regional states for intended future cooperation in the South China Sea. Mutual understanding has been cultivated since then.[24] In 1999, the Philippines

---

[21] http://www.sina.com.cn, 7 October 2003.
[22] http://www.yn.xinhuanet.com, Chronicle of the China-ASEAN Affairs.
[23] Hu Guang-hui and Ouyang Hui-ran, eds., *China-ASEAN Free Competition Zone and Hainan's Economy* (Hainan Publishers, October 2003), pp. 30–36.
[24] http://www.yn.xinhuanet.com, Chronicle of the China-ASEAN Affairs.

brought out the idea of fishery cooperation with China, proposing that it be gradually developed into a treaty among all the littoral states. The Philippines emphasized that the fishery agreement should be for the purpose of conserving resources, not linked with any sovereignty claims. After that, in 2000, China signed a maritime boundary delimitation agreement, and a fishery cooperation agreement on the Beibu Gulf with Vietnam. In 2001, China reached with Indonesia a memorandum of understandings on fishery cooperation matters, initiating projects in the fields of fishing activities, fish products processing, education, harbor construction, and fishing vessel building between the two countries.[25] Among the ASEAN states, a working group for ASEAN ocean and marine environment affairs, an ASEAN fisheries coordination group, and a Center for South East Asian Fishery Development, have been established over the past several years. On 4 November 2002, China signed a Declaration on the Conduct of the Parties in the South China Sea with the ASEAN states in Phnom Penh, Cambodia. In this first multilateral agreement, all the parties in the region promised to settle their disputes with peaceful means. They also agreed to exercise self-restraint, avoiding escalating conflicts and complicating the disputes, before disputes are eventually settled. With such spirits of cooperation and understanding, all the regional parties agreed further to exchange information, and strive to seek cooperation in fields like marine environmental protection, marine scientific research, navigation and transportation safety, and search and rescue.[26]

## International Conventions — Legal Basis of Regional Scheme

Two international treaties have to be taken into account for establishing the South China Sea regional arrangements. The first is the 1982 United Nations Convention on the Law of the Sea (UNCLOS), which provides the basic principles for fishery conservation and cooperation. The second is the 1995 Agreement for the Implementation of the Provisions of the United Nations Convention on the Law of the Sea of 10 December 1982 Relating to the Conservation and Management of Straddling Fish Stocks and Highly Migratory Fish Stocks (Fish Stocks Agreement), which gives much more detailed arrangements on how to conserve the straddling fish stocks in the high seas through international coordination and joint enforcement measures.

---

[25] Wu Shi-chun and Guo Wen-lu, "A Preliminary Study on the Regional Cooperation and Common Conservation in the South China Sea", presented at the South China Sea Resources and Cross-Strait Cooperation, Boao, Hainan, 12 January 2004.

[26] Hu Guang-hui and Ouyang Hui-ran, eds., *China-ASEAN Free Competition Zone and Hainan's Economy* (Hainan Publishers, October 2003), pp. 293–294.

The UNCLOS has been in force since 16 September 1994. All the littoral states in the South China Sea have access to this multilateral treaty. (Taiwan as part of China has no access to it, but has adopted most of the rules in its relevant domestic laws and regulations.) The Fish Stocks Agreement has also been implemented since 12 November 2001. China, Philippines and Indonesia signed this agreement. But none has ratified it.[27]

From the provisions of the two multilateral treaties, a vague shape of the regional scheme for conservation of the fishery resources can be drawn.

*Provisions on Regional Coordination and Cooperation in the UNCLOS*

The South China Sea is a semi-enclosed sea as defined in Article 122 of the UNCLOS. Article 123 of the UNCLOS provides that states bordering an enclosed or semi-enclosed sea should cooperate with each other in the exercise of their rights and in the performance of their duties under this convention. To this end, they shall endeavor, directly or through an appropriate regional organization:

1. to coordinate the management, conservation, exploration and exploitation of the living resources of the sea;
2. to coordinate the implementation of their rights and duties with respect to the protection and preservation of the marine environment;
3. to coordinate their scientific research policies and undertake where appropriate joint programmes of scientific research in the area; and
4. to invite, as appropriate, other interested states or international organizations to cooperate with them in furtherance of the provisions of this article.

Therefore, the future South China Sea fishery resources conservation scheme should include all the littoral states and impose the obligation of coordination to all of them. If necessary, they may also create an international organization to fulfill this obligation.

*Provisions on Regional Cooperation and Conservation in the Fish Stocks Agreement*

According to Article 5 of the agreement, in order to conserve and manage straddling fish stocks and highly migratory fish stocks, coastal states and states fishing on the

---

[27] Up to 3 November 2004, a total of 52 states have either ratified or accessed the Fish Stocks Agreement. But none of the eight South China Sea littoral states are included. See http://www.un.org.table, recapitulating the status of the convention and related agreements, and http://www.un.Org/Depts/los/reference_files/chronological_lists_of_ratifications.htm#Agreement%.

high seas shall, in giving effect to their duty to cooperate in accordance with the convention:

1. adopt measures to ensure long-term sustainability of straddling fish stocks and highly migratory fish stocks and promote the objective of their optimum utilization;
2. ensure that such measures are based on the best scientific evidence available and are designed to maintain or restore stocks at levels capable of producing maximum sustainable yield, as qualified by relevant environmental and economic factors, including the special requirements of developing states, and taking into account fishing patterns, the interdependence of stocks and any generally recommended international minimum standards, whether subregional, regional or global;
3. apply the precautionary approach in accordance with Article 6;
4. assess the impacts of fishing, other human activities and environmental factors on target stocks and species belonging to the same ecosystem or associated with or dependent upon the target stocks;
5. adopt, where necessary, conservation and management measures for species belonging to the same ecosystem or associated with or dependent upon the target stocks, with a view to maintaining or restoring populations of such species above levels at which their reproduction may become seriously threatened;
6. minimize pollution, waste, discards, catch by lost or abandoned gear, catch of non-target species, both fish and non-fish species (hereinafter referred to as non-target species) and impacts on associated or dependent species, in particular endangered species, through measures including, to the extent practicable, the development and use of selective, environmentally safe and cost-effective fishing gear and techniques;
7. protect biodiversity in the marine environment;
8. take measures to prevent or eliminate overfishing and excess fishing capacity and to ensure that levels of fishing effort do not exceed those commensurate with the sustainable use of fishery resources;
9. take into account the interests of artisanal and subsistence fishers;
10. collect and share, in a timely manner, complete and accurate data concerning fishing activities on, *inter alia*, vessel position, catch of target and non-target species and fishing effort, as set out in Annex I, as well as information from national and international research programmes;
11. promote and conduct scientific research and develop appropriate technologies in support of fishery conservation and management; and

12. implement and enforce conservation and management measures through effective monitoring, control and surveillance.

Article 6 of the Fish Stocks Agreement provides further on the precautionary approach that:

1. States shall apply the precautionary approach widely to conservation, management and exploitation of straddling fish stocks and highly-migratory fish stocks in order to protect the living marine resources and preserve the marine environment.
2. States shall be more cautious when information is uncertain, unreliable or inadequate. The absence of adequate scientific information shall not be used as a reason for postponing or failing to take conservation and management measures.
3. In implementing the precautionary approach, states shall:
   (a) improve decision-making for fishery resource conservation and management by obtaining and sharing the best scientific information available and implementing improved techniques for dealing with risk and uncertainty;
   (b) apply the guidelines set out in Annex II and determine, on the basis of the best scientific information available, stock-specific reference points and the action to be taken if they are exceeded;
   (c) take into account, *inter alia*, uncertainties relating to the size and productivity of the stocks, reference points, stock condition in relation to such reference points, levels and distribution of fishing mortality and the impact of fishing activities on non-target and associated or dependent species, as well as existing and predicted oceanic, environmental and socio-economic conditions; and
   (d) develop data collection and research programmes to assess the impact of fishing on non-target and associated or dependent species and their environment, and adopt plans which are necessary to ensure the conservation of such species and to protect habitats of special concern.
4. The states shall take measures to ensure that, when reference points are approached, they will not be exceeded. In the event that they are exceeded, the states shall, without delay, take the action determined under paragraph 3 (b) to restore the stocks.
5. Where the status of target stocks or non-target or associated or dependent species is of concern, the states shall subject such stocks and species to enhanced monitoring in order to review their status and the efficacy of conservation and management measures. They shall revise those measures regularly in the light of new information.

6. For new or exploratory fisheries, the states shall adopt as soon as possible cautious conservation and management measures, including, *inter alia*, catch limits and effort limits. Such measures shall remain in force until there are sufficient data to allow assessment of the impact of the fisheries on the long-term sustainability of the stocks, whereupon conservation and management measures based on that assessment shall be implemented. The latter measures shall, if appropriate, allow for the gradual development of the fisheries.
7. If a natural phenomenon has a significant adverse impact on the status of straddling fish stocks or highly migratory fish stocks, the states shall adopt conservation and management measures on an emergency basis to ensure that fishing activity does not exacerbate such adverse impact. The states shall also adopt such measures on an emergency basis where fishing activity presents a serious threat to the sustainability of such stocks. Measures taken on an emergency basis shall be temporary and shall be based on the best scientific evidence available.

For purposes of regulating cooperation for conservation and management, Article 8 of the Fish Stocks Agreement provides that:

1. Coastal states and states fishing on the high seas shall, in accordance with the convention, pursue cooperation in relation to straddling fish stocks and highly-migratory fish stocks, either directly or through appropriate sub-regional or regional fisheries management organizations or arrangements, taking into account the specific characteristics of the sub-region or region, to ensure effective conservation and management of such stocks.
2. The states shall enter into consultation in good faith and without delay, particularly where there is evidence that the straddling fish stocks and highly migratory fish stocks concerned may be under threat of over-exploitation or where a new fishery is being developed for such stocks. To this end, consultation may be initiated at the request of any interested state with a view to establishing appropriate arrangements to ensure conservation and management of the stocks. Pending agreement on such arrangements, the states shall observe the provisions of this agreement and shall act in good faith and with due regard to the rights, interests and duties of other states.
3. Where a sub-regional or regional fisheries management organization or arrangement has the competence to establish conservation and management measures for particular straddling fish stocks or highly-migratory fish stocks, states fishing for the stocks on the high seas and relevant coastal states shall give effect to their duty to cooperate by becoming members of such organization or participants in such arrangement, or by agreeing to apply the conservation and

management measures established by such organization or arrangement. The states, having a real interest in the fisheries concerned, may become members of such organization or participants in such arrangement. The terms of participation in such organization or arrangement shall not preclude such states from membership or participation; nor shall they be applied in a manner which discriminates against any state or group of states having a real interest in the fisheries concerned.
4. Only those states which are members of such an organization or participants in such an arrangement, or which agree to apply the conservation and management measures established by such organization or arrangement, shall have access to the fishery resources to which those measures apply.
5. Where there is no sub-regional or regional fisheries management organization or arrangement to establish conservation and management measures for a particular straddling fish stock or highly-migratory fish stock, relevant coastal states and states fishing on the high seas for such stock in the sub-region or region shall cooperate to establish such an organization or enter into other appropriate arrangements to ensure conservation and management of such stock and shall participate in the work of the organization or arrangement.
6. Any state intending to propose that action be taken by an inter-governmental organization having competence with respect to living resources should, where such action would have a significant effect on conservation and management measures already established by a competent sub-regional or regional fisheries management organization or arrangement, consult through that organization or arrangement with its members or participants. To the extent practicable, such consultation should take place prior to the submission of the proposal to the inter-governmental organization.

## Development of Techniques for Fishery Cooperation

Ever since the 1990s, the international community has welcomed many new techniques for fishery management, and many new regional entities for fishery cooperation. Their practices have offered valuable experience for the future South China Sea cooperation scheme.

1. The most powerful initiator of new techniques for international fishery management is the Food and Agriculture Organization (FAO). After the UNCLOS was adopted in 1982, FAO has constantly promoted many new ideas and new methods for fishery resources conservation and management.

(a) *Fishing capability control techniques*

In 1991, The Committee on Fisheries (COFI) of the FAO proposed the concept of "responsible fishery".[28] During the 28th meeting of the FAO in 1995, a "Code of Conducts for Responsible Fishery (Code of Conducts)" was adopted. This provided that the method of TAC should be applied for conservation of fishery resources. Also, based on the Code of Conducts, FAO adopted in 1999 an "International Action Plan for Management of Fishing Capabilities". FAO further asked the various fishing states and regional conservation organizations to adopt respectively their own "action plans" in accordance with the said international action plan.

(b) *Precautionary approach*

The Code of Conducts requires the various fishing states and sub-regional or regional fishery organizations to decide, on basis of "the best scientific evidence available", the "target reference point" for specific species and the actions to be taken when such a point is exceeded. More than this, the Code of Conducts further requires that the "limit reference point" and the actions to be taken when such point is exceeded, should be decided as well. Annex II of the 1995 Fish Stocks Agreement has already provided the detailed rules and applicable methods for deciding these reference points.[29]

(c) *Transparent operational management control*

In 1993, FAO adopted an "Agreement for Promoting Obedience of International Conservation and Management Measures at High Seas by the Fishing Vessels Operating on the High Seas". This agreement ascertained that the flag states should exercise jurisdiction and control measures over their fishing vessels. Moreover, on 2 March 2001, the Committee on Fisheries (COFI) of the FAO held its 24th meeting, and adopted the "International Action Plan for Preventing, Stopping and Eliminating the Illegal, Unreported and Unregulated (IUU) Fishing Activities". This action plan intended to make the fishing operations more transparent.

---

[28] The Committee on Fisheries (COFI), a subsidiary body of the FAO Council, was established by the FAO Conference at its 13th Session in 1965. The Committee presently constitutes the only global inter-governmental forum where major international fisheries and aquaculture problems and issues are examined and recommendations addressed to governments, regional fishery bodies, NGOs, fishery workers, FAO and international community, periodically on a worldwide basis. COFI has also been used as a forum where global agreements and non-binding instruments were negotiated.

[29] Kuen-chen Fu, "Precautionary Approaches and Precautionary Principles", *Fa-Ling-Yue-Kan (Law Monthly)*, Vol. 54, No. 9 (Taiwan: September 2003).

(d) *Ecological-social system management methods*

In 1999, the FAO adopted its "Rome Declaration on Implementation of the Code of Conduct". The declaration itself encouraged a more precise route of ecological approach for fishery conservation and management, and emphasized the relevant trade and environmental factors for regulating fishing and aquaculture.[30]

2. While the fishery resources conservation and management rules were improved, the regional cooperation arrangements for fishery resources conservation were expanding.

Up to now, according to incomplete statistics, the number of various kinds of regional fishery conservation cooperation entities in the world has reached 40.[31] After the adoption of the Code of Conducts and the Fish Stocks Agreement in 1995, the international community has responded positively with new agreements or new amendments of previous agreements concerning establishment of the various regional fishery cooperation schemes. These cooperation schemes may be classified into two types — cooperation through the all-member states' assembly, and cooperation through a standing committee.

(a) *Cooperation through the all-member states' assembly*

Two or more states reached a marine living resources preservation agreement by negotiations, and decided to hold regular or irregular assembly meetings with all the member states attending. The assembly will directly make the policies on data collection, resources assessment, and conservation measures. Member states will then take their own actions to implement the decisions made by the assembly. Two forms may be taken for this type of cooperation scheme. One is in the form of bilateral meeting. The other is in the form of multilateral assembly conference. An example of the former can be found in the joint-meeting system as created between China and Korea on fishery management in the Yellow Sea, or in the East China Sea with Japan. The latter can be observed in the annual conference of the member states of the "Agreement for Conservation and Management of the Cod Fish Resources in the Mid-Bering Sea". Countries adopting this type of cooperation scheme usually tend to emphasize more on the economic factors. Their conservation measures would usually include the TAC, the quota system, and other traditional methods.[32]

---

[30] Stuart M. Kaye, *International Fisheries Management* (Kluwer Law International, 2001) pp. 264–286.
[31] http://www.fao.org/fi/body/rfb/chooserfb.htm.
[32] Stuart M. Kaye, *International Fisheries Management* (Kluwer Law International, 2001) pp. 337–341.

(b) *Cooperation through a standing committee*

By way of negotiation, all the participating states might reach a multilateral agreement, and establish accordingly a standing committee constituting representatives from all the member states. This standing committee will decide on the conservation measures, and carry out marine living resources conservation cooperation projects. For example, the "Agreement on Conservation of the Antarctic Marine Living Resources" has already adopted this type of cooperation. Using it will usually help promote protection of the living resources. Conservation techniques adopted are usually more sophisticated. These would include the precautionary approach, and the ecological-social system management methods. Of course, these would require more complete and precise data as well.[33]

The above discussions have indicated that a regional fishery resources management cooperation scheme in the South China Sea would be very much needed. The question remains as how to establish such a scheme for cooperation.

## Project for Implementation of Regional Cooperation Scheme

### Basic Legal Characterization of the Water Area

Serious disputes over territorial sovereignty of the South China Sea islands have hindered development and management of the regional resources. To create an effective cooperation scheme for fishery resources conservation in this area needs to settle the disputes in advance. Even if it would not be the final resolution, a temporary resolution for the disputes will be essential. For this purpose, from the viewpoint of the Chinese people, based on the UNCLOS and the Fish Stocks Agreement, and sufficient historical evidence, which we cannot elaborate in this paper, a "three-level South China Sea" concept has to be understood.[34] I believe, only with this concept or legal characterization of the water area, could the proposed scheme be established, and be executed legally and reasonably.

According to this legal characterization, the South China Sea should be treated as a water area with three different levels. At each level, different ways and

---

[33] Stuart M. Kaye, *International Fisheries Management* (Kluwer Law International, 2001), pp. 377–385.

[34] As can be reasonably expected, China has much more than sufficient historic evidence to show that it has strong historic rights in this region. In the past several years, this author traveled to all the South China Sea littoral states to reach this conclusion. Many other scholars have also indicated the same fact in their writings. Long before Christ, in the West Han Dynasty and the East Han Dynasty (206 BC~220 AD), Chinese warships and civilian vessels were often sailing in the South China Sea. Abundant records were left in China ever since. For these details, see Kuen-chen Fu, *The Legal Status of the South China Sea* (Taipei: 123 Information Company, 1996).

manners should be applied in dealing with the South China Sea matters.[35] The first level of the South China Sea is the whole semi-enclosed South China Sea. Based on the provision of Article 123 of the UN Convention on the Law of the Sea, all littoral states should coordinate in matters concerning living resources, marine environment, and scientific research activities in this semi-enclosed sea. China, like any other coastal state in this region, should be bound by the convention.

The second level encompasses the historic waters in the South China Sea. The Chinese historic water is not the only one in this region. As the Chinese shall enjoy some historic titles within the Chinese U-shaped boundary line, the Thai people should enjoy some privileges in the Gulf of Siam where historic interests may be evidenced. Likewise, the Indonesian people should enjoy some historic rights in the vicinity of Natuna Island.

The third level is that of the islets in the region and their 12-nautical-mile territorial seas. Over the four groups of islets and/or rocks within the Chinese 1947 U-shaped boundary line and their 12-nautical-mile territorial seas, the Chinese enjoy sovereignty over these territories; just like the Indonesians enjoy sovereignty over the Natuna Island. Those who have occupied some of the Chinese islets should know that international law will never, and should never, allow such illegal attempts be turned into legal titles with sheer lapse of time.

International law has never allowed prescription as a legitimate reason for obtaining territorial sovereignty. It has not, even in the seemingly supporting cases of the Island of Palmas and of the Eastern Greenland. If it does, the world will surely face more, not less, confrontations. In the case of the South China Sea, time is not, and should not be, on the side of the states pursuing exploitation, regardless of the historical evidence, because China has repeatedly and publicly protested against such illegal exploitation activities. More importantly, international law should never support such maneuvers.

For many years, this author has been promoting in the region the concept of a "three-level South China Sea". At least in China (both Mainland and Taiwan) alone, this concept has been accepted and recorded in their respective legal documents. In China Mainland, Article 14 of the 1998 Law of the PRC's Exclusive Economic Zone and Continental Shelves provides that "the provisions in this law shall not affect the

---

[35] Kuen-chen Fu, "The U-shaped Boundary Lines in the South China Sea and the Three Levels of the South China Sea Waters", in *Chinese Society of International Law* (Taiwan: July 1994), pp. 1–18. Also see Kuen-chen Fu, *The Legal Status of the South China Sea* (123 Information Company) (Taiwan: April 1995), pp. 201–211.

historical right that the PRC enjoys".[36] In Taiwan, when the government declared its territorial sea baselines in 1999, the publicized list of coordinates contained a remark for the Spratly Islands that "the islands within the traditional U-shaped line are territories of the ROC, their territorial sea baselines..., will be publicized in the future."[37]

Will the other littoral states in the region accept this concept? I really do not know. But, evidently, since this is an issue of sovereignty and the basis of future cooperation, I would think this basic concept may be pondered with sincerity by the neighboring states in this region. After all, they all understand that these South China Sea islands are historically Chinese islands. And the spirit of the UNCLOS would urge anyone to pay due respects to the vested rights of other states based on historical evidence.

Therefore, it is rather obvious that there should be a demarcation of a joint conservation area (or areas) first, for all the littoral states to give their agreement on fishery resources management in the South China Sea. This area of cooperation might very possibly be the whole area of this semi-enclosed sea (the first level of the South China Sea), because this will be more meaningful in terms of fishing and conservation activities under the regulation of Article 123 of the UNCLOS.

If this is the case, then we will have to design a system and invite all the littoral states, including China, to join it and formally give up their rights of priorities in their historic waters (the second level of the South China Sea) respectively.

As to the sovereignty enjoyed by the littoral states over their own islands and rocks, even low-tide elevations, it should never being influenced. For those Chinese islets in the Spratly, which are occupied by other neighboring states, China will have to choose her own ways for settling these issues by herself. Fortunately, these small islets occupied by the others are of very limited area and impact in terms of fishery management. They are not necessary for anchoring, logistic supply, or any other fishery scientific researches. Thus, the sovereignty disputes over the Spratly Islands might very possibly not create any difficulties for the proposed scheme.

### Goals of Cooperation

The goals should include two points:

1. Maintaining the ecological balance in the South China Sea; and
2. Ensuring all fishing activities in the region are not IUU activities.

---

[36] *Collection of the Laws and Regulations of The People's Republic of China*, 3rd Edition (Office of Policy, Law and Regulation, State Oceanic Administration, 2001), pp. 214–215.
[37] Executive Yuan Order No. 88–Nei–05161, issued on 10 February 1999.

## Stages of Cooperation

### Stage one

At this stage, the main purpose is to make preparations for future cooperation. Within two or three years, all the littoral states should first improve their fishery resources conservation within their own territorial seas. At the same time, resources assessment in their respective waters should be made with effective means and generally-accepted scientific methods. Since this will involve only the third level of the South China Sea, any measures to be taken by the South China Sea community are not supposed to have any influence on the sovereignty claims.

1. Cooperation form: No special international organization is formed. Through regional conferences and agreements, various states should encourage public participation in their own countries.[38] With this, the mission of data collection and resources assessment should be better accomplished. Moreover, future cooperation among neighboring states would also be better supported by the public.
2. Scientific researches: The main aims would be for understanding precisely the allocation of fishery resources and the impact of fishing on the ecology in the South China Sea. A balance for sustainable exploitation of the fishery resources should be established.
3. Techniques of controls: Regional conferences and agreements should be used for promoting "input control methods". A better system of suspending fishing activities should be created and implemented among all these littoral states. IUU fishing activities should be lowered as much as possible in these states. Zero growth of fishing vessels and fishing capabilities should be fulfilled by these states, too.
4. Enactment: Regional cooperation through direct or indirect assistance from each other should be made for amending the domestic laws and regulations on fishing activities. Applicable international criteria on fishing permits, fishing gears, fishing data collection, and law enforcement, should be the main contents of these amendments.
5. Guardianship: Various littoral states should begin cooperation with FAO, to exchange information and help improve transparency of their respective fisheries and their governmental policy-making.

---

[38] Stuart M. Kaye, *International Fisheries Management* (Kluwer Law International, 2001), pp. 287–302.

*Stage two*

At this stage, regional cooperation should be initiated and developed. The main water areas for such cooperation should be carried out in the historic waters, including archipelagic waters, which are substantially historic waters too. Hopefully within five years, multilateral arrangements could be established and begin operating within such waters on the initiative of those states which are claiming such historic waters in the South China Sea. Because these waters are the second level of the South China Sea, those states with historic vested rights should take the initiative and the primary responsibility for conserving the fishery resources within their own historic waters (or archipelagic waters).

1. Cooperation form: Bilateral arrangements, i.e., conferences or agreements, between states with historic vested rights and those whose nationals are fishing in those specific historic waters should be made at the beginning. Later on, gradual enlargement of the circle may be achieved to include all the littoral states in all the historic waters in the South China Sea.
2. Scientific researches: More detailed researches on the fishery resources and their impacts on the environment should be emphasized during this second stage. Using on-board observation system should become a regular practice among the state parties in the various waters described as the second level of the South China Sea. Thus the density of the fishery resources, the allocation of all the target species, and the trend of variances could be forecast with reliable scientific evidence.
3. Techniques of control: The conservation techniques based on the TAC should be the main techniques used in this region now. The maximum sustainable yield (MSY) should be decided as the TAC in the various historic waters, and suspension of fishing activities would be imposed upon all the fishing vessels at the moment when the practical catch reaches that TAC.
4. Enactment: The two or more states fishing in the historic waters should help each other to enact uniform laws and regulations on their fisheries, making allocation of quotas, exchange of fishery information, and coordination of law enforcement, a constitutional and feasibly effective mechanism for conservation in those historic waters (or archipelagic waters).
5. Guardianship: Bilateral or multilateral conferences should be held on a regular basis, to review periodically the real situation of national enactment and implementation of their respective conservation endeavors. Exceptional session of these conferences may also be necessary when emergency cases occur.
6. Dispute settlement: At this second stage, each of the above-mentioned bilateral or multilateral conferences may set up its own dispute settlement scheme,

adopting the "negative consensus" rule for settling disputes in specific historic waters.[39] Under the assembly, technical and legal working groups may be established to offer necessary suggestions for any dispute settlement procedures.

*Stage three*

This should be the stage of full-scale cooperation in the region. The water area for these conservation efforts should be the whole semi-enclosed sea, i.e., the first level of the South China Sea. Within five or six years, hopefully, all the littoral states will become members of a single regional cooperation organization to implement all the requirements provided in the UNCLOS and the Fish Stocks Agreement. According to Article 123 of the UNCLOS, these littoral states should see it as their duty and rights to coordinate the living resources conservation and management measures in the semi-enclosed South China Sea.

1. Organization form: Taking the "ASEAN+1" forum as a platform, and eventually creating a standing organization, which should consist of an all-member assembly, and several special committees for the various routine conservation works.
2. Scientific researches: The purpose is to set up and maintain a whole regional cooperation project for fishery resources conservation and environmental preservation. A complete and reliable data bank on the fishery resources is essential for this purpose in the first place. All these researches should be able to serve the need of the following control measures.
3. Techniques of control: While implementing the control measures based on the TAC, the littoral states should adopt gradually the conservation techniques with "precautionary approaches" and "ecological-economical system methods". Limit (or preservation) reference points and target (or management) reference points should be scientifically determined, and enforced in accordance with the articles and Annexes of the Fish Stocks Agreement.[40] Specific TAC for specific species and quotas for specific fishing states should also be determined with the best scientific evidence available and in the spirit of cooperation in the region.
4. Enactment: Uniform laws and regulations concerning national TAC, quotas, fishing grounds, fishing gears, fishing seasons, law enforcement coordination, authorization for release of national information, exchange of data, allocation of budget for the regional organization, division of labor, settlement of disputes,

---

[39] Annex 2 of the Marrakesh Agreement Establishing the World Trade Organization, and Article 16 of the Understanding on Rules and Procedures Governing the Settlement of Disputes.
[40] Annex 2 of the 1995 Fish Stocks Agreement.

etc., should be enacted by the member states for implementing of the regional agreement.
5. Guardianship: A system of routine reviewing conferences should be organized by the international organization. This review process should be carried out by the organization's designated technical committees and the audit branch of the organization. FAO, IMO, and other competent international organizations may be invited to participate in the review procedure, in accordance with Article 123 of the UNCLOS, if member states feel that it is necessary. A contingent plan for the emergency cases in the region, pertaining to living resources conservation, should also be adopted by the organization's assembly.
6. Dispute settlement: A dispute settlement body similar to the DSB of the WTO, may be created as an arbitral body of the international organization.

## Cross-Strait Fishery Cooperation as an Antecedent Arrangement

When discussing the issue of regional fishery resources conservation and management project in the South China Sea, the cross-strait relationship between China and Taiwan has to be treated with some priority. It is not only because Taiwan is an active fishing entity in the region and a military occupant of the largest island, i.e., the Taiping Island or Itu Aba Island, but also because the relationship between China and Taiwan, if not treated properly, might lead to military conflicts and constitute a real setback for the peaceful environment in the region, which is essential for any resources conservation and environment protection endeavors.

### *Feasibility of Establishing a Cross-strait Fishery Cooperation Scheme*

The above scheme for regional cooperation on fishery conservation is preconditioned by the fact that all the littoral states (or non-state parties) in the South China Sea area are willing to work together for that purpose. Today, such cooperation between China and Taiwan appears to be impossible. However, we believe that such cooperation between the two sides of the Taiwan Strait is possible for a number of reasons.

The first reason is that after China and Taiwan entered the WTO,[41] Taiwan has actually admitted its legal status as a non-state entity in the international community. Therefore, for such fishery resources conservation matters, Taiwan should be expected to be willing to participate without insisting on any "statehood".

---

[41] China was admitted to WTO on 11 December 2001 as a "state"; and Taiwan was admitted to WTO on 1 January 2002 as a "separate customs territory", not as a "state".

Secondly, Taiwan is now very much dependant on her economic relationship with China. According to some official statistics, by the end of 2002, the total direct investment from Taiwan in China exceeded US$3,970,640,000.[42] According to Lin Yifu, the former Minister of Economic Affairs of Taiwan, China has already become the largest market for Taiwanese products, the third largest source of Taiwan's imported goods, and the largest source of Taiwan's trade surplus.[43]

The third reason is that, in terms of fishery, China and Taiwan always need each other. For many years, fishery cooperation in other parts of the world, between private sectors from both Taiwan and China, has being ongoing with positive results. A win-win arrangement for the two parties in the South China Sea area should be a realistic proposition. Therefore, I am rather confident that, with some probable incentives from the Chinese government, such cooperation on fishery resources conservation in the South China Sea should be realized with ease.

## *Main Technical Issues to be Taken Care of by the Two Parties*

From the technical aspect, to realize such cooperation, we believe the following main issues have to be taken care of by the two parties first.[44]

1. Creating a mechanism for resources assessment by the Chinese people in the historic waters within the U-shaped line demarcated by the Chinese government since 1947. This should be an essential basis for any further cooperation.
2. Offering administrative and economic incentives for any cross-strait joint ventures on fishery cooperation by the private sectors in the South China Sea area.
3. Exploring new methods to ensure protection of the rights and interests of China's fishing industry.
4. Promoting Chinese fishermen's cultural level by offering adequate education, for the purpose of collecting precise and reliable data from the fishing vessels.
5. Establishing an emergency search-and-rescue plan for protecting the Chinese fishermen in the region.
6. Establishing logistic support bases in the region for the Chinese fishermen to utilize.
7. Amending fishery laws and regulations in both China and Taiwan on the basis of the "precautionary principle".

---

[42] *China Yearbook of Statistics*, 2003.
[43] News report from Taipei by Chen Shunxie of The Central News Agency (27 October 2003).
[44] This author, Kuen-chen Fu, has been a member of the board of directors of the (Taiwan) Overseas Fishery Development Council, and the chairman of its Legal Affairs Committee, for 18 years.

8. Establishing a more detailed and more reasonable scale of fishery suspension system. The current system is not only being implemented unilaterally, i.e., not enforceable on Taiwanese fishing vessels, but is also not reasonably effective.
9. Creating a set of uniform criteria for law enforcement in the region.
10. Constructing a joint force for coordination, protection, and law enforcement at sea.

*Practical Procedures for Implementing Cross-strait Fishery Cooperation*

The fishery resources conservation activities to be carried out by the two sides of the Taiwan Strait may be realized together with the construction of a Southern China economic integrated entity. Such an entity shall not be a Free Trade Area as defined in the GATT and WTO rules. It might, however, be a CEPA-typed gradual, mild integration among the four Chinese parties in the WTO, i.e., China, Hong Kong, Macao and "Chinese Taipei".[45] Basically, such a procedure should be initiated by the private sectors with governmental support. Fishery is a part of the cross-strait economic relationship. It should be cultivated and protected like any other business operated on land. Once a cross-strait fishery cooperation scheme is established, it will certainly help not only the preservation of the historic rights of the Chinese people in the South China Sea, but also the conservation of living resources for the benefit of all the people in the region.

For realizing such a scheme, three steps may be necessary.

*Step one*

At the beginning, joint ventures on South China Sea fishery activities should be encouraged by at least the PRC government. The PRC government may assist by: offering administrative and economic incentives for any cross-strait joint ventures on fishery cooperation by the private sectors in the South China Sea; exploring new methods to ensure protection of the rights and interests of China's fishing industry; and establishing a emergency search-and-rescue plan and some logistic support bases in the region for the Chinese fishermen to utilize.

All these are to facilitate the creation of joint ventures by private fishing companies in both China and Taiwan.

---

[45] For a detailed discussion of the issue, see Kuen-chen Fu, "One China Four Seats in the WTO — A proposed Integrated Economic Zone in Southern China", proceedings of the Joint Conference of the Chinese Society of International Economic Law and Xiamen University Institute of International Economic Law, Xiamen, China, 4–5 November 2004.

*Step two*

Other than doing business together in the South China Sea, the two parties might begin doing scientific researches, and to have the living resources assessment successfully done with assistance from the educated Chinese fishermen in the region. The main purpose of this step is to make the background for fishery conservation feasible in the region for Chinese fishermen.

*Step three*

Then the two sides of the Taiwan Strait might begin amending their respective laws and regulations for fishery resources conservation, concluding a bilateral agreement on a reasonably effective system of fishery suspension, and begin joint operations in the Chinese historic waters for effective law enforcement.

Expectably, when Chinese fishing activities in the region are operating smoothly, with reasonable profits, but not with any IUU fishing business in existence, then all the other littoral states will become more eager to cooperate with the Chinese for fishery resources conservation in the region. And the above proposed regional cooperation scheme for that purpose shall be realized for common interests.

## Chapter 13

## Prospects for Joint Development in the South China Sea

*ZOU Keyuan*

## Introduction

Mineral reserves including oil and gas have a huge potential in the South China Sea. Estimated by scientists, there are five sedimentary basins in the northern part of the South China Sea with an area of 420,000 km². As of 1997, 56 oilfields or structures had been discovered, containing 700 million tons of oil and 310 billion m³ of gas.[1] In the Spratly area, there are eight sedimentary basins with an area of 410,000 km², and 260,000 km² are within China's unilaterally-claimed U-shaped line.[2] Incomplete statistics from China show that these eight sedimentary basins contain 34.97 billion tons of petroleum reserves, including the discovered 1.182 billion tons of oil and 8,000 billion m³ of gas. There are also oily sedimentary basins existing around the Paracel Islands, though no proven hydrocarbon resources have been discovered. Thus the South China Sea is sometimes called a second "Persian Gulf". Unlike the Chinese estimation, the general view outside China is less optimistic. A 1993/1994 figure provided by the US Geological Survey estimated the sum total of discovered oil reserves and undiscovered resources in the offshore basins of the South China Sea at 28 billion barrels.[3] However, if this figure is added with the reserves of natural gas which, according to the US Geological Survey, accounts for 60% to 70% of the total potential hydrocarbon resources in

---

[1] See Zhao Huanting *et al.*, *Geomorphology and Environment of the South China Coast and the South China Sea Islands* (Beijing: Science Press, 1999) (in Chinese), p. 484.
[2] For details of this line and its legal implications, see Zou Keyuan, "The Chinese Traditional Maritime Boundary Line in the South China Sea and Its Legal Consequences for the Resolution of the Dispute over the Spratly Islands", *International Journal of Marine and Coastal Law*, Vol. 14 (1), pp. 27–55 (1999).
[3] See "South China Sea Region", *Country Analysis Briefs*, September 2003, http://www.eia.doe.gov/emeu/cabs/schina.html (accessed 4 October 2004).

this region,[4] the overall picture of petroleum exploration and development, though within the conservative calculation, is still very encouraging. Someone even used the Natuna gas field as an indicator to presume that the potential in the South China Sea could be very significant.

The political situation in the South China Sea is complicated, as it contains potential for conflict with different national interests. The dispute over the Spratly Islands is most complicated since it has been lingering on for a long time and involves as many as five states, i.e., China including Taiwan, Malaysia, Vietnam, the Philippines, and Brunei. It is not usual in the history of international relations that so many countries make claims over such small islets, either as a whole or in part, of the Spratly Islands and their surrounding maritime zones (see Figure 13.1). The complicated political and legal situation in the South China Sea does not stop adjacent countries concerned from unilaterally exploring and exploiting oil and gas in undisputed or disputed offshore areas (see Tables 13.2–13.3).

Indonesia is the only OPEC member in East Asia and important to the world energy market. The Natuna Islands in the South China Sea and surrounding seas are a major petroleum producing site. The 400-mile Natuna pipeline is one of the longest undersea gas pipelines in the world, transporting gas to Singapore; new pipeline proposals from East Natuna to the Philippines are under consideration.[5] Though there is no territorial dispute with other South China Sea claimants, China considered Indonesia to have invaded its claimed sea areas by extracting oil and gas within the Chinese claimed U-shaped line.

The Philippines reached a crude oil production averaging 23,512 barrels per day (bbl/d) in 2002 from the 2001 figure of only 1,000 bbl/d. This dramatic increase was primarily due to the development of deep-sea oil deposits in the Malampaya field.[6] In February 2004, the Philippines unilaterally announced an international bidding for its oil development in the South China Sea, near its offshore Malampaya gas field close to the southern island of Palawan. It is reported that 16 foreign firms had submitted bids.[7] China expressed its concern about this unilateral move.

Malaysia produced monthly crude oil of between 650,000 bbl/d and 730,000 bbl/d from 1996 till mid-2003 and its domestic production is primarily

---

[4] Ibid.
[5] "Indonesia", *Country Analysis Briefs*, July 2004, http://www.eia.doe.gov/emeu/cabs/indonesa.html (accessed 4 October 2004).
[6] "Philippines", *Country Analysis Briefs*, July 2004, http://www.eia.doe.gov/emeu/cabs/philippi.html (accessed 4 October 2004).
[7] See "Sixteen Firms Want to Hunt For Oil in Philippines", Reuters, 3 March 2004, http://www.forbes.com/business/newswire/2004/03/03/rtr1284135.html (accessed 26 March 2004).

Prospects for Joint Development in the South China Sea  247

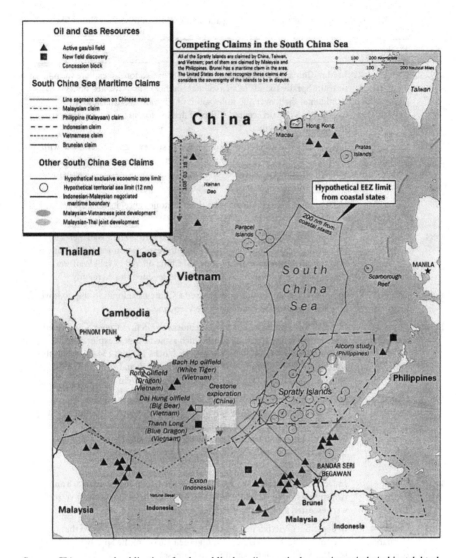

Source: CIA maps and publications for the public, http://www.eia.doe.gov/emeu/cabs/schinatab.html (accessed 4 October 2004).

Figure 13.1   Competing Claims in the South China Sea

Table 13.1  Disputes over Oil and Gas Drilling and Exploration in the South China Sea

| Date | Countries | Disputes |
|---|---|---|
| 1992 | China, Vietnam | In May, China signed a contract with US firm Crestone to explore for oil near the Spratly Islands in an area that Vietnam says is located on its continental shelf, over 600 miles south of China's Hainan Island. In September, Vietnam accused China of drilling for oil in Vietnamese waters in the Gulf of Tonkin. |
| 1993 | China, Vietnam | In May, Vietnam accused a Chinese seismic survey ship of interfering with British Petroleum's exploration work in Vietnamese waters. The Chinese ship left Vietnamese block 06 following the appearance of two Vietnamese naval ships. |
| 1993 | China, Vietnam | In December, Vietnam demanded that Crestone cancel its offshore oil development in nearby waters. |
| 1994 | China, Vietnam | Crestone joined with a Chinese partner to explore China's Wan' Bei–21 (WAB-21 block). Vietnam protested that the exploration was in Vietnamese waters in their blocks 133, 134 and 135. China offered to split Wan' Bei production with Vietnam, as long as China retained all sovereignty. |
| 1994 | China, Vietnam | In August, Vietnamese gunboats forced a Chinese exploration ship to leave an oilfield in a region claimed by the Vietnamese. |
| 1996 | China, Vietnam | In April, Vietnam leased exploration blocks to US firm Conoco, and ruled out cooperation with US oil firms which signed Chinese exploration contracts in disputed waters. Vietnamese blocks 133 and 134 cover half the zone leased to Crestone by China. China protested, and reaffirmed a national law claiming the South China Sea as its own in May. |
| 1997 | China, Vietnam | In March, Vietnamese issued a protest after the Chinese Kantan-3 oil rig drills near Spratly Islands in March. The drilling occurred offshore Da Nang, in an area Vietnam calls Block 113. The block is located 64 nautical miles off Chan May cape in Vietnam, and 71 nautical miles off China's Hainan Island. The diplomatic protests were followed by the departure of the Chinese rig. |
| 1997 | China, Vietnam | In December, Vietnamese protested after the Exploration Ship No. 8 and two supply ships entered the Wan' Bei exploration block. All three vessels were escorted away by the Vietnamese navy. |
| 1998 | China, Vietnam | In September, Vietnamese protested after a Chinese report stated that Crestone and China were continuing their survey of the Spratly Islands and the Tu Chinh region (Wan'an Bei in Chinese). |
| 2003 | Malaysia, Brunei | In May 2003, a patrol boat from Brunei acted to prevent TotalFinaElf from undertaking exploration activities in an area offshore from Northern Borneo disputed by the two countries. |

Source: Energy Information Administration, http://www.eia.doe.gov/emeu/cabs/schinatab.html (accessed 4 October 2004).

Table 13.2  Oil and Gas in the South China Sea Region

|  | Proven Oil Reserves (Billion Barrels) | Proven Gas Reserves (Trillion Cubic Feet) | Oil Production (Barrels/Day) | Gas Production (Billion Cubic Feet) |
|---|---|---|---|---|
| Brunei | 1.4 | 13.8 | 189,000 | 366 |
| Cambodia | 0 | 0 | 0 | 0 |
| China* | 1 (est.) | 5.0 (est.) | ~350,000 (est.) | ~200 (est.) |
| Indonesia* | 0.1 (est.) | 32.0 (est.) | ~323,000 (est.) | ~50 (est.) |
| Malaysia | 3.0 | 75.0 | 751,973 | 1,895 |
| Philippines | 0.2 | 3.8 | 24,512 | <1 |
| Singapore | 0 | 0 | 0 | 0 |
| Taiwan | <0.01 | 2.7 | 3,300 | 26 |
| Thailand | 0.6 | 13.3 | 193,162 | 661 |
| Vietnam | 0.6 | 6.8 | 339,595 | 46 |
| Total | Est. 7.0 | Est. 150.3 | 2,174,542 | 3,244 |

Source: Energy Information Administration, http://www.eia.doe.gov/emeu/cabs/schinatab.html (accessed 4 October 2004).
*Only the regions around the South China Sea are included. Note: There are no proved reserves for the Spratly and Paracel Islands. Proved oil and natural gas reserves are as of 1/1/2003. Oil production is a 2002 average. Oil supply includes crude oil, natural gas plant liquids, and other liquids. Natural gas production is the 2001 average. All figures cited for China and Indonesia are estimates of offshore reserves and production in the South China Sea only.

Table 13.3  Oil and Gas in the South China Sea — Comparison with Other Regions

|  | Proven Oil Reserves (Billion Barrels) | Proven Gas Reserves (Trillion Cubic Feet) | Oil Production (Million Barrels/Day) | Gas Production (Trillion Cubic Feet/Year) |
|---|---|---|---|---|
| Caspian Sea Region | 17.2–32.8 | 232 | 1.6 | 4.5 |
| North Sea Region | 16.8 | 178.7 | 6.4 | 9.4 |
| Persian Gulf | 674.0 | 1,923.0 | 19.3 | 8.0 |
| South China Sea | Est. 7.0 | Est. 150.3 | 2.2 | 3.2 |

Source: Energy Information Administration, http://www.eia.doe.gov/emeu/cabs/schinatab.html (accessed 4 October 2004).
*Proved reserves as of 1/1/2003. Oil production as of 2002. Oil supply includes crude oil, natural gas plant liquids, and other liquids. Natural gas production as of 2001.

from offshore oilfields near Peninsular Malaysia.[8] In June 2003, Petronas concluded an agreement with Petrovietnam and Pertamina of Indonesia for joint exploration of Block SK 305, offshore from Sarawak. Commercial production of natural gas at Bintang in the South China Sea began in 2003.[9] Since April 2003, Malaysia has sent

---

[8] "Malaysia", Country Analysis Briefs, July 2004, http://www.eia.doe.gov/emeu/cabs/malaysia.html (accessed 4 October 2004).
[9] Ibid.

four cruises with 11 survey vessels to Nantong Jiao (Louisa Reef) in the Spratly area for petroleum-prospecting activities.[10]

Brunei is a small but wealthy country in East Asia, thanks to profits generated from crude oil and natural gas. It produced 196,000 bbl/d in 2003.[11] Since Brunei is adjacent to the South China Sea, the majority of its oilfields are located offshore.

Vietnam has a very long coastline and offshore oil reserves are very promising. As recorded, its crude oil production in 2003 was 352,507 bbl/d on average. In January 2003, Petrovietnam and Petronas Carigali Overseas Company of Malaysia signed a contract to explore and exploit oil and gas in lot 01-02-97, located on Vietnam's continental shelf.[12]

China (PRC) is thirsty for oil and gas for its economic growth. It was the second largest consumer of petroleum products in 2003, surpassing Japan for the first time, with total demand of 5.56 million bbl/d.[13] The China National Offshore Oil Corporation (CNOOC) — the third largest state-owned oil company in China — is responsible for offshore oil and gas exploration and production. It has been operating petroleum activities in the South China Sea, mainly adjacent to Hainan Island and Hong Kong, such as Weizhou, Huizhou, Lufeng, Dongfang oil/gas fields. CNOOC has attempted to increase its production to 20% of China's total petroleum production by 2010.[14] It is to be noted that except for the frustrating Crestone deal, China does not have any actual oil and gas exploration and exploitation around the Spratly Islands.

However, unilateral exploration activities often cause conflicts with interested countries. The Crestone Concession case is an example (see Table 13.1). In addition, conflicts over resources in the South China Sea also occur between other claimants, such as Brunei and Malaysia on their conflicting EEZ claims. The reason behind this is simple: extraction by a single state can deplete the fluid deposit in which other adjacent states are entitled to share.[15] For that reason, joint development seems only feasible for petroleum resources in the disputed areas.

---

[10] "More than Ten Countries Carve Up South China Sea Resources and the Chinese Government Can Do Nothing", *Liaowan Dongfang*, 15 January 2004, http://peacehall.com (accessed 16 January 2004).

[11] "Brunei", *Country Analysis Briefs*, August 2004, http://www.eia.doe.gov/emeu/cabs/brunei.html (accessed 4 October 2004).

[12] "Vietnam", *Country Analysis Briefs*, June 2004, http://www.eia.doe.gov/emeu/cabs/vietnam.html (accessed 4 October 2004).

[13] "China", *Country Analysis Briefs*, July 2004, http://www.eia.doe.gov/emeu/cabs/china.html (accessed 4 October 2004).

[14] For details, see Zou Keyuan, "China's Governance over Offshore Oil and Gas Development and Management", *Ocean Development and International Law*, Vol. 35 (4), pp. 339–364 (2004).

[15] Masahiro Miyoshi, "The Joint Development of Offshore Oil and Gas in Relation to Maritime Boundary Delimitation", *Maritime Briefing*, Vol. 2 (5), p. 5 (1999).

## Joint Development in International Law and State Practice

Joint development is defined as "an agreement between two states to develop so as to share jointly in agreed proportions by inter-state cooperation and national measures the offshore oil and gas in a designated zone of the seabed and subsoil of the continental shelf to which both or either of the participating states are entitled in international law".[16] This definition is broad enough to cover all relevant situations which a joint development is needed and/or required. Generally speaking, the concept of joint development contains several characteristics: (1) it is an arrangement between two countries; (2) it is usually concerned with an overlapping maritime area; (3) it can be used as a provisional arrangement pending the settlement of the boundary delimitation disputes between the countries concerned; (4) it is designed to jointly develop the mineral resources in the disputed area or a defined area shared by two countries. In this sense, joint development is a most feasible mechanism to shelf the dispute so as to pave the way of cooperation, pending the settlement of the territorial and/or maritime disputes over a certain sea area due to their overlapping claims.

It is necessary to differentiate the concept of joint development from similar concepts such as unitization which straddles the same structure licensed to two oil companies, or joint venture which is established between, say, a Chinese enterprise and a foreign investor, though the detailed implementation of joint development will be related to unitization and/or joint venture. Furthermore, joint development carries special meaning and should not be misunderstood simply as something equivalent to joint cooperation either. Finally, while joint development is mainly applicable to the use of mineral resources, it is not excluded in the area of marine living resources management, as manifested in some existing cases, such as the Colombia-Jamaica Treaty of 12 November 1993 which set up a "Joint Regime Area" for joint management, control, exploration and exploitation of the living and non-living resources,[17] and the Guinea-Bissau-Senegal Agreement of 14 October 1993 which covers not only oil and gas but also living marine resources.

The LOS Convention provides a legal basis for joint development in disputed maritime areas pending the settlement of the maritime boundary delimitation. Articles 74(3) and 83(3) provide that pending agreement reached between them on the delimitation of the EEZ and continental shelf, the states concerned, in a spirit of understanding and cooperation, are required to "make every effort to enter into

---

[16] British Institute of International and Comparative Law, *Joint Development of Offshore Oil and Gas: A Model Agreement for States for Joint Development with Explanatory Commentary* (London: British Institute of International and Comparative Law, 1989), p. 45.

[17] See Miyoshi (1999), *supra* note 15, p. 23.

provisional arrangements of a practical nature and, during this transitional period, not to jeopardize or hamper the reaching of the final agreement". This legal norm has been reflected in many existing cases and can apply to the South China Sea. However, the LOS Convention leaves other options open to the states concerned about whether they make joint development after the definite maritime boundary delimitation. It may be recalled that as early as 1969, the International Court of Justice (ICJ) touched this issue in the North Sea Continental Shelf cases by stating that "if... the delimitation leaves to the parties areas that overlap, these are to be divided between them in agreed proportions or failing agreement, equally, unless they decide on a regime of joint jurisdiction, use, or exploitation for the zone of overlap or any part of them".[18] Finally, it should be noted that peaceful resolution of international disputes is one of the fundamental principles in international law, as stipulated both in the Charter of the United Nations and the LOS Convention. Joint development is one of the options for states to reach a peaceful resolution in the maritime domain.

In state practice, there are many such precedents setting forth the joint development arrangements. The first of this kind was made between Kuwait and Saudi Arabia in 1922, the earliest example of a joint development regime. Based on the 1922 Aqeer Agreement, the two countries were co-tenants in the Neutral Zone, holding shares equally and jointly in a condominium and later on they consented to joint development by their concessionaires.[19] It is summarized that there are generally three types of joint development schemes: (1) to be devised with the maritime boundary delimited, such as the Bahrain-Saudi Arabia Agreement concerning the Delimitation of the Continental Shelf of 22 February 1958, the France-Spain Convention on the Delimitation of the Continental Shelves of the Two States in the Bay of Biscay of 29 January 1974, the Sudan-Saudi Arabia Agreement Relating to the Joint Exploration and Exploitation of the Natural Resources of the Seabed and Subsoil of the Red Sea in a Defined Area of the Two Countries in the Red Sea of 16 May 1974, and the Iceland-Norway Agreement on the Continental Shelf between Iceland and Jan Mayen of 22 October 1981; (2) for the purpose of unitizing hydrocarbon deposits which straddle the boundary line, such as the Norway-United Kingdom Agreement Relating to the Delimitation of the Continental Shelf between the Two Countries of 10 March 1965 and its subsequent agreements for the exploitation of Frigg Field gas and Statfjord Field and Murchison Field petroleum respectively;

---

[18] North Sea Continental Shelf Cases, Judgement of 20 February 1969, ICJ Report, 1969, p. 63.

[19] See Ibrahim F.I. Shihata and William T. Onorato, "Joint Development of International Petroleum Resources in Undefined and Disputed Areas", in G.H. Blake, M.A. Pratt & C.H. Schofield, eds., *Boundaries and Energy: Problems and Prospects* (London: Kluwer Law International, 1998), pp. 436–437.

and (3) to be worked out with the issue of boundary delimitation shelved or kept unresolved, such as the Japan-Korea Agreement Concerning Joint Development of the Southern Part of the Continental Shelf Adjacent to the Two Countries of 30 January 1974, the Malaysia-Thailand Memorandum of Understanding on the Establishment of a Joint Authority for the Exploitation of Resources of the Seabed in a Defined Area of the Two Countries in the Gulf of Thailand of 21 February 1979, and the Australia-Indonesia Treaty on the Zone of Cooperation in an Area between the Indonesian Province of East Timor and Northern Australia of 11 December 1989.[20] The third type is linked to the provisional arrangements stipulated under the LOS Convention.

Some existing joint development models are conducive to the initiation and development of a similar one for the South China Sea. They actually exist all over the world from the North Sea Model to the Arabic Sea, and from Asia to Latin America.[21] In the East Asian region, joint development is also a mode of bilateral cooperation between the countries concerned, such as the Japan-South Korean Arrangement in the Sea of Japan and the East China Sea in the 1970s, the Malaysia-Thailand Joint Development Area in the Gulf of Thailand and the Australian-Indonesia Joint Development Zone for the Timor Gap.[22]

## Future Prospects

Though difficulties in realizing joint development in the South China Sea exist and it will take a long time to reach such an agreement among the interested countries, there are some positive signs that favor a possible joint development in the disputed areas there, as the complicated situation in the South China Sea indicates that joint development may be the only feasible means for regional cooperation for offshore oil and gas development.[23]

The first significant development regarding the South China Sea is the signing by China and all ASEAN member states of the Declaration on the Conduct of

---

[20] Masahiro Miyoshi, "Is Joint Development Possible in the South China Sea?" in Mochtar Kusuma-Atmadja, Thomas A. Mensah and Bernard H. Oxman, eds., *Sustainable Development and Preservation of the Oceans: The Challenges of UNCLOS and Agenda 21* (Honolulu: the Law of the Sea Institute, University of Hawaii, 1997), pp. 613–614.

[21] For details of some existing cases, see Miyoshi (1999), *supra* note 15, pp. 7–41.

[22] For details, see Shihata and Onorato, *supra* note 19, pp. 438–441.

[23] In a document released in 2002 by the Philippines military, the Philippines realized that it could develop oil deposits in the Kalayaan Island group because "no sensible foreign investor would come in because the government cannot guarantee a climate of security to underwrite their investments". See "China Accused Over Disputed Spratly Islands", *Energy Compass*, 18 July 2002, http://80-proquest.umi.com.libproxy1.nus.edu.sg (accessed 24 February 2004).

Parties in the South China Sea in December 2002, pledging to explore or undertake cooperative activities in the South China Sea including: "marine environmental protection; marine scientific research; safety of navigation and communication at sea; search and rescue operation; and combating transnational crime, including but not limited to trafficking in illicit drugs, piracy and armed robbery at sea, and illegal traffic in arms, pending a comprehensive and durable settlement of the disputes in the South China Sea".[24] The modalities, scope and locations, in respect of bilateral and multilateral cooperation, should be agreed upon by the parties concerned prior to their actual implementation.[25] They promise to resolve their territorial and jurisdictional disputes by peaceful means, without resorting to the threat or use of force.

It is worth mentioning that the informal Workshop on Managing Potential Conflict in the South China Sea, sponsored by Canada and Indonesia and involving all claimants, had discussed the need for joint development or joint cooperation in the disputed area at the very beginning since 1990. At its second meeting in Bandung in 1991, the Workshop stated that "in areas where conflicting territorial claims exist, the claimant states may consider the possibility of undertaking cooperation for mutual benefit, including ... joint development".[26] The Technical Working Group on Resources Assessment and Ways of Development established within the Workshop framework found the need "to define clearly the area that could be subjected to joint effort in the multiple claims area around the Spratly Islands groups, without prejudice to the various territorial or jurisdictional claims in the area", and "to collect and analyze as much as possible the various concepts of joint development that are already existing, particularly in the region, and to use them, as far as practicable, as models for developing joint efforts regarding the multiple claims area".[27] In June 1998, a meeting on non-living resources was convened in Vientianne, where various models of joint development and possibilities were discussed. They included the Malaysia-Thailand and Malaysia-Vietnam agreements, the Indonesian-Australian agreement, the China-Japan fisheries agreement in the East China Sea, the Argentine-United Kingdom agreement in the South West Atlantic, and the Antarctic Treaty.[28] Those models indicate that once agreement is reached on the need, it will take a lot of time and efforts to conclude the

---

[24] Text of the Declaration on the Conduct of Parties in the South China Sea is available at http://www.aseansec.org/13163.htm (accessed 2 July 2003).
[25] *Ibid.*
[26] Hasjim Djalal, "The Relevance of the Concept of Joint Development to Maritime Disputes in the South China Sea", *Indonesian Quarterly*, Vol. 27 (3), pp. 180–181 (1999).
[27] Cited in Djalal, *ibid.*, p. 182.
[28] Djalal, *ibid.*, p. 183.

joint development arrangement.[29] The Vientianne meeting also observed that the concept of joint development should not be limited to the development of non-living resources. There were attempts in 1993–1994 to find out an "area of a zone" for joint development. After several consultations, some claimants were happy with the suggestion, some had reservations but were willing to discuss it, and some rejected it immediately.[30] After all these constructive discussions, it appears possible that the "area" could start with a relatively small area involving not so many parties,[31] based on a non-official arrangement, including private commercial companies. Joint cooperation in least controversial matters, such as marine scientific research, marine environmental protection and ecotourism, can lead to joint resources exploration and exploitation.[32] However, through the discussions at the Workshop, some difficulties in joint cooperation and development exist, such as the difficulty in clearly defining the disputed area since it is difficult to precisely know which areas are being claimed by certain countries, and the difficulty in interpreting the concept of joint development, for there is a strong tendency among the claimants that joint development should not be attempted in an area which a claimant believes to be its own, and that the concept should only be used in an area claimed by others or for an area outside of its claims.[33] Despite the existing difficulties, the discussions in the Workshop have provided some basis for further discussions among the interested parties regarding the possibility of formulating a joint development arrangement in the South China Sea.

Secondly, China's position on joint development is very favorable. Encouraged or triggered by relevant developments in East Asia, China also put forward the idea of joint development in the disputed sea areas. As early as 1980s, Deng Xiaoping, the former paramount Chinese leader, made the famous statement regarding China's policy towards disputed areas in China's adjacent seas by joint development. Deng regarded "joint development" as one of the two most important peaceful means for international dispute resolution.[34] Since then, China has been pushing very actively to realize the goal of joint development and reiterated on many international occasions its proposal. In 1990, when former Chinese Prime Minister Li Peng visited Malaysia, he expressly put forward the joint development proposal

---

[29] Djalal, *ibid.*, p. 183.
[30] Djalal, *ibid.*, p. 185.
[31] Participants in the joint development plan should be those which are directly interested parties and which are maintaining presence in the area. Djalal, *ibid.*, p. 185.
[32] See Djalal, *ibid.*, p. 185.
[33] See Djalal, *ibid.*, p. 186.
[34] The other is "one country, two systems". See *Selected Works of Deng Xiaoping*, Vol. 3, p. 87.

as "shelving the disputes and developing jointly" (*gezhi zhengyi, gongtong kaifa*).[35] Since then the Chinese government has reiterated this proposal on many occasions when the Spratly issue was raised and it has remained unchanged to date. For example, when Wu Bangguo, Chairman of the National People's Congress, visited the Philippines in August 2003, he proposed to his Philippine counterpart to jointly develop petroleum in the South China Sea.[36]

China's efforts proved not to be wasted in the end. On 11 November 2003, the CNOOC and the Philippine National Oil Company agreed to jointly explore oil and gas in the South China Sea through a letter of intent between the two sides. A joint committee will be set up to help select exploration areas in the South China Sea. They also agreed to a program to "review, assess and evaluate relevant geographical, geophysical and other technical data available to determine the oil and gas potential in the area".[37] As stated by the Philippine side, the joint exploration will be conducted in the Northwest Palawan offshore area, "not even close enough to the Spratlys".[38] While there is an initial agreement between the two countries, there is still some likelihood that such an agreement might not be executed in practice or would fail in the end. It may be recalled that China and Russia signed an agreement to build an oil pipeline from Angarsk in East Siberia to Daqing in Northeast China, but the Russian side changed its original position by cancelling this project.[39] We just have to wait and see whether the agreement between China and the Philippines will encounter the same fate as the proposed Angarsk-Daqing project in the end.

Thirdly, in addition to the joint development arrangements addressed above, East Asian countries also have concluded bilateral agreements concerning maritime boundary delimitation and fishery management, which can provide some basis for possible joint development arrangements. Taking China as an example, it has concluded three fishery agreements with Japan (1997), South Korea (2000) and Vietnam (2000), as well as a maritime boundary agreement with Vietnam (2000). All three fisheries agreements have established joint fishery management zones in the Yellow Sea (with South Korea), the East China Sea (with Japan) and the Gulf of

---

[35] See *People's Daily* (in Chinese), 14 December 1990.
[36] "Wu Bangguo Proposes a Multiple Cooperation for Oil in the Spratly Islands", *Lianhe Zaobao*, 1 September 2003.
[37] "Chinese, Philippine Firms Join Forces to Look For Oil in South China Sea", *Agence France Presse*, 13 November 2003.
[38] "RP-China Oil Exploration", *BusinessWorld*, Manila, 12 November 2003.
[39] See Xie Ye, "Crude-Oil Quandary Causes Concern", *China Daily*, 24 February 2004.

Tonkin (with Vietnam).[40] The cooperative experiences accumulated from fishery resource management will no doubt enhance joint cooperation in the management of non-living resources. It may be recalled that the initial step for the joint development arrangement between Australia and Indonesia in the Timor Sea was the conclusion of the MOU for provisional fisheries surveillance and enforcement in October 1981.[41] In this sense, the bilateral fisheries arrangements can become the basis for countries concerned to make joint development arrangements.

More significant are agreements concerning maritime boundary delimitation, since joint development is usually invoked in state practice as a provisional measure pending the settlement of boundary delimitation. In this sense, the 2000 Agreement on the Maritime Boundary Delimitation in the Gulf of Tonkin between China and Vietnam is noteworthy. In addition to the settlement of the maritime boundary issue in the Gulf of Tonkin permanently, it reflects the idea of joint development of mineral resources in the gulf as it provides that:

> "In case any single geophysical structure of oil and gas or other mineral deposits straddles the demarcation line as provided in Article 2 of this Agreement, the parties shall, through friendly consultation, reach an agreement on developing the structure or deposit in the most effective way, as well as on equal sharing of the profits resulting from the development.[42]"

It indicates that, although for the time being there is no imminent prospect for the two sides to carry forward any type of joint development of petroleum resources in the Gulf, they leave the opportunity for cooperation in the future through the boundary agreement. It may be recalled that the early Sino-Vietnamese boundary negotiations in the 1970s initiated by Vietnam was actually triggered by the prospect for oil and gas in the Gulf when Vietnam intended to grant the exploration right for some sedimentary blocks in the Gulf to an Italian oil company. In the 1980s, Vietnam proposed that the two countries undertake a joint development program

---

[40] For relevant details, see Zou Keyuan, "Sino-Vietnamese Fishery Agreement for the Gulf of Tonkin", *International Journal of Marine and Coastal Law*, Vol. 17 (1), pp. 127–148 (2002); and "Sino-Japanese Joint Fishery Management in the East China Sea", *Marine Policy*, Vol. 27 (2), pp. 125–142 (2003).
[41] Text is reprinted in Jonathan I. Charney and Lewis M. Alexander, eds., *International Maritime Boundaries*, Vol. 1 (Dordrecht: Martinus Nijhoff Publishers, 1993), pp. 1238–1239.
[42] Article 7 of the 2000 Boundary Agreement. An unofficial English version of this agreement is attached to Zou Keyuan, "Sino-Vietnamese Agreement on the Maritime Boundary Delimitation in the Gulf of Tonkin", *Ocean Development and International Law*, Vol. 36, pp. 13–24 (2005).

in the Gulf of Tonkin.[43] With the implementation of the agreement from 1 July 2004, a joint development arrangement for non-living resources in the Gulf of Tonkin can be made at any time when the two sides consider it necessary, based on the above provision.

There are also further concerns and considerations in relation to the realization of the joint development idea. Firstly, China's behavior sometimes worries Southeast Asian countries. Whether or not the smaller claimant countries "will have the confidence to negotiate a settlement of the disputes, enter into provisional cooperative arrangements, or feel the need instead to beef up their defense capabilities and unilaterally strengthen their respective positions, depends in large part on what they see as China's policy and posture towards the issue". China's previous insistence in bilateral talks on the Spratlys issue in fact has blocked the way of seeking any possible joint development on a multilateral basis. However, the recent change of China's posture from sticking to bilateral talks to accommodating the whole ASEAN region has paved the way for reaching a joint development arrangement in future. Though with some reluctance, China has realized that regionalization of the Spratlys issue is inevitable and may help to break the deadlock in the resolution of the disputes.

Secondly, joint development is mainly used as an interim measure, pending the settlement of territorial and/or maritime disputes. Joint development can help to stabilize the status quo of the disputed area, and may finally lead to a permanent dispute resolution. Unless such a resolution comes true, the disputes will still be there. It may, on the other hand, affect from time to time the effective implementation of any joint development arrangement.

Thirdly, it is to be noted that all the claimant states surrounding the South China Sea are developing countries with rapid economic growth, which will be accompanied by an increasing demand for energy. It is estimated that oil demand for Asian nations will increase from about 14.5 million barrels per day in 2000 to nearly 29.8 million barrels per day by 2025.[44] The demand for energy and increased energy consumption will definitely make the claimants more actively involved in oil and gas exploration and exploitation in the South China Sea. This on the one hand may cause more conflicts amongst the claimants if any of them conduct unilateral petroleum activities in the disputed maritime area; but on the other, it could create

---

[43] See Epsey Cooke Farrell, *The Socialist Republic of Vietnam and the Law of the Sea: An Analysis of Vietnamese Behaviour within the Emerging International Oceans Regime* (The Hague: Martinus Nijhoff, 1998), p. 251.
[44] See "South China Sea Region", *Country Analysis Briefs*, September 2003, http://www.eia.doe.gov/emeu/cabs/schina.html (accessed 4 October 2004).

a window of opportunity for joint development when the claimants have realized that the unilateral act becomes a costly, disturbing and unfeasible option.

Fourthly, it is clear that joint development between two states in the areas with multiple claims causes problems. It may be recalled that when Japan and South Korea signed the joint development agreement for the East China Sea, it invited furious protests from China. The recent move between China and the Philippines regarding the possible joint development in the South China Sea met with protests at least from Vietnam. For that reason, joint development in the South China Sea launched by only two claimant countries is not a wise option. On the other hand, joint development proposals at the bilateral level may not always be welcomed by relevant claimants. It may be recalled that China once proposed a joint development with Vietnam for the Vanguard Bank where part of the area was a concession given to Crestone by China, but Vietnam rejected China's proposal on the ground that the area is close to Vietnamese coastlines and within the limits of its continental shelf.

A new model for potential joint development in the South China Sea can thus be designed, one which comprises China including Taiwan as one party and ASEAN including all ten member states as the other party. This model is based on the existing formula of the Declaration on the Conduct of the Parties in the South China Sea, which has been signed by China and ASEAN members. If this can stand, then a bilateral but multiple-party cooperation between China and ASEAN in the South China Sea can be formulated. Of course, some scholars have already expressed their concerns about whether the interests and rights of China including Taiwan can be guaranteed in the future joint development scheme, as Hurng-yu Chen, a Taiwanese scholar, warned that "[w]hat Chinese should remember before they enter into any joint development agreement is that Taipei and Peking occupy far fewer islands than either Vietnam or the Philippines, and joint development with these countries might well prejudice Chinese rights and interests". However, with a cooperative framework between China and ASEAN, the interests of all the parties concerned may well be assured. The bilateral scheme of joint development is foreseeable and backed up by a number of favorable factors.

On the Chinese side, both mainland China and Taiwan have taken similar political and legal positions on the South China Sea issue based on the same historical evidence and reasons on most occasions, and never launched any challenge against each other regarding the territorial sovereignty over the South China Sea islands. The presence of mainland China (PRC) in the Spratly Islands is largely due to the preceding presence of Taiwan (ROC) there by having occupied the largest island Taiping within the Spratlys. On the other hand, cooperation between the two sides across the Taiwan Strait regarding the petroleum exploration and exploitation has already begun. As early as July 1996, the Taiwan Chinese Petroleum Corporation and the mainland China Offshore Oil Corporation signed the agreement on the exploration for the Tainan Basin and Zhaoshan sunken area, and this was finally approved by both sides in April 1998.[45] As to the utilization of the mineral resources in the South China Sea, the two sides can make some form of cooperation and consultation based on their previous joint development experience in the Taiwan Strait as well as in the sea areas around the Pratas Islands.

On the ASEAN side, a concerted and unified policy towards the South China Sea has been formulated since 1992, when ASEAN member states issued the Declaration on the South China Sea, which expressed their resolve to explore the possibility of cooperation in the South China Sea and to establish a code of international conduct there.[46] In 1995, the ASEAN foreign ministers reiterated the letter and spirit of the 1992 Declaration.[47] ASEAN as a whole signed the 2002 Declaration with China. Following this trend, joint development such as between China and ASEAN could be at least one of the feasible options for the South China Sea claimants to consider seriously.

Finally, even if there is a chance of joint development in the South China Sea, difficulties still remain to be tackled. How to create such a mechanism and how to reflect the representation of each claimant in the joint management authority remains a problem at the technical level. Questions such as whether the authority should include an equal number of representatives from each claimant state, or whether the allocation of rights and profits should be based on the number of islands each nation occupies, or whether they should be divided equally, have already been posed in the South China Sea literature. It is pointed out that the "process of identifying the common interests of the claimant states is an essential prerequisite to the consideration of JDZ (joint development zone) proposals".

## Conclusion

It has been predicted that some kind of joint development, no matter which form it takes, can be arranged for the disputed area in the South China Sea, provided that all the interested parties have such an intention and goodwill so that they can exert necessary efforts in reaching the agreement. The existing state practices in East Asia concerning joint development, in particular the tripartite one in the

---

[45] See *People's Daily* (in Chinese), 1 June 1998.
[46] "ASEAN Declaration on the South China Sea", Manila, Philippines, 22 July 1992, http://www.aseansec.org/5233.htm (accessed 27 October 2004).
[47] "Recent Developments in the South China Sea", 18 March 1995, http://www.aseansec.org/5232.htm (accessed 27 October 2004).

Gulf of Thailand, are good examples to show that joint development is not totally alien in East Asia. While there are a number of difficulties and issues, the window of opportunity always remains. Based on the recently signed Declaration between China and ASEAN countries, the window is now even wider. Once a form of joint development has been reached amongst countries concerned, long-term peace and security in the South China Sea can definitely be guaranteed, and this may positively lead to the final settlement of the territorial and maritime disputes in the area.

## Chapter 14

## Maintaining Maritime Safety in Southeast Asia: Regional Cooperation

*CHEN Haibo**

## Introduction

The ocean space of Southeast Asia, as one scholar said, "has become three-dimensional. Not only is it now used for surface navigation but the sea is also vital for its resources, an area of pollution, and a field of movement for submarines; the sea-bed has also vast resources of energy and minerals and is suitable as a base for sophisticated naval devices."[1] Nobody will deny the essential role oceans and seas of Southeast Asia play in economic developments, civilian activities, social affairs, national securities, and cultural evolutions of all ASEAN[2] states and China.

However, according to the analysis by the International Maritime Organization (IMO), in Southeast Asia, including maritime zones of ASEAN countries, South China Sea, the Straits of Malacca and Singapore (the Straits), Hong Kong SAR (HK), and southern part of China, there were at least 35 very serious casualties of ocean vessels, excluding fishing vessels, in the year 1999,[3] 41 in 2000,[4] and 38 in 2001.[5] Only one very serious casualty of fishing vessels occurred in both 2000[6] and 2001.[7] Among these casualties, some vessels were totally lost, some polluted

---

*The author would like to thank Dr. Zou Keyuan, senior research fellow, East Asian Institute, National University of Singapore, Singapore, for his invaluable comments made on an earlier version of this paper. However, all the errors and unclear explanations should be the author's responsibilities.
[1]Lee Yong Leng, *Southeast Asia and the Law of the Sea: Some Preliminary Observations on the Political Geography of Southeast Asian Seas*, Revised Edition (Singapore: Singapore University Press, 1980), p. 69.
[2]The Association of Southeast Asian Nations, which was formed in 1967 by Indonesia, Malaysia, Philippines, Thailand and Singapore was joined by Brunei in 1984, and by Cambodia, Laos, Myanmar and Vietnam in the 1990s.
[3]http://www.imo.org/includes/blastDataOnly.asp/data_id%3D5397/2.pdf (accessed 10 October 2004).
[4]http://www.imo.org/includes/blastDataOnly.asp/data_id%3D5118/3.pdf (accessed 10 October 2004).
[5]http://www.imo.org/includes/blastDataOnly.asp/data_id%3D8934/4.pdf (accessed 10 October 2004).
[6]http://www.imo.org/includes/blastDataOnly.asp/data_id%3D8933/3.pdf (accessed 10 October 2004).
[7]http://www.imo.org/includes/blastDataOnly.asp/data_id%3D9119/4.pdf (accessed 10 October 2004).

the marine environment, while personnel in some vessels were lost or injured. All these numbers remind us to pay much more attention to maritime safety in Southeast Asia. The theme "Safe, Secure and Efficient Shipping on Clean Oceans" was highlighted as a message of the World Maritime Day 2002 by the Secretary-General of the IMO. This theme also emphasized the objectives that we should pursue for maritime activities in Southeast Asia.

Maritime safety, which we are eager to pursue in Southeast Asia, is composed of various aspects, including but not limited to safety of navigation by vessels of various sizes and types, avoidance of damages or injuries to vessels, properties and persons at sea, efficient search and rescue of persons in distress, successful assistance to vessels in distress, prevention and control of marine environmental pollution, and so on.

It is submitted that maritime safety could be realized only by cooperation among all interested parties. We ratified the United Nations Convention on the Law of the Sea, 1982 (UNCLOS) and numerous IMO conventions, guidelines and directions, with the purpose of constructing a basic global cooperation framework. In the next section, the author examines the status of ASEAN states and China (including Hong Kong) in this global cooperation framework and finds out four disadvantages of the current situation that hinder maritime safety cooperation in Southeast Asia. The third section analyzes three forms of regional cooperation on maritime safety issues in Southeast Asia. Cooperation between China and ASEAN is expected to be rational and operational. In the fourth section, the author provides the foundation for ASEAN-China Regional Maritime Safety Cooperations. In the fifth section, she discusses the process of maritime safety cooperation between ASEAN and China. The sixth section proposes some maritime safety cooperation measures or projects in the Southeast Asian area that should be given priorities. Conclusions are given finally.

## Current Global Cooperation Framework and Its Disadvantages to Regional Maritime Safety in Southeast Asia

### UNCLOS and IMO Conventions: Basis of Global Cooperation

"UNCLOS is acknowledged to be an 'umbrella convention' because most of its provisions, being of a general kind, can be implemented only through specific operative regulations in other international treaties".[8] IMO Conventions (see Table 14.1) are

---

[8] Agustín Blanco-Bazán, "IMO interface with the Law of the Sea Convention (2000)", http://www.imo.org/InfoResource/mainframe.asp?topic_id=406&doc_id=1077 (accessed 10 October 2004).

specific operative regulations to this extent.[9] The IMO Assembly, IMO Maritime Safety Committee (MSC) and IMO Marine Environment Protection Committee (MEPC) also adopt recommendations for the same purposes.[10] Accordingly, UNCLOS and IMO instruments fulfill two aspects of the same maritime safety objective. One provides the general rules,[11] the other works on implementation or operational rules (see Table 14.2). They mutually support each other.[12]

Nevertheless, four disadvantages of the current situation substantially limit the benefits of global cooperation in maritime safety in Southeast Asia.

## Disadvantage One

As mentioned above, numerous basic rules were stipulated by UNCLOS. Among them, rights and responsibilities of flag states of vessels and port states controlling maritime safety in certain areas are essential. Although UNCLOS is broadly acknowledged in the world, two ASEAN countries, Cambodia and Thailand, which could be flag states and port states as defined by UNCLOS, have not ratified UNCLOS by 16 July 2004. That is to say, rules on maritime safety under UNCLOS, especially those concerning rights and responsibilities of flag states and port states, are not uniformly applicable to all ASEAN countries.

Laos, a land-locked country, is not an IMO member and does not ratify any IMO conventions accordingly.[13] As for other ASEAN countries and China, including Hong Kong,[14] their ratification of IMO instruments on maritime safety are varied

---

[9] IMO conventions and recommendations usually stipulate UNCLOS as their basis, and/or that there should be no overlap, inconsistent, incompatible, or prejudicial provisions to the law of the sea established by UNCLOS.

[10] In several cases, recommendations by IMO are incorporated into IMO treaties at a later stage. They should show the effect of convention provisions consequently. For instance, SOLAS and MARPOL incorporated the International Code for the Construction and Equipment of Ships carrying Dangerous Chemicals in Bulk (IBC Code), made it one effective part of them.

[11] UNCLOS stipulates numerous basic provisions on maritime safety issues, which construct a basic framework for maritime safety. Those provisions include, but are not limited to, Articles 21, 22, 23, 24, 39, 41, 42, 43, 44, 53, 60, 80, 91, 92, 94, 98, 100, 108, 142, 163, 197, 198, 199, 200, 201, 202, 203, 204, 205, 206, 208, 210, 211, 212, 214, 216, 217, 218, 219, 220, 221, 222, 223, 226, 228, 230, 231, 233, 235, 237, 242, 243, 244, 261, 262, 268, 269, 270, 271, 272, 275, 276, 277, etc.

[12] "Implications of the United Nations' Convention on the Law of the Sea for the International Maritime Organization", Document of IMO, LEG/MISC/3/Rev.1, 6 January 2003.

[13] Notably, according to IMO statistics, many land-locked countries are IMO members. Mongolia and Luxemburg could be examples. They had ratified various IMO conventions on maritime safety.

[14] Hong Kong SAR is a member of IMO, and a ratified party to almost all IMO conventions.

Table 14.1   IMO International Conventions Dealing with Maritime Safety

1. International Convention for the Safety of Life at Sea, 1974, as amended (SOLAS (amended) 1974), and its protocol 1978 (SOLAS Protocol 1978)
2. International Convention on Load Lines, 1966 (LL 1966)
3. International Convention on Tonnage Measurement of Ships, 1969 (TONNAGE 1969)
4. Convention on the International Regulations for Preventing Collisions at Sea, 1972, as amended (COLREG (amended) 1972)
5. International Convention on Standards of Training, Certification and Watchkeeping for Seafarers, 1978, as amended (STCW (amended) 1978)
6. International Convention on Standards of Training, Certification and Watchkeeping for Fishing Vessel Personnel, 1995 (STCW-F 1995)
7. International Convention on Maritime Search and Rescue, 1979 (SAR 1979)
8. International Convention for the Prevention of Pollution from Ships, 1973, as modified by the Protocol of 1978 relating thereto (MARPOL (amended) 73/78) (Annex I/II, Annex III, Annex IV, Annex V), and MARPOL Protocol 97 (Annex VI)
9. International Convention relating to Intervention on the High Seas in Cases of Oil Pollution Casualties, 1969 (INTERVENTION 1969), and its protocol 1973 (INTERVENTION Protocol 1973)
10. International Convention on Civil Liability for Oil Pollution Damage, 1969 (CLC 1969), and its protocols 1976, 1992
11. International Convention on the Establishment of an International Fund for Compensation for Oil Pollution Damage, 1971 (FUND 1971), and its protocols 1976, 1992, 2003
12. Convention relating to Civil Liability in the Field of Maritime Carriage of Nuclear Material, 1971 (NUCLEAR 1971)
13. Convention for the Suppression of Unlawful Acts against the Safety of Maritime Navigation (SUA 1988), and its protocol 1988 (SUA Protocol 1988)
14. International Convention on Salvage, 1989 (SALVAGE 1989)
15. International Convention on Oil Pollution Preparedness, Response and Cooperation, 1990 (OPRC 1990)
16. International Convention on Liability and Compensation for Damage in connection with the Carriage of Hazardous and Noxious Substances by Sea, 1996 (HNS 1996)
17. 2000 Protocol on Preparedness, Response and Cooperation to Pollution Incidents by Hazardous and Noxious Substances (OPRC/HNS PROT 2000)
18. International Convention on Civil Liability for Bunker Oil Pollution Damage, 2001 (BUNKERS 2001)
19. International Convention on the Control of Harmful Anti-Fouling Systems on Ships, 2001 (ANTI FOULING 2001)

Source: Website of IMO, http://www.imo.org.

(See Table 14.3). Taking the COLREG 1972 as an example, the Philippines is not a party to this essential convention regulating navigation safety. As for Cambodia and Thailand, both did not ratify SAR 1979, INTERVENTION 1969 and its protocol, SUA 1988 and its protocol, SALVAGE 1989, etc. Thailand is not a party to the MARPOL 1973/78 and all its Annexes. Cambodia is yet to become a party of OPRC 1990.

Table 14.2  IMO Instruments in which Certain Provisions can be Used as Major Implementation or Operative Directions of UNCLOS on Maritime Safety Issues

1. SOLAS (amended) 1974 and its protocol
2. LL 1966
3. COLREG (amended) 1972
4. STCW (amended) 1978
5. MARPOL (amended) 73/78
6. Code of Safety for Nuclear Merchant Ships
7. IMO/IAEA Safety Recommendations on the Use of Ports by Nuclear Merchant Ships
8. The International Code for the Transport of Dangerous Goods by Sea (IMDG Code)
9. The International Code for the Construction and Equipment of Ships Carrying Dangerous Chemicals in Bulk (IBC Code)
10. The International Code for the Construction and Equipment of Ships Carrying Liquified Gases in Bulk (IGC Code)
11. The Code for the Safe Carriage of Irradiated Nuclear Fuel, Plutonium and High-Level Radioactive Wastes in Flasks on Board Ships (INF Code)
12. SUA 1988 and its protocol 1988
13. The Convention on the Prevention of Marine Pollution by Dumping of Wastes and Other Matter, 1972 (LC or London Convention 1972)
14. INTERVENTION 1969 and its protocol 1973
15. The Torremolinos International Convention for the Safety of Fishing Vessels, 1977 (SFV 1977)
16. The International Code of Signals (Code of Signals)
17. SAR 1979
18. The Global Maritime Distress and Safety System (GMDSS)
19. Convention on Facilitation of International Maritime Traffic, 1965, as amended (FAL (amended) 1965)
20. OPRC 1990 and its protocol 2000
21. International Convention on the Control of Harmful Anti-fouling Systems (AFS 2001)
22. CLC 1969 and its protocols
23. FUND 1971 and its protocols
24. HNS 1996
25. BUNKERS 2001

Sources: "Implications of the United Nations' Convention on the Law of the Sea for the International Maritime Organization", Document of IMO, LEG/MISC/3/Rev.1, 6 January 2003.

Due to the above situation, international cooperation in maritime safety in Southeast Asia could not be fulfilled substantially, thoroughly and comprehensively based on the UNCLOS and IMO maritime safety platforms.

*Disadvantage Two*

Some IMO conventions on maritime safety have not been enforced until now, such as MARPOL 1973/78 Annexes IV and VI, the 1996 LC Protocol, SFV 1977 and its protocol 1993, International Convention on Standards of Training, Certification and Watchkeeping for Fishing Vessel Personnel, 1995 (STCW-F),

Table 14.3  State(s) and/or Area which is/are not a Party/Parties to Certain Major IMO Conventions Relating to Maritime Safety

| Essential IMO Conventions Related | Non-Party States or Area |
|---|---|
| SOLAS 1974 | Laos |
| SOLAS Protocol 1978 | Laos, Philippines, Thailand |
| LOAD LINES 1966 | Laos |
| TONNAGE 1969 | Laos |
| COLREG 1972 | Laos, Philippines |
| STCW 1978 | Laos |
| SAR 1979 | Laos, Brunei, Cambodia, Indonesia, Malaysia, Myanmar, Philippines, Thailand, Vietnam |
| MARPOL 73/78 (Annex I/II) | Laos, Thailand |
| MARPOL 73/78 (Annex III) | Laos, Brunei, Indonesia, Malaysia, Myanmar, Thailand, Vietnam |
| MARPOL 73/78 (Annex IV) | Laos, China, Indonesia, Malaysia, Myanmar, Singapore, Thailand, Vietnam, Hong Kong SAR |
| MARPOL 73/78 (Annex V) | Laos, Brunei, Indonesia, Myanmar, Thailand, Vietnam |
| MARPOL Protocol 1997 (Annex VI) | Laos, Brunei, Cambodia, China, Indonesia, Malaysia, Myanmar, Philippines, Thailand, Vietnam, Hong Kong SAR |
| INTERVENTION 1969 | Laos, Brunei, Cambodia, Indonesia, Malaysia, Myanmar, Philippines, Singapore, Thailand, Vietnam |
| INTERVENTION Protocol 1973 | Laos, Brunei, Cambodia, Indonesia, Malaysia, Myanmar, Philippines, Singapore, Thailand, Vietnam |
| OPRC 1990 | Laos, Brunei, Cambodia, Indonesia, Myanmar, Philippines, Vietnam, Hong Kong SAR |
| CLC 1969 | Laos, Brunei, China (Denunciation), Myanmar, Philippines, Singapore, Thailand, Vietnam |
| CLC Protocol 1976 | Laos, China (D), Indonesia, Malaysia, Myanmar, Philippines, Thailand, Vietnam, Hong Kong SAR (D) |
| CLC Protocol 1992 | Laos, Myanmar, Thailand |
| FUND 1971 | Laos, Brunei, Cambodia, China (D), Indonesia, Myanmar, Philippines, Singapore, Thailand, Vietnam |
| FUND Protocol 1976 | Laos, Brunei, Cambodia, China, Indonesia, Malaysia, Myanmar, Philippines, Singapore, Thailand, Vietnam, Hong Kong SAR (D) |
| FUND Protocol 1992 | Laos, China, Indonesia, Malaysia, Myanmar, Thailand, Vietnam |
| FUND Protocol 2003 | Laos, Brunei, Cambodia, China, Indonesia, Malaysia, Myanmar, Philippines, Singapore, Thailand, Vietnam, Hong Kong SAR |
| NUCLEAR 1971 | Laos, Brunei, Cambodia, China, Indonesia, Malaysia, Myanmar, Philippines, Singapore, Thailand, Vietnam, Hong Kong SAR |
| SUA 1988 | Laos, Cambodia, Indonesia, Malaysia, Thailand, Hong Kong SAR |
| SUA Protocol 1988 | Laos, Cambodia, Indonesia, Malaysia, Singapore, Thailand, Hong Kong SAR |
| SALVAGE 1989 | Laos, Brunei, Cambodia, Indonesia, Malaysia, Myanmar, Philippines, Singapore, Thailand, Vietnam |
| HNS 1996 | Laos, Brunei, Cambodia, China, Indonesia, Malaysia, Myanmar, Philippines, Singapore, Thailand, Vietnam, Hong Kong SAR |
| OPRC/HNS 2000 | Laos, Brunei, Cambodia, China, Indonesia, Malaysia, Myanmar, Philippines, Thailand, Vietnam, Hong Kong SAR |
| BUNKERS 2001 | Laos, Brunei, Cambodia, China, Indonesia, Malaysia, Myanmar, Philippines, Singapore, Thailand, Vietnam, Hong Kong SAR |
| ANTI FOULING 2001 | Laos, Brunei, Cambodia, China, Indonesia, Malaysia, Myanmar, Philippines, Singapore, Thailand, Vietnam, Hong Kong SAR |

Source: Website of IMO, http://www.imo.org.

HNS 1996, OPRC/HNS PROT 2000, BUNKER 2001, AFS 2001 and so on. As a result, the implementation and efficiency of UNCLOS and IMO maritime safety framework could be decreased to a large extent.

Notably, civilians in Southeast Asia depend a lot on fishery for a very long time. However, under the UNCLOS and IMO framework, SFV 1977, its protocol 1993 and STCW-F are the only three international conventions specifically regulating safety, proper manning and personnel of fishing vessel. Non-enforcement of these treaties until now is bad news for fishing safety in particular and maritime safety in general.

## Disadvantage Three

Certain issues had emerged or are being highlighted in recent years. However, since an international regime requires concessions by numerous state parties with different interests, IMO has difficulties passing new resolutions at the very beginning. UNCLOS and existing IMO conventions could not be modified promptly for the same reason.

In this regard, regional resolutions are being hoped for, with their advantages of prompt attention to contending issues. EU could be a good example. After the "ERIKA" oil spill accident in 1999, EU enacted a series of measures to increase safety of ships and marine pollution prevention in European waters. The Maritime Safety Agency was set up to enact additional regulations on marine pollution prevention. After the "Prestige" oil spill disaster, EU adopted Regulation (EC) No. 1726/2003 of the European Parliament and of the Council of 22 July 2003 amending Regulation (EC) No. 417/2002 on the accelerated phasing-in of double-hull or equivalent design requirements for single-hull oil tankers. By this amendment, phasing out of one-hull oil tankers is to be speeded up.

## Disadvantage Four

UNCLOS and IMO conventions seek compromise by numerous member states with different interests from different parts of the world. They write down the most basic principles and set minimum standards. Undoubtedly, interests of EU on maritime safety could be different from those of other parts of the world, and vice versa. Those most basic principles and minimum standards might not be able to actualize the interests of particular member states or certain regions. This is the fourth disadvantage of the current UNCLOS and IMO framework.

## Models to Overcome these Disadvantages

There are two models to overcome these disadvantages. One is the USA model. After the "EXXON VALDEZ" catastrophe in 1989, instead of joining CLC Convention

1969 and its protocols, and Fund Convention 1971 and its protocols, the United States enacted OPA 1990, which set up higher and more comprehensive coverage for marine pollution in US waters.

Another is the EU model, which is submitted to be proper for Southeast Asia maritime safety. After the "Prestige" accident occurred, single-hull tankers had the opportunities to be still in business for years more under IMO. However, someone from EU was blatant in setting up a regional timetable to phase out single-hull tankers, not considering the functions and documents of the IMO within a worldwide construction. After negotiations, EU decided to respect the functions of the IMO in phasing out single-hull tankers issues within a worldwide framework, but it will do more on its part to protect maritime safety as much as possible. As mentioned above, on the basis of IMO instruments, EU adopted the Regulation (EC) No. 1726/2003 of the European Parliament and of the Council of 22 July 2003 amending Regulation (EC) No. 417/2002 on the accelerated phasing-in of double-hull or equivalent design requirements for single-hull oil tankers, and established a European liability and compensation regime, featuring a fund of 1 billion euros to top up the IOPC Fund under the Fund Convention 1971 and its protocols, with respect to oil spills in European waters.[15] Meanwhile, after considering the suggestions raised by EU, IMO decided to fulfill its functions concerning maritime safety much more actively. Protocol 1978 of MARPOL 1973/78 was amended to speed up phasing out of single-hull tankers internationally. Regional maritime safety and international maritime safety could benefit a lot from positive interactions between regional regime and international regime.

## Arrangements of Regional Maritime Safety Cooperation in Southeast Asia

Maritime safety is of public interest. It needs compromise by sovereign states and implementation by authorized administrations. There are three alternative arrangements for regional maritime safety cooperation. However, which is a proper one should be considered.

### Multilateral Agreements Among All Interested Countries?

By contracting and implementing a multilateral agreement on regional maritime safety cooperation by all countries in the region, every interested party could fulfill

---

[15]TETLEY, William, *International Maritime and Admiralty Law* (Quebec: Les Editions Yvon Blais Inc., 2002), p. 464.

rights and undertake responsibilities provided by the agreements, to get a win-win benefit. In this regard, multilateral agreements could be a rational way. However, compromise by all interested countries takes a long time, and is a heavy drain on the economy and manpower. Multi-compromise could help to set up a basic framework but not a detailed and technical one. Even a special organ like the European Maritime Safety Agency might be needed to implement a multilateral agreement. Countries in this region have difficulty setting up an efficient multilateral organ dealing with maritime safety issues in a short time.

### Bilateral Agreements between Interested Countries?

Compared to multilateral agreement, bilateral agreement could be concluded easily, quickly, and economically. Cooperation between administrations could be efficient also. There are numerous bilateral agreements in various fields between ASEAN countries, as well as between individual ASEAN countries and China. Until now, communications and cooperation between China and ASEAN countries are mostly based on concluding and implementing bilateral agreements before the year 2002. To define rights and responsibilities of all these countries in maintaining regional maritime safety, China should conclude bilateral agreements with all ASEAN countries separately, and vice versa. In addition, all these bilateral agreements should be effected, and not conflict with each other. Otherwise, the aim of regional maritime safety could hardly be fulfilled.

As a matter of fact, China has concluded bilateral maritime transport agreements with five ASEAN countries, i.e., Singapore, Malaysia, Thailand, Indonesia and Vietnam. They are: Agreement on Maritime Transport between the Government of the People's Republic of China and the Government of Republic of Singapore, 1989; Agreement on Maritime Transport between the Government of the People's Republic of China and the Government of Malaysia, 1987; Agreement on Maritime Transport between the Government of the People's Republic of China and the Government of the Kingdom of Thailand, 1979; Agreement on Maritime Transport between the Government of the People's Republic of China and the Government of the Republic of Indonesia, 1965; and Agreement on Maritime Transport between the Government of the People's Republic of China and the Government of the Socialist Republic of Vietnam, 1962. No bilateral maritime agreements were concluded between China and other ASEAN countries. In every bilateral maritime transport agreement concluded, there is one article stipulating rescue operations of one contracting party for vessel of the other contracting party, distressed in the ports, internal waters, territorial sea or maritime areas within the sovereignty and jurisdiction of the rescuing contracting party, as well as crew members, passengers and cargoes of the distressed vessel.[16] Nothing more is provided on maritime safety issue.

*Cooperation between ASEAN and China*

Cooperation between ASEAN and China on maritime safety could be a proper arrangement based on the combination of advantages of the ASEAN grouping itself and above-mentioned arrangements.

Firstly, ASEAN is a platform for cooperation among member countries on matters of common interest.[17] Also, ASEAN represents the collective will of the nations of Southeast Asia.[18] It is a platform for collaboration between ASEAN states and other countries, particularly China, on matters of maritime safety.

Secondly, negotiations between ASEAN and China could be hoped to represent multilateral interests among China and Southeast Asian countries. Rights and responsibilities of every interested party could be defined accordingly.

Thirdly, for the negotiation partnership and process, they could be similar to those of bilateral negotiations. One part is ASEAN, representing ASEAN member states. Another part is China. Separate negotiations between China and every Southeast Asian country, and vice versa, would not be needed. Compromise could be achieved much more easily, quickly, economically and effectively. In addition, potential conflicts between bilateral agreements could never happen. ASEAN can authorize its certain professional organ, or establish one under ASEAN framework, to cooperate with that of China on maritime safety issue. It will also be much easier and more effective than establishing a multilateral organ to implement multilateral agreements.

---

[16] See Article 10(1) of the Agreement on Maritime Transport between the Government of the People's Republic of China and the Government of the Republic of Singapore, 1989; Article 9(1) of the Agreement on Maritime Transport between the Government of the People's Republic of China and the Government of Malaysia, 1987; Article 13(1) of the Agreement on Maritime Transport between the Government of the People's Republic of China and the Government of the Kingdom of Thailand, 1979; Article 5 of the Agreement on Maritime Transport between the Government of the People's Republic of China and the Government of the Republic of Indonesia, 1965; Article 9 of the Agreement on Maritime Transport between the Government of the People's Republic of China and the Government of the Socialist Republic of Vietnam, 1962.

[17] According to Article 2(3) of the ASEAN Declaration, one of the aims and purposes of ASEAN is to promote active collaboration and mutual assistance, on matters of common interest in the economic, social, cultural, technical, scientific and administrative fields.

[18] Article 5 provides that, "the Association represents the collective will of the nations of Southeast Asia to bind themselves together in friendship and cooperation and, through joint efforts and sacrifices, secure for their peoples and for posterity the blessings of peace, freedom and prosperity."

Fourthly, ASEAN has concluded various agreements or memorandums with certain regions or countries outside of ASEAN. However, ASEAN is not authorized to sign them by itself. Those agreements or memorandums are signed by ASEAN countries and those regions or countries. Therefore, these are still multilateral agreements or memorandums, which are applicable to all ratified parties.

## Foundation of ASEAN-China Regional Maritime Safety Cooperation

When establishing a regional maritime safety cooperation system between ASEAN and China, ASEAN's experience in dealing with maritime safety issues within its own framework, as well as bilateral maritime agreements between China and five ASEAN member states, as mentioned above, could be considered as a baseline.

### *Agreement for the Facilitation of Search of Ships in Distress and Rescue of Survivors of Ship Accidents, 1975*

In order to implement the Regulations contained in the Safety of Navigation, Chapter V annex to the *International Convention for the Safety of Life at Sea, 1960*, original member states of ASEAN, i.e., Indonesia, Malaysia, Philippines, Singapore and Thailand, adopted the Agreement for the Facilitation of Search of Ships in Distress and Rescue of Survivors of Ship Accidents in 1975.

Under the agreement, contracting parties undertake to provide necessary measures of assistance to ship in distress in their territories as they may find practicable, and to permit the owners of the ship or authorities of the state in which the ship is registered to provide such measures of assistance as may be necessitated by the circumstance.[19]

According to Article 3 of the agreement, contracting parties should permit immediate entry of aircraft, vessels, equipment and personnel necessary to search for ship in distress, rescue survivors of ship accidents (search and rescue persons), into any areas other than those prohibited, and provide the necessary facilities. Contracting parties should, through its Rescue Coordination Center (RCC) and those of other contracting parties, undertake necessary information exchange required by the Agreement.[20]

The agreement is submitted to be an essential inter-ASEAN legal instrument dealing with maritime safety within the ASEAN framework.

---

[19]See Article 1 of the Agreement for the Facilitation of Search of Ships in Distress and Rescue of Survivors of Ship Accidents, 1975.
[20]See Articles 5, 7 of the Agreement for the Facilitation of Search of Ships in Distress and Rescue of Survivors of Ship Accidents, 1975.

As mentioned above, bilateral maritime agreements between China and certain ASEAN members provide search and rescue undertakings between China and other contracting party. Generally, these undertakings provided by bilateral maritime agreements are similar to those under the Agreement for the Facilitation of Search of Ships in Distress and Rescue of Survivors of Ship Accidents, 1975. Rights and responsibilities under this agreement are much more detailed. It is submitted that search and rescue issues and related rights and responsibilities could be one of the basic foundations for ASEAN-China maritime safety cooperation.

### *Singapore Declaration 1992 and Framework Agreement on Enhancing ASEAN Economic Cooperation 1992*

By the Singapore Declaration 1992, and the Framework Agreement on Enhancing ASEAN Economic Cooperation 1992, ASEAN countries agreed to further enhance regional cooperation to provide safe, efficient and innovative transportation.

### *ASEAN Plan of Action in Transport and Communications (1994–1996)*

To help achieve those goals raised by the Singapore Declaration 1992 and the Framework Agreement on Enhancing ASEAN Economic Cooperation 1992, the ASEAN Senior Economic Officials Meeting (SEOM) convened a working group on transport and communications, and determined priority theme issues, key objectives, projects, and activities for ASEAN transport and communication cooperation. Those points of agreement were included in the ASEAN Plan of Action in Transport and Communications (1994–1996). "Development of ASEAN rules and regulations for carriage of dangerous goods and industrial wastes on land and by sea" was raised as the fifth initiative. The plan was to establish a response system to chemical spills, and harmonize maritime instruments on port state control procedures in ASEAN.

### *Ministerial Understanding on ASEAN Cooperation in Transportation, 1996*

During the First ASEAN Transport Ministers' Meeting, ASEAN adopted the Ministerial Understanding on ASEAN Cooperation in Transportation in 1996, to express its desire for maritime safety. According to this understanding, ASEAN is to establish and develop a harmonized and integrated regional transportation system in order to provide a safe, efficient and innovative transportation infrastructure, and to establish a mechanism to coordinate and supervise cooperation projects and activities in the transport sector.[21]

---

[21] See Article 1(a) and (c) of the Ministerial Understanding on ASEAN Cooperation in Transportation, 1996.

Subject to Article 2, member sates shall strengthen and enhance existing cooperation efforts, formulate action programs in areas not covered by existing cooperation agreements through, *inter alia*, policy coordination, development of infrastructure, exchange of information, transfer of technology, strengthening of regional institutional capacities, and developing ASEAN common stands on international issues in areas of common interest.

Emphasis was given to specific tasks, such as harmonization of laws, rules and regulations; enhancing capacity and efficiency of transport facilities; development of institutional and human resource capability; and promotion of an environment-friendly and safe transport system.[22]

Article 4 provides that ASEAN member states shall endeavor to warrant a thorough implementation of the ASEAN Plan of Action in Transport and Communications, especially on safety of maritime transport and prevention of pollution from ships.

During the first formal meeting of the ASEAN Transport Ministers in 1996, the implementation timeframe of the Plan of Action was revised from 1994–1996 to 1996–1998.

### Integrated Implementation Program for the ASEAN Plan of Action in Transport and Communications, 1997

During the third SEOM, the Integrated Implementation Program for the ASEAN Plan of Action in Transport and Communications, 1997, was endorsed and adopted by the second ASEAN Transport Ministers' (ATM) Meeting in 1997.

Among seven programs under this Integrated Implementation Program, Program 5 deals with maritime safety. Its aim is to ensure that ASEAN ships shall be of good quality, well-maintained and manned by competent seafarers along international standards, and in the collective implementation of regional harmonized standards for the safe operation of ships and the prevention of marine pollution. Accordingly, member states were urged to provide measures for the eventual ratification and/or implementation of IMO conventions, protocols and agreements. According to the program, six projects and activities on maritime safety should be implemented. They are: implementation of maritime conventions; harmonization of maritime instruments in respect of port state control procedures for ASEAN countries; establishment of an ASEAN ship-reporting system; accession to the Agreement for the Facilitation of Search for Ships in Distress and Rescue of Survivors by Ship Accidents, 1975; trans-boundary ship-borne pollution; and common ASEAN near-coastal voyages.

---

[22] See Article 3(a), (b), (c) and (d) of the Ministerial Understanding on ASEAN Cooperation in Transportation, 1996.

## ASEAN Framework Agreement on Facilitation of Goods in Transit, 1998

In 1998, the ASEAN Framework Agreement on the Facilitation of Goods in Transit was adopted, giving a general consideration on maritime safety of goods in transit arrangements. ASEAN has concluded the arrangement and is concluding implementation protocols under this framework agreement. Maritime safety issues were dealt with by some protocols, e.g., Protocol 9 on Dangerous Goods.[23]

## ASEAN Transport Action Plan 2005–2010

During the tenth ATM in 2004, the ASEAN Transport Action Plan 2005–2010 was adopted, and the further development of an integrated, harmonized, safe and secure regional transportation system was stressed.

The plan summarized lessons in ASEAN Transport Cooperation during 1999–2003, ascertained the development of trans-ASEAN transportation and inter-ASEAN cooperation. As provided, cooperative partnerships had been forged with ASEAN dialogue partners like China, India and Japan, to further enhance regional transport programs and to support the rising trend towards inter-regional free-trade arrangements.[24] Specifically, the plan made intensifying cooperative bonds with dialogue partners like China and some others, by way of policy consultation and joint programs and activities, as one of the policy directions for enhanced cooperation in the ASEAN transport sector for 2005–2010.[25]

Under the goals of achieving globally-acceptable standards in maritime safety, security and protection of marine environment, the plan made improving maritime safety and security and protection of the marine environment, by enhancing cooperation amongst ASEAN member states to facilitate the acceptance and implementation of IMO conventions, one of the strategic thrusts.

According to the plan, proposed actions on maritime safety and related issues are as follows:[26] enhance the activities of the ASEAN Forum on IMO Conventions to facilitate the accession and implementation of relevant IMO conventions by ASEAN member countries; strengthen the institutional capacity, human resource base and cooperation linkages of ASEAN member countries for achieving improved maritime safety and security and preventing marine pollution (e.g.,

---

[23] Protocol 9 on Dangerous Goods was signed by transport ministers of ASEAN in 2002.

[24] Paragraph 5 of the ASEAN Transport Action Plan 2005–2010, http:// www.aseansec.org/16596.htm (accessed 30 November 2004).

[25] Paragraph 14(h) of the ASEAN Transport Action Plan 2005–2010, http://www.aseansec.org/16596.htm (accessed 30 November 2004).

[26] See Annex A of the ASEAN Transport Action Plan 2005–2010, http://www.aseansec.org/16596.htm (accessed 30 November 2004).

ISPS Code, STCW trainers' training, etc.); pursue the "ASEAN Clean Seas Strategy"; intensify maritime transport security through capacity building and IT-based programs; pursue the Handling of Dangerous Goods in ASEAN Ports' Project and other APA-initiated mutually-beneficial projects; enhance regional capacity for maritime search and rescue (SAR) operations; strengthen maritime transport human resource capacity; and regular exchange of information and best practices in maritime transport policy and development programs.

The ASEAN Senior Transport Officials' Meeting (STOM) was required to have the overall responsibility in the supervision, coordination and implementation of the Plan 2005–2010.

## Process of ASEAN-China Regional Maritime Safety Cooperation

China was accorded full Dialogue Partner status at the 29th ASEAM Ministerial Meeting (AMM) in 1996. Since then, cooperation between ASEAN and China has been broadened and deepened. Maritime safety cooperation was considered or raised by several documents adopted by ASEAN[27] and China.

### Joint Statement of Meeting of Heads of State/Government of Member States of ASEAN and President of the People's Republic of China, 1997

The Joint Statement of the Meeting of the Heads of State/Government of the Member States of ASEAN and President of the People's Republic of China, 1997, opened the doors to all-dimensional cooperation between ASEAN and China. In their joint statement, both sides undertook to heighten cooperation at bilateral and multilateral levels in promoting economic growth, sustainable development and social progress, based on the principles of equal mutual benefit and shared responsibility in the interest of achieving national and regional prosperity in the 21st century.[28] They also agreed not to allow existing differences to hamper the development of friendly relations and cooperation.[29] This is the basis for maritime safety cooperation between ASEAN and China.

---

[27] As mentioned above, these documents were signed by China and ASEAN member states, which were referred to as "ASEAN".

[28] Paragraph 5 of the Joint Statement of the Meeting of the Heads of State/Government of the Member States of ASEAN and President of the People's Republic of China, 1997.

[29] See paragraph 8 of the Joint Statement of the Meeting of the Heads of State/Government of the Member States of ASEAN and President of the People's Republic of China, 1997.

*Declaration on the Conduct of Parties in the South China Sea, 2002*

Being cognizant of the need to promote a peaceful, friendly and harmonious environment in the South China Sea between ASEAN and China for the enhancement of peace, stability, economic growth and prosperity in the region, the Declaration on the Conduct (DOC) of Parties in the South China Sea was adopted in 2002, as a commitment to enhancing the principles and objectives of the 1997 joint statement.

Both sides agreed to explore or undertake the following maritime safety, science and security cooperative activities, pending a comprehensive and durable settlement of the disputes:[30] marine environmental protection; marine scientific research; safety of navigation and communication at sea; search and rescue operation; and combating transnational crime, including but not limited to trafficking of illicit drugs, piracy and armed robbery at sea, and illegal traffic in arms.

As required by the DOC,[31] both sides carried out the ASEAN-China Senior Officials' Meeting on the Implementation of the Declaration on the Conduct of Parties in the South China Sea in 2004.[32] At the meeting, both sides adopted the Terms of Reference of the ASEAN-China Joint Working Group on the Implementation of the Declaration on the Conduct of Parties in the South China Sea (the Terms of Reference). The ASEAN-China Joint Working Group (ASEAN-China JWG) was set up by both sides to study and recommend measures to translate the provisions of the DOC into concrete cooperative activities that will enhance mutual understanding and trust.[33]

According to the Terms of Reference, the ASEAN-China JWG is tasked to formulate recommendations on:[34] guidelines and action plan for the implementation of the DOC; specific cooperative activities in the South China Sea, particularly in the areas of marine environmental protection, marine scientific research, safety of navigation and communication at sea, search and rescue operation, and combating transnational crime; a register of experts and eminent persons who may provide

---

[30] See paragraph 6 of the Declaration on the Conduct of Parties in the South China Sea, 2002.

[31] Paragraph 6 of the Declaration on the Conduct of Parties in the South China Sea, 2002, provided that the parties concerned prior to their actual implementation should agree upon the modalities, scope and locations, in respect of bilateral and multilateral cooperation.

[32] ASEAN-China Senior Officials' Meeting on the Implementation of the Declaration on the Conduct of Parties in the South China Sea, Kuala Lumpur, 7 December 2004, http://www.aseansec.org/16888.htm (accessed 1 March 2005).

[33] See the Terms of Reference of the ASEAN-China Joint Working Group on the Implementation of the Declaration on the Conduct of Parties in the South China Sea, 2004, http://www.aseansec.org/16885.htm (accessed 1 March 2005).

[34] Paragraph 7 of the Terms of Reference of the ASEAN-China Joint Working Group on the Implementation of the Declaration on the Conduct of Parties in the South China Sea, 2004.

technical inputs, non-binding and professional views or policy recommendations to the ASEAN-China JWG; and the convening of workshops, as the need arises.

*Joint Declaration of ASEAN and China on Cooperation in the Field of Non-traditional Security Issues, 2002, and Memorandum of Understanding between the Governments of Member Countries of ASEAN and the Government of China on Cooperation in the Field of Non-traditional Security Issues, 2004*

Some non-traditional security issues, such as sea piracy, terrorism, and arms-smuggling, are also threats to maritime safety. Aiming to deal with non-traditional security issues cooperatively, ASEAN and China signed the Joint Declaration of ASEAN and China on Cooperation in the Field of Non-traditional Security Issues in 2002 (Joint Declaration on Non-traditional Security Issues). Desiring to strengthen the friendly cooperation among the parties, ASEAN and China adopted the Memorandum of Understanding between the Governments of the Member Countries of the Association of Southeast Asian Nations (ASEAN) and the Government of the People's Republic of China on Cooperation in the Field of Non-traditional Security Issues, 2004 (MOU on Non-traditional Security Issues) to deepen cooperation in the field of non-traditional security issues among the parties.

According to the Joint Declaration on Non-traditional Security Issues and the MOU on Non-traditional Security Issues, both sides, on the basis of deepening their existing multilateral and bilateral cooperation, are to:[35] strengthen information exchange; strengthen personnel exchange and training, and enhance capacity-building; strengthen practical cooperation on non-traditional security issues; strengthen joint research on non-traditional security issues; and explore other areas and modalities of cooperation. Mid- and long-term cooperation would be information exchange cooperation, personnel exchange and training cooperation, law enforcement cooperation, and joint research.[36]

The implementation agencies for the MOU on Non-traditional Security Issues are the ASEAN Secretariat, in coordination with the relevant national agencies of the ASEAN member countries, and the Ministry of Public Security of the People's Republic of China.[37]

---

[35] Part II "Priority and Form of Cooperation" of the Joint Declaration of ASEAN and China on Cooperation in the Field of Non-traditional Security Issues, 2002.

[36] See Article 2 of the Memorandum of Understanding between the Governments of the Member Countries of the Association of Southeast Asian Nations (ASEAN) and the Government of the People's Republic of China on Cooperation in the Field of Non-traditional Security Issues, 2004.

[37] See Article 3 of the Memorandum of Understanding between the Governments of the Member Countries of the Association of Southeast Asian Nations (ASEAN) and the Government of the People's Republic of China on Cooperation in the Field of Non-traditional Security Issues, 2004.

*Joint Declaration of the Heads of State/Government of ASEAN and China on Strategic Partnership for Peace and Prosperity, 2003, and Plan of Action to Implement the Joint Declaration on ASEAN-China Strategic Partnership for Peace and Prosperity, 2005*

In 2003, ASEAN and China signed the Joint Declaration of the Heads of State/Government of the Association of Southeast Asian Nations and the People's Republic of China on Strategic Partnership for Peace and Prosperity (Joint Declaration on Strategic Partnership), declaring to establish "a strategic partnership for peace and prosperity", to foster friendly relations, mutually beneficial cooperation and good neighborliness between ASEAN and China by deepening and expanding ASEAN-China cooperative relations in a comprehensive manner in the 21st century, thereby contributing further to the region's long-term peace, development and cooperation.

Security cooperation is one of the five cooperation areas provided in the Joint Declaration on Strategic Partnership, which comprises three components, namely: to expedite the implementation of the Joint Statement on Non-Traditional Security Issues and to actively expand and deepen cooperation in such areas; to hold, whenever appropriate, ASEAN-China security-related dialogue to enhance mutual understanding and promote peace and security in the region; and to implement the DOC, discuss and plan areas and projects for follow-up actions.

In 2005, ASEAN and China, pursuant to the Joint Declaration on Strategic Partnership, concluded a plan for deepening and broadening ASEAN-China relations and cooperation in a comprehensive and mutually-beneficial manner for the period of 2005–2010. The Plan of Action to Implement the Joint Declaration on ASEAN-China Strategic Partnership for Peace and Prosperity, 2005 (Plan of Joint Declaration on Strategic Partnership) listed detailed joint actions and measures in six major aspects[38] to pursue in the period, some of which will be noted as follows, including maritime safety.

As to DOC issues, the Plan of Joint Declaration on Strategic Partnership stressed the effective implementation of DOC and related mechanism, the promotion of joint cooperation and dialogue in areas such as marine scientific research, protection of the marine environment, safety of navigation and communication at sea, search and rescue operation, humane treatment of all persons in danger or distress, fight against transnational crimes, as well as cooperation among military officials and adherence to the terminologies used in the UNCLOS and IMO instruments.

---

[38] They are political and security cooperation, economic cooperation, functional cooperation, cooperation in international and regional, funding, and institutional arrangements.

As to non-traditional security issues, the implementation measures and activities under the Joint Declaration on Non-traditional Security Issues 2002 and its MOU 2004, were emphasized.

The plan addressed the conclusion and implementation of the Memorandum of Understanding between ASEAN-China on Transport Cooperation;[39] pursued the formalization of a consultation mechanism for maritime transport cooperation; emphasized the carrying out of mutually-beneficial projects in such areas as maritime safety and security, search and rescue at sea, marine environmental protection against pollution from ships, ship ballast water management, port state control and international shipping; aimed for developing and signing an ASEAN-China regional maritime transport cooperation framework agreement; and sought to strengthen capacity building in port management, navigation-channel technology, transportation infrastructure design and construction, and other areas of common interest in the transport sector.[40]

*Memorandum of Understanding between the Governments of the Member Countries of ASEAN and the Government of People's Republic of China on Transport Cooperation, 2004*

Common consensus of ASEAN and China were reached on furthering ASEAN-China transport cooperation in 2002. Both sides adopted the Memorandum of Understanding between the Governments of the Member Countries of ASEAN and the Government of the People's Republic of China on Transport Cooperation, 2004 (MOU on Transport Cooperation), in the spirit of laying solid foundations for lasting partnership and cooperation in the transport sector.[41]

---

[39] The Memorandum of Understanding between the Governments of the Member Countries of ASEAN and the Government of the People's Republic of China on Transport Cooperation, signed in Vientiane, 27 November 2004.

[40] See 2.6 "Transport Cooperation" of the Plan of Action to Implement the Joint Declaration on ASEAN-China Strategic Partnership for Peace and Prosperity, 2005.

[41] The Memorandum of Understanding between the Governments of the Member Countries of ASEAN and the Government of the People's Republic of China on Transport Cooperation, 2004, is a result of maneuvering for a further and comprehensive fulfillment of the best mutual interest of the parties. China initiated to strengthen the ASEAN-China transport cooperation at the Fourth ASEAN-China Summit in 2000, and proposed to establish the ASEAN-China Transport Ministers' Meeting mechanism at the Fifth ASEAN-China Summit in 2001, so as to strengthen communication and coordination, and promote cooperation in the transport sector between the parties. In 2002, the two parties signed the Framework Agreement on the Comprehensive Economic Cooperation Agreement, aimed at strengthening and enhancing economic, trade and investment cooperation, as well as in the liberalization of trade in goods and services. Common consensus on furthering the ASEAN-China transport cooperation was reached at the First ASEAN-China Transport Ministers' Meeting in 2002. See the foreword of the Memorandum of Understanding between the Governments of the Member Countries of ASEAN and the Government of the People's Republic of China on Transport Cooperation, 2004.

Article II of the MOU on Transport Cooperation laid down various areas, which should be promoted for medium- and long-term cooperation. Parties of the MOU were asked to establish consultative mechanism, in consultation with relevant ASEAN fora, to discuss, identify and implement certain mutually-beneficial projects, namely maritime safety, search and rescue at sea, marine environmental protection against pollution, port state control, and maritime security. The parties shall cooperate: in the implementation of relevant IMO conventions on safety of navigation at sea, to which the parties have agreed; on search and rescue at sea in accordance with national laws and regulations and international laws; on the management of ship's ballast water and marine environmental protection against pollution from ships, within the IMO framework; closely on port state control, in line with the Tokyo Memorandum of Understanding on Port State Control in the Asia-Pacific Region (1993–1994), in so far as the respective parties have agreed to; and by exchanging information and sharing of experiences on the implementation of the IMO International Ship and Port Facility Security (ISPS) Code, in so far as it binds the respective parties.

It was provided that the parties shall cooperate in organizing workshops or seminars on subjects of common interest in the transport sector. Subjects may cover, but shall not be limited to port management, maritime transport security, techniques of navigation-channel regulation, maritime safety administration, and management on ship's ballast water, etc.[42]

The MOU on Transport Cooperation an implementation framework with specific functions of the Ministry of Communications of China and the ASEAN Secretariat, the ASEAN-China Senior Transport Officials' Meeting, the ASEAN-China Transport Ministers' Meeting, Working groups or ad hoc expert task forces pursuant to be formed as appropriate, etc.[43]

---

[42] See 5 "Human Resources Development" in Article II "Areas of Cooperation" of the Memorandum of Understanding between the Governments of the Member Countries of ASEAN and the Government of the People's Republic of China on Transport Cooperation, 2004.

[43] Article III "Implementation" of the Memorandum of Understanding between the Governments of the Member Countries of ASEAN and the Government of the People's Republic of China on Transport Cooperation, 2004. This article stipulates that: "(1) The Ministry of Communications of the People's Republic of China and the ASEAN Secretariat shall be the agencies responsible for the identification, coordination, implementation and monitoring of projects and activities conducted pursuant to this Memorandum of Understanding; (2) The identification, implementation, monitoring and appraisal of the joint cooperation projects and activities shall be undertaken through the ASEAN-China Senior Transport Officials' Meeting, for final consideration and approval by the ASEAN-China Transport Ministers' Meeting; (3) Working groups or ad hoc expert task forces may be formed, as appropriate, to expedite the implementation of specific areas of cooperation under this Memorandum of Understanding; (4) The specific tasks, obligations and conditions relating to the cooperative activities under this Memorandum of Understanding, including the responsibility for cost payment, shall be discussed and

To sum up, the process of ASEAN-China regional maritime safety cooperation could be considered under three time frames.

The first period is from 1996 to 2002. During this period, the Joint Statement of the Meeting of the Heads of State/Government of the Member States of ASEAN and President of the People's Republic of China, 1997, is significant for its declaration of cooperation bilaterally and multilaterally, which is the basis for ASEAN-China cooperations thereafter.

The second period is from 2002 to 2004. The Declaration on the Conduct of Parties in the South China Sea, 2002, is the most important cooperation agreement of this period on maritime safety issues. It gave priority to maritime safety issues. However, no detailed cooperation activities, measures, and detailed implementation framework were provided, except some basic rules, by the declaration and the Joint Declaration of ASEAN and China on Cooperation in the Field of Non-traditional Security Issues, 2002, and the Joint Declaration of the Heads of State/Government of the Association of Southeast Asian Nations and the People's Republic of China on Strategic Partnership for Peace and Prosperity, 2003.

The third period is from 2004 until now, while specific plans of action, terms of reference and memorandums of understanding were concluded for the effective and comprehensive fulfillment and implementation of those agreements adopted in the first two periods. These instruments are: Terms of Reference of the ASEAN-China Joint Working Group on the Implementation of the Declaration on the Conduct of Parties in the South China Sea, 2004; Memorandum of Understanding between the Governments of the Member Countries of the Association of Southeast Asian Nations (ASEAN) and the Government of the People's Republic of China on Cooperation in the Field of Non-traditional Security Issues, 2004; Memorandum of Understanding between the Governments of the Member Countries of ASEAN and the Government of the People's Republic of China on Transport Cooperation, 2004; and Plan of Action to Implement the Joint Declaration on ASEAN-China Strategic Partnership for Peace and Prosperity, 2005. Provisions on maritime safety issues in these instruments are much more detailed, comprehensive, and could be estimated to implement effectively with satisfaction. It is difficult to define which of them is the most important, since they define regional maritime safety cooperation issues from different aspects, and the purposes of those provisions are the same, that is to realize regional maritime safety. However, we can give priorities to strengthening and deepening cooperation in certain issues.

---

agreed to by the relevant parties prior to the implementation of such activities, and be subject to the availability of funds and technical personnel of the parties."

## Highlights of ASEAN-China Regional Maritime Safety Cooperation

In 2004 and earlier, there were only three ASEAN-China cooperation projects relating to maritime safety or environment protection.[44] We need more cooperation dialogues and projects on maritime safety.

Certain issues should be raised to specify, deepen or implement maritime safety in Southeast Asia, subject to the common understanding forged by UNCLOS, IMO instruments, ASEAN-China cooperation documents, and other bilateral or multilateral documents. Some of them were provided in specific ASEAN-China cooperation documents.

### Ratification and Implementation of UNCLOS and IMO Conventions

Maritime safety cooperation in Southeast Asia can be achieved only by common understanding and operations. As mentioned above, international and regional framework are interactive. If international common sense is difficult to be realized, many issues will have to be dealt with within the regional framework. Regional cooperation will be carried out easily if there is an international common sense. Therefore, I should say that ASEAN and China need international conventions, and international conventions need recognition and support by ASEAN and China.

The first priority is to promote and facilitate the ratification and implementation of UNCLOS and IMO conventions by ASEAN and China. This is especially true for Cambodia, Thailand and Laos. Cambodia is a flag-of-convenience state, being the flag state of many vessels. Thailand is usually a port state. As mentioned above, provisions of UNCLOS on flag states and port states form the basis of various important IMO conventions. Ratifying and implementing UNCLOS by these two countries as well as IMO conventions by all ASEAN member states and China will promote ASEAN-China maritime safety cooperation.

However, the ratification and implementation of UNCLOS and IMO conventions need the support of relative infrastructure, human resources, and technology, as well as monetary fund. ASEAN-China cooperation should be carried out to help certain countries prepare for the ratification and implementation of UNCLOS and IMO conventions in these aspects, to a certain extent.

---

[44] They are: Project "China-ASEAN Workshop on Ships' Ballast Water Management" (Year 2004) (Project ID: TRN/MRT/04/001/REG); Project"China-ASEAN Environmental Policy Dialogue" (Year 2004) (Project ID: ENV/EVN/04/002/REG); Project "ACCF-China-ASEAN Training Project for Vessel Masters Chief Engineers and Safety Administrative Personnel on the Lancang-Mekong River (Phase One)" (Year 2001) (Project ID: TRN/01/003), http://www.aseansec.org/14402.htm, and http://www.aseansec.org/14470.htm (accessed 1 March 2005).

## Construction of Infrastructure, Training of Human Resources and Technical Assistance

Good construction of infrastructure can help to avoid collision and grounding of vessels. It can bring about efficient disaster response and minimize the damages. Cooperation in the building of infrastructure would need investment and technology. ASEAN members and China could cooperate to fund these infrastructure, and to share their experiences and technology.

Proper undertaking of their duties by seafarers and staff of companies or administrations could reduce some maritime accidents, or prevent ships from encountering accidents under certain circumstances. There are several branches of the World Maritime University (WMU), and professional institutions in ASEAN and China. Cooperation among these institutions could improve the capability of seafarers and staff to a large extent.

Technology of sea affairs, management, monitoring, etc., is essential to maritime safety. However, the levels of technology in this region are different. We need assistance among ASEAN member states and China to improve the technical level as a whole.

## Exchanges of Information and Data

Exchange of information and data has been emphasized in several ASEAN-China cooperation documents. This manifests the importance of such exchanges. Almost all operations or implementations of maritime safety rules should be based on necessary information on related ships, ports and seaways. Exchange of information, especially that on ship reporting, is necessary for implementation of the Port State Control. In order to effectively exchange information, ASEAN and China should build up a joint information exchange center. The center should perform several functions concerning information, such as information analysis, reporting information necessary to specific international institutions, regional institutions, some related countries, and certain private sectors to draw up solutions to specific maritime safety problems.

## Maritime Science and Research

We need various scientific data to analyze the maritime safety situation, as well as the pollution situation of Southeast Asia, and to conclude proper measures and operations. Data is also essential to determine the wrong doer, the liability of certain party, the compensation amount the victim could claim, and so on. While carrying out scientific research for exploration and exploitation of resources, we could do scientific research for maritime safety purpose.

*Construction of Accident/Disaster Report, Search and Rescue, Response and Investigation Cooperative Framework*

Accident/disaster report, search and rescue, response and investigation are a series of essential measures to be taken after the accident/disaster occurs or such a threat appears. They should be prompt, quick and effective, and accordingly ask for a high level of cooperation in the region.

The "Prestige" accident[45] prompted EU to set up the Maritime Safety Agency (MSA) to deal with maritime safety issues in its jurisdictional maritime areas. The MSA has many tasks concerning maritime safety. One important role of the MSA is acting as the platform for ship reporting, safety of navigation, accident investigation, cooperation with EU member states, liability and compensation, oil pollution response, etc. Meanwhile, IMO adopted the Maritime Assistance Services (MAS) in 2003.[46]

It is submitted to establish the joint ASEAN-China MAS to assist vessels in distress, and to undertake functions of accident/disaster report, search and rescue,

---

[45] The "Prestige" oil spill accident was one of the biggest disasters in the marine environment, and caused great concern in the whole world. On 13 November 2002, the tanker "Prestige", with a full load of heavy oil, had a large linkage on her hull, due to heavy weather. She lost her navigational capability, facing a risk of sinking, and floated toward the Spain coast. One day later, the Spain government refused to let the "Prestige" enter any ports of Spain. The distressed tanker had to be towed outward to the open sea, where the weather prevented the tanker from being salvaged smoothly. Cargo onboard the tanker could not be transfer from her either. On 18 November, the Portugal government refused to let the tanker enter Portuguese ports also. One day later, the Prestige broke into two parts and sunk finally. At least 10,000 tons of heavy oil spilt into the sea, causing immeasurable damages to the coasts of Spain, Portugal and France. The marine environment of Western Europe was damaged in definitely and could hardly be recovered for a long period in the future.

The "Prestige" accident is a flashpoint since several issues of maritime safety were raised or highlighted accordingly, such as: (1) whether or not the coastal states or other countries should offer a place of refuge for the distressed tanker, and how it can be done; (2) whether or not Bahamas, the convenient flag state of the Prestige, was unable to fulfill her obligations; (3) whether or not the loading port, in Latvia as the port state, was unable to fulfill her obligations; (4) whether or not the ABS, the classification society of the Prestige, was unable to fulfill its obligations, and how it can be blamed; (5) should the speed of phasing out single-hull tankers established by the IMO be much quicker; (6) should the compensation for damages caused by the accident be much higher; (7) how the relationship between the IMO, as a global institution, and EU, a regional organization, should be established. Some of these issues are discussed in the IMO framework and/or within certain regions. Common sense has been reached to some extent on some issues.

[46] IMO Resolution A.950(23) "Maritime Assistance Services (MAS)".

response and investigation in this region, promptly and effectively. Experiences of EU MSA should be considered for ASEAN-China maritime safety cooperation. Preparedness programs and response operations for various maritime incidents, as well as technical cooperation by related institutions of ASEAN member states and China will also be necessary.[47]

## Port State Control and Place of Refuge

To supplement the functions of flag states, the UNCLOS and IMO conventions confirm the port state control functions of port states. Port state control could help to diminish risks of maritime accidents.[48] Port state control monitoring cooperation should be emphasized for ASEAN-China maritime safety cooperation. Also, it is necessary to conclude certain procedures for various port state control stages, such as reporting procedures on port state control detentions, analysis and evaluation of reports, measures review procedure, and so on. Certainly, port state control requires prompt information exchange, efficient infrastructure, qualified personnel, quick and stable technology.

Providing places of refuge is believed to be one of the Coastal States' fundamental contributions to maritime safety. As stated by the IMO Resolution A.949(23), "Guidelines on Place of Refuge for Ships in Need of Assistance",[49] 2004, "[w]hen

---

[47] IMO made great efforts to improve the response system and technology. The International Ship and Port Facility Security Code ("ISPS Code") stipulated certain articles providing a consistent and standardized framework for related countries to make suitable response to certain circumstances, and is submitted to be considered.

[48] According to IMO documents, a total of 17,379 inspections were carried out in 2001 (16,034 in 2000) under the Asia-Pacific MOU Annual Report 2001 (FSI 11/INF.4), leading to 1,349 detentions (1,101 in 2000) for an overall detention rate of 7.76% (three-year rolling average of 7.11% in 2000). It is indicated that technical cooperation programs were developed. A total of 5,520 inspections were carried out in 2001 (4,949 in 2000) under the Asia-Pacific MOU Annual Report 2001 (FSI 11/INF.4), and leading to 291 detentions (336 in 2000) for an overall detention rate of 5.27% (6.8% in 2000), that a website had been created and that an IT system for handling the region's inspection data was under development. See the "Port State Control: An Update on IMO's Work in 2003" on IMO's website.

[49] The IMO and the Committee of Maritime International (CMI) made a lot of efforts to identify places of refuge issue within a global construction. On 5 March 2004, the IMO assembly adopted Resolution A.949 (23), "Guidelines on Place of Refuge for Ships in Need of Assistance" (the Guidelines). However, CMI discussed topics relevant to places of refuge in a report, identifying that there was no international conventions establishing the rights and obligations of a coastal state when it was faced with a request to provide a place of refuge. Consequently, the IMO agreed that the issues raised in the report needed further study before making any decision. See Circular of Legal Committee (LEG) of IMO, http://www.imo.org/Newsroom/mainframe.asp?topic_id=280&doc_id=3662 (accessed 31 October 2004).

a ship has suffered an incident, the best way of preventing damage or pollution from its progressive deterioration would be to lighten its cargo and bunkers; and to repair the damage. Such an operation is best carried out in a place of refuge."[50] It is also recognized that to bring a distressed ship into a place of refuge near a coast may endanger the coastal state economically and environmentally, and local authorities and civilians may object to the operation strongly. However, in some circumstances, "the longer a damaged ship is forced to remain at the mercy of the elements in the open sea, the greater the risk of the vessel's conditions deteriorating or the sea, weather or environmental situation changing for the worse, thereby causing the ship to become a greater potential hazard".[51] Therefore, the guidelines strengthen the importance of case-by-case consideration, to get a balance between the advantage for the affected ship and the environment if the ship is accepted into a place of refuge, and the risk to the environment resulting from that ship being near the coast.[52] However, EU had adopted a ship reporting and monitoring directive (2002/59/EC), enclosing plans to accommodate vessels in distress in some places. All EU coastal countries and representatives from Norway and Iceland participated in it.

It is submitted that ASEAN and China could draw up agreements on cooperative environmental protection and port reception facilities. In the long term, cooperation on providing a place of refuge for ships in need of assistance could protect marine environment of Southeast Asia from damages much better than not doing so. IMO's documents, such as IMO Resolution A.949(23) ("Guidelines on Places of Refuge for Ships in Need of Assistance") and the regional cooperation framework, practices and considerations by EU, are being suggested for our region. For the purpose of preventing damages to the coastal state providing a place of refuge as much as possible, it is proposed that certain regular places of refuge be provided, each with the most prompt information exchange, the most sufficient and efficient infrastructure, the most qualified personnel, and the fastest and most stable technology.

## Conclusions

China and ASEAN are closely connected by the South China Sea. Maritime safety is one of the essential topics in Southeast Asia. China and ASEAN have many

---

[50] IMO Guidelines on Place of Refuge for Ships in Need of Assistance, Article 1.3.
[51] IMO Guidelines on Place of Refuge for Ships in Need of Assistance, Article 1.6.
[52] IMO Guidelines on Place of Refuge for Ships in Need of Assistance, Article 1.7.

common interests in maritime safety. Based on their common interests and interdependence, we can have complete understanding of specific circumstances of each other, and cooperate comprehensively and deeply.

As stated,[53] if the need for cooperation is generally accepted, we have to deal with: who we cooperate with; how widely we cooperate; over what areas and ways we cooperate; and the mechanisms we need to establish to achieve this cooperation.

---

[53] Keynote address by Efthimios E. Mitropoulos, Secretary-General, IMO, at the ASEAN Regional Forum Conference "Regional Cooperation in Maritime Security", 2–4 March 2005, http://www.midef.gov.sg/display.asp?number=2383 (accessed 12 March 2005).

# Part VII

# China-ASEAN Relations in Regional Perspectives

# Chapter 15

# China-ASEAN FTA and Korean FTA Policies

*Moon-Soo CHUNG**

## Introduction

Until 1989, the idea of a regional economic integration had been almost an anathema to the East Asian region and multilateralism was taken for granted as being in the best interests of the region. APEC was formed in 1989 but with its declared policy of "open regionalism" and its wide dispersion of membership across the Pacific, it did not really measure up as a regional economic entity. The East Asian Economic Caucus was proposed by Malaysia but it proved too politicized and narrow to win acceptance as a practical proposition. In 1992, ASEAN agreed to form an FTA but still at a very shallow level.

The defining moment for the future of the East Asian economic integration came when the region was hit by an unprecedented financial and economic crisis that began in 1997. This not only prompted worldwide re-evaluation of the institutional underpinnings of Asian economic development models, but led directly to the inauguration of regular summit meetings between leaders of this region. The stalemate of multilateral DDA on the one hand and the active bilateralism of the United States on the other also called for the region's awakening. Added to this was the Chinese entry into the WTO in 2001 that signified China's status as a global producer and trader with direct impact on the rest of the region.

All of a sudden, the idea of a regional integration became the buzzword and everybody rushed into it. A spate of formulae were put forward and pursued aggressively: ASEAN plus Three (China, Japan, Korea); ASEAN plus Five (three plus Australia and New Zealand); ASEAN plus China; ASEAN plus Japan; ASEAN

---

*Preparation of this paper was aided by an Inha University Research Grant. The author is thankful to fellow professors Dong Chun Suh, Young Il Park and Jung Taik Hyun for their valuable comments on the earlier draft.

plus Korea; China-Japan-Korea; Japan-Korea; China-Korea; Japan versus individual members of ASEAN (Singapore, Thailand, Philippines, Malaysia); China versus individual ASEAN members; etc.

The first regional FTA to be agreed on was the Japan–Singapore EPA (Economic Partnership Agreement) in 2002, but given the special feature of Singapore, not much significance may be attached to it. At the end of 2004, FTAs under official negotiations in the region were the ASEAN-China FTA, Japan-Korea FTA, Japan-Thailand FTA and Korea-Singapore FTA. Korea and ASEAN agreed to negotiate one in 2005.

Among those under negotiations, the ASEAN-China FTA is apparently making the fastest progress and is expected to be completed soon. It was first proposed by China in 2000 and agreed to in 2001, and negotiated from 2002 to 2004. Once concluded, it will have wide and significant impact and ramifications not only for China and ASEAN member countries but all other countries in the region.

Given the likelihood of successful conclusion of the China-ASEAN FTA, this paper tries to answer the following three key questions:

1. What likely economic impact would the FTA have on Korea?
2. What would be the political reaction of Korea and other regional countries, Japan in particular, in terms of regional integration, i.e., whether they will try to join the group or to form a competing group?
3. What would be the desirable and feasible policy options available to Korea in these FTA matters?

## Impact of China-ASEAN FTA on Korea

The China-ASEAN FTA will significantly affect the Korean economy. While it is difficult to assess its likely impact in the absence of concrete details on the agreement, it will certainly help strengthen international competitiveness of both Chinese and ASEAN economies, particularly the latter's. The FTA will accelerate the pace of industrial restructuring and the realization of scale of economy in China, which are ongoing since the Chinese entry into WTO. The free access to the vast Chinese market will induce more FDI to ASEAN economies and faster industrial restructuring.

In 2003, China and ASEAN markets represented 18.1% and 10.4% of Korean exports respectively. Duty-free imports of Chinese goods will replace Korean and Japanese goods in ASEAN markets, and similarly ASEAN goods in Chinese market. This outcome will be avoided if Korean and Japanese companies are already in the market, i.e., manufacturing inside the FTA through FDI. In this regard, Japan has a much stronger market presence than Korea, which means less severe impact.

Table 15.1  Exports and Imports of Korea

(in million US$)

| Year | | Total | Japan | China | ASEAN |
|---|---|---|---|---|---|
| 1995 | Exports | 125,058 | 17,049 (13.6%) | 9,144 (7.3%) | 17,979 (14.4%) |
| | Imports | 135,119 | 32,606 (24.1%) | 7,401 (5.5%) | 10,137 (7.5%) |
| 2000 | Exports | 172,268 | 20,466 (11.9%) | 18,455 (10.7%) | 20,134 (11.7%) |
| | Imports | 160,481 | 31,828 (19.8%) | 12,799 (8.0%) | 18,173 (11.3%) |
| 2003 | Exports | 193,817 | 17,276 (8.9%) | 35,110 (18.1%) | 20,253 (10.4%) |
| | Imports | 178,827 | 36.313 (20.3%) | 21,909 (12.3%) | 18,458 (10.3%) |

Source: Homepage of Korea International Trade Association, http://www.kita.net/.

Cheong and Oh (2003) estimates that the China-ASEAN FTA will have a negative impact on Korea: Korean exports to China will decrease by around $0.3 billion and those to ASEAN by about $1.1 billion, or a total of $1.4 billion. This deficit is largely attributable to the trade diversion effect. If the FTA helps ASEAN economies realize the scale of economy that leads to increased imports from Korea, such deficit would be substantially reduced, according to the same study.

These figures are not so significant when compared with the overall trade figures but should, however, be taken as representing only a part of the impact only. The strengthened competitiveness of Chinese and ASEAN companies will negatively affect Korean businesses in the global market as well.

## Domino Effect and Spaghetti Bowl Effect

Of more concern and interest, however, is how Korea, Japan and other regional countries will react to the above scenario. It is not easy to predict but there seem to be several theories and projections. Brief reference will be made to the domino effect and the spaghetti bowl effect theories here.

Regional economic integration, once started, is very hard to resist. Once one group of countries begins to liberalize preferentially, others find themselves compelled to react by either joining the pioneer group, or by forming their own counter balancing group. Thus, a domino effect comes in (Baldwin, 1993). China's announcement of closer relations with ASEAN had shaken up government ministries throughout the region, especially in Tokyo and Seoul. While regionalism is not yet a reality in East Asia, it is noted that there is a good chance that the domino effect will soon be triggered (Baldwin, 2004). According to Baldwin, the key to the logic driving the domino effect is the exclusion or the threat of exclusion, i.e., the extent to which various nations would care about being excluded from various trade blocks, which he terms "exclusion indices". Table 15.2 shows a number of

Table 15.2 Exclusion Indices for Various Trade Blocks

| ex. Sing | Asean ex. Singapore | Asean + China ex. Sing | Japan + Korea FTA | China + Japan + Korea FTA | China + Japan + Korea + Asean |
|---|---|---|---|---|---|
| Indonesia | 9% | 14% | 26% | 31% | 39% |
| Malaysia | 11% | 18% | 30% | 37% | 48% |
| Philippines | 11% | 21% | 33% | 43% | 54% |
| Thailand | 13% | 21% | 38% | 46% | 59% |
| China | 9% | 14% | 41% | 47% | 56% |
| Japan | 17% | 36% | 7% | 25% | 42% |
| Korea | 11% | 22% | 27% | 38% | 49% |

Source: Baldwin, Richard, "Asian Regionalism: Promises and Pitfalls", mimeo prepared for KIEP conference in Seoul Korea, October 2002, reprinted in Baldwin (2004).

possibilities. For the country listed in each row, the figures show the percentage of that nation's exports that go to the markets indicated in the columns. This analysis shows that a well-implemented China-ASEAN FTA or a Korea-Japan FTA is likely to trigger a domino effect. Following this logic, Korea is best advised to join the China-ASEAN FTA, negotiate bilateral FTAs with China or ASEAN, or form a counter balancing group.

We should, however, not discount the possibilities of quite an opposite development: instead of integrating into a single regional entity, there emerge competing groups with respective FTA arrangements with outside non-regional countries. For example, Japan goes its own way with multiple FTA arrangements with Mexico, Chile, Malaysia and other members of ASEAN; China with ASEAN, India, Australia, etc.; Korea with Chile, United States and other APEC countries. As more and more bilateral FTAs are negotiated, they may result in what Bhagwati and Panagariya (1996) refer to as "spaghetti-bowl" effect which may arise due to the differential tariff rates for FTA member and non-member countries that cancel out each other. Thus, the spokes or "spaghetti strands" may emanate out in many different and overlapping directions, with consequential negative welfare effects. Then, even if a single regional entity may ultimately emerge true to the domino theory, it may well take a long period of this spaghetti regionalism.[1]

---

[1] For discussions on spaghetti regionalism, Freund, C.L., "Spaghetti Regionalism", *Board of Governors of the Federal Reserve System, International Finance Discussion Papers*, Vol. 680 (September 2000). Also, Ibarra-Yunez, Alejandro, "Spaghetti Regionalism or Strategic Foreign Trade: Some Evidence for Mexico", *Journal of Development Economics* No. 72, pp. 567–584 (2003).

Another possibility suggested is that the United States become the hub and East Asian countries merely its spokes.[2] From the viewpoint of the United States, emergence of a regional entity in East Asia may be seen as a serious threat to the current global economic order dominated by itself, and it may seek an alternative to engage individual East Asian countries in bilateral FTAs. An FTA with the United States, with its huge market, will certainly be very attractive to East Asian countries, which compete with each other in the US market. With one country joining, the domino effect may work to bring in other countries.

It is not easy to predict which way it will take. Initially, China-ASEAN FTA and Japan-Korea FTA may emerge respectively and later merge into one.[3] China-ASEAN FTA may pull in Korea and Japan through respective bilateral FTAs with ASEAN. China-Korea-Japan FTA may advance and pull in ASEAN members later. Equally likely are, however, the spaghetti bowl scenarios or at least China and Japan forming competing groups.

## Competition and Rivalry

The shape of the regional integration in East Asia will be determined by the play of several factors. Two major ones among them are: rivalry between China and Japan competing for the leadership of the region; and the knotty issue of the agricultural sector for Korea and Japan. In other words, taking the external forces as neutral,[4] in the absence of these two factors, there seem to be no major obstacles to the East Asian region moving to form one grand single FTA under the umbrella of the so-called ASEAN plus Three. I will discuss the first factor here.

Banking on the world's highest growth rate for the last two decades, absorbing the largest share of the global FDI, and having finally acquired membership of the WTO, China is emerging as an active player in the international arena. In the integration of the region, China has taken Japan and Korea by surprise by agreeing and negotiating the China-ASEAN FTA on a fast track. Behind this active involvement

---

[2] Park (2004).
[3] This seems like what Baldwin (2004) predicts.
[4] Given the reality of the United States being the undeniable superpower in the current world order, no serious opposition from the United States would be a necessary condition to any solid regional integration. Regarding the options open to the United States against possible East Asian integration, Park (2004) lists (1) trying to maintain the status quo under the umbrella of multilateralism; (2) trying to recharge the APEC vehicle through implementation of the 1994 Bogor Declaration; and (3) trying to dampen East Asian initiatives through bilateral FTAs with individual countries. All of them are similar in posing a serious obstacle to the regional integration in East Asia. Japan and Korea, which are dependent on the US for their security, will be particularly vulnerable to the negative pressure from the US. Discussions in the text are limited to consideration of internal factors only.

in the regional integration in East Asia lies the intention to build a China-centered (Sinocentric) "East-Asianism" (Cho, 2004). Taiwan's Li Ying-Ming (2001) claims that in East Asia in the post-Cold War era, a few regional groupings have been growing while restraining each other; that is, China-centered and/or Japan-centered East–Asianism, which China and Japan are pursuing for their respective interests, coexists with Asia–Pacificism centered around the United States, and ASEANism formed by the ten ASEAN member countries.[5] China is also approaching the FTA from the security perspective as well. Through its pursuit of economic integration within the region, China seems to be seeking to change security order within the region. As a matter of fact, China had already made individual security cooperation arrangements with all of the 10 ASEAN countries prior to the formal proposal of the FTA in November 2000 (Cho, 2004). On the part of ASEAN, the FTA may be viewed as mitigating their concerns about the so-called "China Threat".

Prior to the latest emergence of China, it was Japan which appeared to have become the regional leader. The 10-year-long economic recession, which hit Japan since the mid-1990s, has however sharply decreased the weight and influence of Japan in the region. Japan Economic Research Center states in a report "China to Change Asia — Japan's Survival Strategy", published in December 2002, that Japan's political and economic influence is rapidly decreasing within East Asia due to the rise of China, which is expediting changes in regional economic structure while changing the regional industrial map. It is noted that Japan's economic and trading status in the region is deteriorating while China's status is on the rise.

Recent discussions and movements toward an FTA within the region, aside from the AFTA, began with the Korea-Japan Summit Meeting in October 1998 where both countries agreed to jointly study the idea of a Korea-Japan FTA. A two-year-long report concluded that a Japan-Korea FTA is overall economically desirable. That is, although Korea may lose from trade liberalization, the FTA could provide a lot of dynamic gains to Korea by realizing economies of scale, enhancing competitive business environment, increasing inflows of investments, and expanding strategic alliances among companies. In the late 2001, the Korea-Japan FTA Business Forum was established and it endorsed the bilateral FTA.[6] Official negotiations commenced early this year. Japan's push for this Japan-Korea FTA, as well as others with Singapore, other members of ASEAN, Mexico and Chile,

---

[5] Li, Ying-Ming, *Taiwan and Cross-strait Relations under Information Era* (Taipei: Sheng-Zhi Publishing House, 2001).

[6] It is interesting to note that while the Korean business community in general, other than small and medium-sized companies, had been reported as favoring the FTA in 2002, the mood has lately changed to one of caution, according to newspaper reports.

must have alarmed China, which was until then engrossed in negotiations for its entry into the WTO. China reacted by agreeing to negotiate the China-ASEAN FTA and speeding it up to overtake the Japan-Korea FTA. While China made overtures for a China-Japan-Korea FTA, Japan appears not to favor it, citing China's political system and inferior economic system as major obstacles to the FTA. It is claimed that an FTA with China poses risks of uncertainty for its partner country and the risks could be mitigated and shared if the partner is a group of countries; the partner should be large enough and comparable to China'a potential; and Japan and Korea together under an FTA could be the candidate.[7] According to Igawa (2004), the best strategy for promoting an East Asian FTA is having the Japan-Korea FTA first and extending it to ASEAN and finally to China.[8] This idea, of course, does not appeal to China. China has proposed a China-Japan-Korea FTA, citing the bigger economic benefits, or bilateral FTAs with Korea and Japan. Japan is, however, shying away from this and China is not on its priority list of FTA candidates.

It is not yet clear whether China-ASEAN FTA and Japan-Korea FTA will materialize soon, which of the two will be first and in what form. As noted earlier, Korea is to start negotiations with ASEAN in 2005. How smooth and fast it will progress will certainly affect the ongoing Korea-Japan negotiations. Japan is also expected to open its negotiation with ASEAN. It is too early to predict which one will come first. In the event that two competing groups will result, the next step will be to try to combine the two into one single FTA as soon as possible. In this regard, it is very important that the two FTAs are structured with low entry barriers. Mitigating this play of rivalry between China and Japan and brokering the two into a single entity is the role to be played by Korea and ASEAN.

## Agriculture

Agriculture is the single most important factor behind the reticence of the Korean and Japanese governments in pursuing regional integration and FTA, particularly with China and ASEAN. The agricultural sectors of China, Korea and Japan share the same characteristics of small-scale labor-intensive cultivation and have similar composition of produces. Compared to the Chinese, Korean and Japanese agricultural sectors lack competitiveness and it is feared that once the market is open, it will be a matter of time before they are swept away.

---

[7] Igawa, Kazuhiro, "Japan's FTA Policy and Position for a Japan-Korea FTA", a paper presented at the conference on FTA Policies of Korea and Japan and Policy Implications for a Bilateral FTA, Jeongsuk Institute for Logistics and International Trade, Inha University, Incheon, May 2004.
[8] *Ibid.*

The total agricultural/fishery production of China, Korea and Japan stands above $300 billion as of 2001 and its share of GDP is about 5.4%, higher than that of advanced countries but lower than the average of developing countries. Average farm size per household is 0.54 ha in China, 1.4 ha in Korea and 1.6 ha in Japan, all well below the size in the US (176 ha), Australia (400 ha) and the average of the 15 EU members (18 ha). The number of farmers has decreased in China, Korea and Japan, but the number of farm households has increased in China, while it has decreased in Korea and Japan. The proportion of farmers over the total population is about 8% in Korea and Japan but still over 60% in the case of China.

For both Korea and Japan, agriculture is the heavily protected sector and indeed a very sensitive one. While most non-tariff barriers have been removed during the Uruguay Round, tariff rates for many agricultural produces are still prohibitively high and rice was taken out from tarriffication altogether. Pressure is mounting for import liberalization at the ongoing DDA but the Korean and Japanese governments seem to be trying hard to fend off the pressure rather than

Table 15.3 Agricultural Indices of Korea, China and Japan

(Unit: US$ hundred million and %)

|  | Korea | China | Japan | Total |
|---|---|---|---|---|
| GDP | 4,574 | 11,558 | 41,760 |  |
| (per capita) | (8,918) | (909) | (32,524) | 57,892 |
| Agri/Fishery GDP | 199 | 2,253 | 668 |  |
| (Growth Rate) | (1.9) | (6.1) | (0.4) | 3,120 |
| Weight of Agri/Fishery (%) | 4.4 | 22.2 | 1.6 | 5.4 |
| Population (million) | 47.3 | 1,276.3 | 126.9 | 1,450.5 |
| - Farm Population (Ratio) | 3.9 (8.3%) | 795.6 (62.3) | 10.2 (8.1) | 809.7 (55.8) |
| Land (1,000 km$^2$) | 99.5 | 9,597 | 378 | 10,075 |
| - Agri (Ratio) | 18.8 (18.9%) | 1,300 (13.5) | 48 (12.7) | 1,367 (13.6) |
| - per Household | 1.4 ha | 0.54 | 1.6 | 0.56 |
| Exports | 1,504 | 2,492 | 4,035 | 8,031 |
| (Agricultural) | (15.8) | (154.5) | (24.9) | (195.2) |
| Imports | 1,411 | 2,251 | 3,491 | 7,153 |
| (Agricultural) | (84.6) | (97.6) | (354) | (536.7) |
| Trade Balance | 93 | 241 | 544 | 878 |
| (Agriculture) | (−68.8) | (57) | (−329) | (−341) |

Source: Korea Ministry of Agriculture/Forestry, *Major Statistics of Agriculture and Forestry* (2003); Japan Ministry of Finance, *International Trade Statistics* (2002); People's Bank of China, *Statistical Yearbook* (2001).

facing the issue directly.[9] Admittedly, the Korean and Japanese agricultural industry is so inefficient that it will be wiped out once the imports flood in. Despite its relatively insignificant weight in terms of GDP, it raises the so-called non-trade concerns of food security and devastation of rural community and culture. Politically, if not economically, the issue does raise almost insurmountable obstacle. The political leadership is neither strong enough to carry it through nor seems to be convinced of the need for liberalization.

Lately, rethinking of the issue seems to have started. The Japanese government's White Paper for 2004 cautiously suggests the possibility of liberalizing the farm sector and exporting farm products.[10] Although export possibilities are extremely limited at present, it is argued, a number of products are making a splash despite their relatively high prices because consumers put a higher premium on health and safety concerns.[11] It is easier said than done, but it seems to suggest a viable alternative. Consumers around the world, not to mention the Japanese and Koreans, are more and more health-conscious and would not mind paying three or four times higher for organic products that are free of chemicals and pesticides. Expenditure on foodstuffs out of the total household expenditure is insignificant already. This raises the possibility of organic farming. The critical question to which no satisfactory answer has yet been found is how such organic products can be differentiated from the same products grown with chemical fertilizers and pesticides but sold at much lower prices.

It is very clear that Korea and Japan cannot continue to fend off the pressure for liberalizing their agricultural sector. Japan has lately agreed to the tariffication but Korea is trying to defer the issue again. In view of the inevitability of the market opening sooner or later, deferring the issue another five or ten years would not help, in view of the rather poor record for the past ten years since the Uruguay Round. As noted above, opening up of the sector would not necessarily lead to the disastrous collapse of the sector either. With adequate packages for restructuring the sector and compensation for damages, the issue can be addressed. In summary, the agricultural issue should not be allowed to block regional integration and taking the whole liberalization as hostage.

---

[9] Japan subsequently agreed to the tariffication in 2001 but Korea is still pursuing it in its DDA negotiations. It is reported that Korea is close to an agreement with China and USA to extend the tariffication by another ten years on condition that minimum market access report be raised up to 8%.
[10] "Making the Farm Sector Competitive", *The Japan Times*, 28 July 2004.
[11] *Ibid*.

## Korean Policies on FTA

Korea cannot be said to have been active in pursuing FTAs so far. While it has become a trading economy heavily dependent on international trade, it sought multilateralism under the WTO regime rather than pursuing bilateral or regional arrangements. In fact, until the 1990s, Korea had little reason to even consider regional arrangements because its multilateral approach had worked well. Korea's largest trading partner, the United States, was not a proponent of regionalism until 1990s and Korea's neighbors did not show much interest either. I will first briefly go through the recent developments in FTA areas in Korea before adding my comments.

In the 1990s, the world experienced a sudden increase in regionalism. Almost all countries became members of regional trade agreements of one sort or another, and exceptions to basic multilateral rules of the GATT/WTO have become the norm in global trade. In the face of mushrooming FTAs and suddenly realizing that it might be left out alone, Korea started looking for possible FTAs in the aftermath of the 1997 East Asian economic crisis. Korea chose Chile as its first partner and started negotiation in 1999.[12] The negotiation took three years and ratification another two years, taking effect only in 2004. The lapse was by and large attributable to the Korean side, which was not fully ready to absorb an FTA. Throughout the negotiation and ratification, its government was dragging its feet, wary of the angry farmers. While actual damages to the farmers were limited as the negotiated draft per se did contain many exceptions and waivers in respect of most products sensitive to Korean farmers, Korean farmers and NGOs took it as the litmus test of the opening of the agricultural sector and took to the highways and streets in protest until the last minute. One might say that the successful ratification of the FTA with Chile was an epochal turning point for Korea's trade policy, as it paved way for further FTAs with major trading partners in the future.[13]

With its first exercise finally behind it, Korea is now actively moving further. First of all, it has begun official negotiations with Japan and Singapore. As for Japan, there were six conferences regarding FTAs after the joint research group of industry, government and academia was launched in July 2002. Formal negotiation started early 2004. As for Singapore, the joint research group was launched in March 2003

---

[12] One might argue that the Korean government was deliberate in choosing Chile as its first candidate for FTA. Chile is located oceans away and has the least impact in terms of volume and nature of trade.

[13] Cheong, Inkyo, "Korea's FTA Policy and Position for a Japan-Korea FTA", a paper presented at the international conference on FTA Policies of Korea and Japan and Policy Implications for a Bilateral FTA, Jeongsuk Institute for Logistics and International Trade, Inha University, Incheon, May 2004.

for a period of six months and negotiation began in early 2004. Korea-Singapore FTA has no agricultural issue and negotiation was completed. The FTA with Japan would not be that easy. Japan has to take up agricultural issue. For Korea, there are growing concerns that it will lead to serious setback for many industrial sectors, particularly small and medium-sized companies producing parts and materials that still lack competitiveness.

Korea has also been investigating possible FTAs with ASEAN, USA, Mexico and EFTA. As Mexico serves as gateway to NAFTA, the Korean industry is strongly in favor of an FTA with it. It is at a preliminary stage at the moment, as Mexico had declared a moratorium against further FTAs in November 2003, but both governments agreed to resume it in May 2004. As for ASEAN, ASEAN-Korea FTA was first proposed by ASEAN in November 2001, but Korea was then hesitant in accepting it, again largely due to its concern over agriculture. Korea and ASEAN ministers were reported to have agreed on opening formal negotiations in 2005. For the FTA with the US, the US seems to have changed its hitherto passive attitude and working-level discussions are expected in 2005. It was reported that negotiations with EFTA will start in 2005.

Still at the initial stage of joint researches but with potentially the most formidable developments will be the China-Japan-Korea FTA. The three countries have agreed to study a trilateral FTA at the summit in November 2002. At present, joint research groups on the economic effects are in the works by research institutes of the three countries. As briefly discussed above, this triangular FTA will be of much benefit to the three countries economically and will also contribute to the opening of North Korea and stabilization in the Korean peninsular. Little progress has been made at the moment, however. Sino-Japanese rivalry may partially explain why Japan is apparently not interested (China is only a long-term candidate for FTA in the Japanese calendar). Korea is hesitant again because China brings in the tough issue of agriculture.

Korea has been lacking clear vision and strategy for FTA. Not having done its homework on agriculture, it has been not proactive but only reactive. According to Hyun (2003), the reason why Korea took a step-by-step approach to FTAs was because it wanted to minimize the reaction of opposition groups and to gain political support. Indeed, it is very difficult to get political support for trade liberalization in Korea. Koreans in general and farmers in particular harbor a rather defensive attitude against foreign influences and the Korean government tends to accommodate them than to dissuade them. Spending five years on an FTA with Chile should, however, be viewed not as a success but an example of passive ducking. Yes, Korea has an FTA but what significance does it have in reality? The same may be said of Japan with its EPA with Singapore.

If Korea keeps an extremely slow pace and assumes the position of a follower rather than a leader, it will run the risk of derailing its outward-driven economy. Korea has to move faster toward larger FTAs. Hyun (2003) argues that FTAs with China, Japan and the United States need to be negotiated simultaneously and I agree. One with ASEAN must also be given priority.

Recently, the Korean government seems to have shifted its stance from being passive to aggressive. Two new developments may be noted. It has declared an ambitious plan of pursuing, in addition to ongoing negotiations, FTAs with the US, Canada, Mexico and EFTA simultaneously. For the Korea-US FTA, the US government showed little interest in the past. Lately, however, the US government seems to have a change of heart. Both governments seem to be interested and serious this time. There are still obstacles such as the thorny issue of screen quotas and agriculture, but if it materializes, it will have significant repercussions on the region as well. It is also reported that joint studies for Korea-Chinese FTA, which take place before official negotiations, will be elevated from ongoing civilian-level institutions to official governmental ones, signaling the Korean government's change of attitude from hitherto lukewarm to earnest.

On the other hand, the Korean government seems to be slowing down its ongoing negotiations with Japan. Inside and outside the government, there are growing concerns over the negative impacts of the FTA on Korean industries, in particular small and medium-sized industries for parts and intermediate materials. The Korean trade deficit with Japan has worsened from around $10 billion in 2000 to more than $20 billion in 2004. It is projected that the Korea-Japan FTA will mean a further $5–6 billion in the red. The latest flare-up of patent squabbles raised by large Japanese companies against Korean companies involving IT technologies is also cooling negotiations. The Korean government has made it known that it would not seek to meet the deadline of the end of 2005 set for completion of negotiations for FTA with Japan.[14]

The above two developments taken together seem to signal a subtle change in Korea's policy; the Korean-Japan FTA will no longer be given priority and negotiations with China, ASEAN and the US will be pursued simultaneously. This may be a mere shift in negotiation strategy but it is more likely a genuine change of priority. It remains to be seen how significant this change of policy will be and how the people on the streets, and in particular the farmers and ever-growing NGOs, will react, but it is certainly a welcome change. The Korean government seems to have finally found the right course, albeit belated.

---

[14] *The Maeil Business Newspaper*, 8 November 2004, p. 5.

In conclusion, Korea need to be more active in promoting FTAs with major trading partners inside and outside the region. In terms of regional FTAs, I think the ASEAN plus Three formula is the best for Korea, and Korea should, in cooperation with ASEAN, take the lead in trying to realize it.

## Challenge for East Asia

East Asia faces an uncertain and turbulent future in this 21st century of globalization. Globalization promises huge economic benefits but at the cost of uneven distribution of "winner-takes-all" and structural instability. To survive in these uncharted turbulent seas, countries, particularly those not big enough on their own, need to align together. East Asia is the only area with no effective region-wide alignment. No wonder the 1997 financial crisis started in this region.

East Asian economic integration is an answer. If China and Japan, the two major economies in the region, will cooperate and show leadership, an East Asian economic entity will not be that far away. Aside from the Sino-Japanese rivalry, they have differences *vis-à-vis* the regional role of the US, which has been dominant in shaping the post-World War law and order in the region. Internally also, they have constraints. Japan is constrained by its heavily US-dependent foreign policy on the one hand and ineffective domestic political system on the other.[15] China has recently been displaying very active leadership, but it has yet to find ways on how to co-opt and build harmony with pre-existing orders and how to quickly reform its domestic systems in full accordance with global norms.

ASEAN and Korea also have their own constraints in mitigating the two. ASEAN is still too weak to be an effective international institution with its consensus decision-making and little binding power to ensure implementation. For Korea, the peaceful solution of the North Korean nuclear crisis and the building of peaceful inter-Korean relations are preempting its diplomatic attention.

Having noted these constraints, however, there is much room for progress if all the parties concerned try to share a common vision and put in joint efforts toward realizing that vision. ASEAN plus Three is a good vision deserving utmost efforts of all the parties, I believe.

Before concluding, however, I would add a quick note of caution easily overlooked in discussions of FTAs. We should not allow regionalism to undermine the importance of the global trading system. After all, an FTA is only the second best option to multilateralism based on MFN. An FTA is taking exceptions from them.

---

[15] Discussions here of constraints to leadership of respective regional countries are quoted from Park (2004), with author's permission.

The benefits of FTA to participants emanating from such favored treatment, in particular those from trade-diversion effect, are basically short-term in nature and gradually decrease as new members join it. In that sense, an FTA in East Asia is to be viewed as a defense against rampant regionalism in the rest of the world, and we should try to promote multilateralism which has proven to be the most beneficial to all in the end.

## References

Cheong, Inkyo and Oh Dong Yoon, "Progress of a China-ASEAN FTA and Its Impacts", *KIEP Policy Study Series No. 03–07* (December 2003).
Cheong, Inkyo, *et al.*, "Need for and Precondition to the Korea-China-Japan FTA", *KIEP Joint Report Series No. 03–01* (2003).
Cheong, Inkyo, *et al.*, "Overall Joint Report: Korea-China-Japan FTA", *KIEP Joint Report Series No. 03–01* (in Korean) 2003.
Baldwin, Richard, "A Domino Theory of Regionalism," NBER WP 4465, in Baldwin, R, P. Haaparanta and J. Kiander, eds., *Expanding European Regionalism: The EU's New Members* (Cambridge University Press, 1995).
Baldwin, Richard, "East Asian Regionalism: A Comparison with Europe", mimeo prepared for the conference on FTA Policies of Korea and Japan and Policy Implications for a Bilateral FTA, Jeongsuk Institute for Logistics and International Trade, Inha University (28 May 2004).
Bhagwati, Jagdish and Arvind Panagariya, "Preferential Trading Areas and Multilateralism — Strangers, Friends, or Foes?" in Jagdish Bhagwati and Arvind Panagariya, eds., *The Economics of Preferential Trade Agreements* (Washington, D.C.: American Enterprise Institute, 1996).
Cho, Hyun-Jun, "China's Approach toward FTAs with East Asian Nations and their Implications for Korea", a paper presented to the international conference on Rising China and the East Asian Economy, KIEP, Seoul (19–20 March 2004).
Japan Economic Research Center, Asia Study Report: "China to Change Asia — Japan's Survival Strategy", Tokyo (2002).
Igawa, Kazuhiro, "Japan's FTA Polilcy and Position for a Japan–Korea FTA", a paper presented at the conference on FTA Policies of Korea and Japan and Policy Implications for a Bilateral FTA, Jeongsuk Institute for Logistics and International Trade, Inha University, Incheon (May 2004).
Cheong, Inkyo, "Korea's FTA Policy and Position for a Japan–Korea FTA", a paper presented at the international conference on FTA Policies of Korea and Japan and Policy Implications for a Bilateral FTA, Jeongsuk Institute for Logistics and International Trade, Inha University, Incheon (May 2004).
Hyun, Jung Taik, "Free Trade Agreement and Korea's Trade Policy", *Journal of International and Area Studies*, Vol. 10, No. 2 (December 2003).
Krueger, Anne O., "Are Preferential Trading Arrangements Trade–Liberalizing or Protectionist?" *Journal of Economic Perspectives*, Vol. 13(4), pp. 105–124 (1999).

Park, Young Il, "Institutionalization of East Asian Economy: Is it Possible?", a paper presented at an international conference on Evaluation and Review of Korea–China Economic Development and Exchange, jointly organized by the Jeongsuk Institute for Logistics and International Trade, Inha University and the Economics Research Center of the Chinese Institute of Social Sciences, Incheon (25 November 2004).

Japan External Trade Organization (JETRO), *Prospects for Free Trade Agreements in East Asia*, Tokyo (January 2003).

## Chapter 16

# China's Ties with Southeast Asia in the Post-Cold War Era: Japan's Response

*LAM Peng Er*

China's growing ties with Southeast Asia in the post-Cold War era have stimulated Japan to seek a closer relationship with the region. Reacting to Beijing's offer of a Free Trade Agreement (FTA) to Southeast Asia and acceptance of ASEAN's (Association of South East Asian Nations) Treaty of Amity and Cooperation (TAC), Tokyo has sought to reinforce its considerable economic, political and strategic ties with Southeast Asia. If China has not stolen a march on Japan by boldly offering an FTA to Southeast Asia in November 2000, Japan would have procrastinated on forging bilateral FTAs with the region. Thus, Southeast Asia is in the enviable position of being wooed by the two great Asian powers of China and Japan.

Japan already has established a good political and economic relationship with the ASEAN states even before China's recent overtures to the region. Tokyo is the largest foreign aid donor to the ASEAN states and has disbursed more aid to Southeast Asia than any other regions in the world. Moreover, Japanese investments[1] in and trade with this region outstrip those of China's (See Appendix 1 and 2). However, the phenomenal economic rise of China coupled with a decade-long economic stagnation of Japan have made many Japanese nervous about a more powerful and assertive Beijing eroding if not displacing Tokyo's influence in Southeast Asia.[2]

---

[1] Between 1995 and 2002, the cumulative amount of Japan's FDI (Foreign Direct Investments) in ASEAN was US$27.97 billion or 15% of ASEAN's total FDI, while China's FDI to ASEAN was only US$861.7 million or 0.5%. The US was number one at US$33.95 billion or 18.2%. See ASEAN-Japan Center, *ASEAN-Japan Statistical Pocketbook 2004*.

[2] Lam Peng Er, "Trading Places? Japan-Southeast Asian Relations: The Leading Goose and the Ascendant Dragon", *Comparative Connections*, CSIS Pacific Forum (April 2002). See also Quansheng Zhao, "Japan's Leadership Role in East Asia: Cooperation and Competition between Japan and China", *Policy and Society*, Vol. 23, No.1 (2004); and Jian Yang, "Sino-Japanese Relations: Implications for Southeast Asia", *Contemporary Southeast Asia*, Vol. 25, No. 2 (August 2003).

Given the electoral clout of farmers in rural Japan and their close links with the ruling Liberal Democratic Party (LDP), it is politically difficult for the Japanese government to forge bilateral FTAs, especially over the sensitive issue of agricultural exports from Southeast Asia. Nevertheless, the China factor was a catalyst to the Japanese government pressing ahead with bilateral FTA negotiations with various ASEAN states while treading gingerly on the agriculture issue. If not for the perceived competition from China, the ruling LDP would have been most reluctant to forge FTAs with the Southeast Asian countries. However, the failure to sign FTAs might well have eroded Japan's position in Southeast Asia. But Japan is not prepared to yield to China for influence in Southeast Asia despite the political reality of offending Japanese farmers, hitherto the most loyal supporters of the LDP.[3]

Indeed, many opinion shapers in Japan are concerned about being outflanked by China in Southeast Asia and some even perceived China's FTA offer as another "threat" to Japan.[4] In November 2004, Nippon Keidanren (the most powerful and prestigious business federation in Japan and key financial contributor to the LDP) sent a delegation to four Southeast Asian countries to study the economic partnership agreements between Japan and these countries. The Japanese media noted that the Keidanren delegation was driven by a sense of crisis towards China's active economic diplomacy in the region.[5]

However, it is not inconceivable that competition between China and Japan for closer economic ties with Southeast Asia could possibly underpin an East Asian Community in the future. A China-Southeast Asia FTA and a Japan-Southeast Asia Economic Partnership could eventually overlap and link the whole East Asian region together. Rather than being sandwiched and marginalized by the two great Asian powers, the ASEAN 10 might find a new role to play in the post-Cold War era: acting as a facilitator and pivot to an East Asian Community. Saddled with the burden of history, China and Japan find it difficult to forge a bilateral FTA and a strategic partnership in the near future. However, the ASEAN 10 can provide a multilateral setting for both China and Japan to mitigate their differences in various

---

[3] The LDP has paid an electoral price for offending the farmers. In the 2004 Upper House Elections, the Democratic Party of Japan (DPJ), the main opposition party, made substantial inroads into rural Japan, hitherto the bastion of support for the LDP. Besides the FTA issue, the Japanese farmers were also unhappy with the decline of public works budget in rural Japan. Moreover, the DPJ, for the first time, campaigned hard in rural Japan and promised direct agricultural subsidies to the farmers. Interestingly, the DPJ did not criticize or propose reneging Japan's FTAs with Southeast Asia.

[4] An analyst noted: "Moreover, as China increases its military might and deepens its economic influence in Southeast Asia, a growing number of Japanese have begun to regard China as a threat". See Keizo Nabeshima, "Strengthen Sino-Japanese ties", *Japan Times*, 25 September 2002.

[5] *Sankei Shimbun*, 7 November 2004.

regional institutions, such as an emerging East Asian Community (though still admittedly nascent) via the ASEAN Plus Three framework and the ASEAN Regional Forum. Simply put, the strategic competition between China and Japan for the hands of Southeast Asia might well pave the way to an East Asian Community which will benefit all in the long run.

This paper first examines why China's relations with Southeast Asia was not perceived to be a serious challenge to Japan's position in the region during the Cold War era. Next, it analyzes the changes in the post-Cold War environment that led to Beijing's offer of an FTA to Southeast Asia and, in turn, triggered Tokyo's bid for a closer partnership with the region. Following that is an examination of Japan's multi-faceted responses to China's growing ties with Southeast Asia. These include not only the counter-offer of an economic partnership but also new Japanese initiatives to play a political role, such as peace-building in areas suffering from ethnic conflicts in Aceh and Mindanao and addressing piracy in the sea lanes of Southeast Asia, especially the Straits of Malacca. The paper concludes with an assessment of Japan's responses to China's diplomatic overtures to Southeast Asia and their implications for Tokyo's quest for a political leadership role in the region.

## The Cold War Era: China as a Non-Challenger to Japan

During the Cold War era, Japan did not perceive China to be a political and economic challenger to its economic pre-eminence and political ambitions in Southeast Asia. After its defeat in the Second World War, Japan lost its important China market and Southeast Asia emerged as an attractive alternative. War reparations and ODA (Official Developmental Assistance), which privileged Japanese big businesses for infrastructural projects, facilitated Japan's economic presence in non-communist Southeast Asia. With the rise of Japan as an economic superpower by the early 1970s, the ASEAN region became an integral part of Japan's economic production network in East Asia.[6]

According to the flying geese model of economic development, Japan, as the leading goose, moved up the ladder of technology and productivity and shed its relatively less high-tech and labor-intensive industries (in the wake of higher labor costs) to the next echelon of geese — Singapore, Taiwan, South Korea and Hong Kong. These four NIEs (Newly Industrializing Economies) would, in turn, upgrade themselves and shed their less hi-tech industries to the next tier of geese: the ASEAN countries of Malaysia, Thailand, Indonesia and the Philippines.

---

[6]Walter Hatch and Kozo Yamamura, *Asia in Japan's Embrace: Building a Regional Production Alliance* (Cambridge: Cambridge University Press, 1996).

Despite residual resentment among some Southeast Asians towards Japan's brutal military occupation of Southeast Asia from 1941–1945, the original ASEAN Five (Indonesia, Malaysia, the Philippines, Singapore and Thailand) pragmatically welcomed Japanese investments, corporations and trading companies to generate economic growth and jobs which help nation-building and domestic political stability. Indeed, Japan and not China led East Asia's economic development during the Cold War era.

Aspiring to play a political role commensurate with its economic preeminence as the world's second largest economic power, then Japanese Prime Minister Fukuda Takeo articulated the Fukuda Doctrine in 1977 that expressed Tokyo's intent to play an active political role in Southeast Asia.[7] Moreover, inspired by the so-called Japanese economic miracle, Singapore and Malaysia viewed Japan favorably as a model of rapid state-led economic development. Singapore also studied Japan's *koban* system (neighborhood police post system), cooperative management-labor practices and work improvement teams (WITS) to enhance productivity. Then Prime Minister Mahathir Mohamad of Malaysia advocated a Look East Policy (to Japan and not China). Moreover, the long-term one-party dominance of Japan's ruling LDP since 1955 was probably very attractive to the ruling parties of Singapore and Malaysia. Indeed, the 1970s and 80s were the halcyon days of Japan's role in Southeast Asia — an era when the Japanese economy appeared to be a juggernaut and Japan was much admired and emulated in the region. That China did not pose a strategic challenge to Japan in Southeast Asia during the Cold War was, in part, due to Japan's considerable economic presence and arguably Japanese "soft" power as a developmental model.

Others reasons also explain why China was not perceived to be a strategic challenger to Japan during the Cold War. Until paramount leader Deng Xiaoping advocated the Open Door Policy in 1978 and embraced various market reforms, China adhered to Marxist-Leninist-Maoist ideology, adopted a planned socialist economy and had only modest economic interaction with Southeast Asia. Even though post-Mao China pursued market reforms in the 1980s, it was not apparent then that it would emerge as an economic powerhouse. Unlike Japan, China during the Cold War was not an attractive model of development to non-communist Southeast Asia.

Moreover, China supported various communist insurgencies in Southeast Asia and was viewed with suspicion by the non-communist ASEAN states. In addition, Southeast Asian states with a sizeable Chinese ethnic minority also feared that Chinese migrants domiciled in their countries might have divided loyalties which

---

[7] Sueo Sudo, *The Fukuda Doctrine and ASEAN: New Dimensions in Japanese Foreign Policy* (Singapore: Institute of Southeast Asian Studies, 1992).

could be exploited by Beijing. However, Deng Xiaoping pragmatically dropped the Chinese Communist Party's support for sister parties in the ASEAN states[8] after he needed ASEAN's support in a United Front[9] against Vietnam — China's erstwhile friend which became an enemy after it forged an alliance with the Soviet Union (China's enemy) and invaded Cambodia (China's friend). Alarmed by the Vietnamese invasion of Cambodia, ASEAN joined hands with Beijing, Washington and Tokyo to oppose Hanoi.

Against the backdrop of the Cold War, Japan was not threatened by the rapprochement between ASEAN and China because it was on the same side of the Cold War divide against Vietnam and the USSR. Moreover, Japan, ASEAN and the US were rather muted when China militarily attacked the northern provinces of Vietnam in 1979 to "teach it a lesson" for the latter's invasion of Cambodia. Given the logic of the Cold War, Japan was also not unduly alarmed when the Chinese navy forcibly seized the Paracels from Vietnam in 1988. Simply put, Japan viewed China as a *de-facto* ally against the Soviet Union and as not a strategic challenger in Southeast Asia during the Cold War.

## Post-Cold War Milieu: China Rising, Japan Stagnating

It is astonishing that China emerged as an economic powerhouse despite the 1989 Tiananmen Incident, subsequent boycott by the West and a more challenging post-Cold War environment in which most communist regimes in the world had collapsed. With the demise of the Soviet Union and Vietnam's withdrawal from Cambodia, the US and Japan no longer regarded China as a strategic ally. In the 1990s, Tokyo became increasingly concerned about Beijing potentially emerging as its geo-strategic competitor in Pacific Asia, as evidenced by China's maritime dispute with Malaysia, Vietnam, Brunei and the Philippines over the Spratlys in the South China Sea, missile tests in the Taiwan Strait (1995 and 1996) and resumption of nuclear tests. In 1995, tension rose in the region when China and the Philippines squabbled over the Mischief Reef in the South China Sea. Then Japanese Prime Minister Murayama Tomiichi sought to discuss the issue with top Chinese leaders

---

[8] For an autobiographic account by the Secretary General of the Malayan Communist Party (MCP) on China's initial support and subsequent betrayal of fraternal parties in Southeast Asia for China's own national interest, see Chin Peng, *My Side of History* (Singapore: Media Masters, 2003). See pp. 457–460 on Chin Peng's meeting with Deng Xiaoping when the latter informed him the rationale of terminating aid to the MCP. Apparently, Lee Kuan Yew was instrumental in persuading Deng to give up support for the CCP's fraternal party in Malaysia.

[9] On China's United Front with Southeast Asia, see Lam Peng Er, "China's United Front Doctrine and Her Foreign Policy Towards Thailand", Honors Year Academic Exercise, Department of Political Science, National University of Singapore (1984).

on behalf of the Philippines, but was rebuffed by Beijing, which felt that Tokyo should not intervene in the Spratlys dispute because it was not a claimant state.[10]

Tokyo's strategic response to a more assertive China was to forge an even tighter alliance with Washington and closer strategic relations with Southeast Asia. The 1997 Defense Guidelines obliged Japan to provide logistical support to US forces in the event of conflict in "areas surrounding Japan". This can be interpreted as Japan's commitment to its US ally if war were to break out in the Korean peninsula or the Taiwan Strait — both theaters critical to China's national interest. It is also not inconceivable that the Guidelines may be extended to the South China Sea if the US chooses to intervene in a hypothetical conflict among claimant states, especially if oil and gas are discovered in the vicinity of the Spratlys. Any conflict in the South China Sea between the claimant states may possibly disrupt Japan's vital sea lanes of communications because at least 70% of its oil tankers traverse in these waters.

While Japan closed an eye to China flexing its military muscles against Vietnam during the Cold War era, it is unlikely to do so in the post-Cold War era because China and Japan are no longer de facto allies. Besides strengthening its alliance with the US, Japan has also reinforced its political and security ties with Southeast Asia. In 1997, then Prime Minister Hashimoto Ryutaro proposed that Japan and the ASEAN states hold annual top-level summits and also bilateral security discussions.

In actuality, the greatest threat to Southeast Asia in the 1990s was not a traditional military one but the Asian Financial Crisis of 1997–1998 that resulted in political change and mass sufferings in East Asia (South Korea, Thailand and Indonesia). Many interpretations abound about the causes of the Asian financial, meltdown, but a key reason was the rapid inflow and outflow of speculative slush funds, which destabilized regional currency markets and triggered the collapse of the Thai baht followed by other Asian currencies, especially the Indonesian rupiah.[11]

Initially, Japan suggested the establishment of an Asian Monetary Fund (AMF) but the scheme was shot down by the US, which apparently feared that it would lead to the formation of a yen bloc in East Asia at the expense of the US dollar as the global currency. Tokyo then offered a rescue package of US$ 63 billion to needy East Asian countries to stabilize their economies. Nevertheless, Japan was roundly criticized in Southeast Asia for caving in to the US veto over the AMF proposal and not disbursing its package fast enough to rescue the regional economies,

---

[10] Lam Peng Er, "Japan and the Spratlys Dispute", *Asian Survey*, Vol. 36, No. 10 (October 1996).
[11] Lam Peng Er, "The Asian Financial Crisis and Its Impact on Regional Order: Opening Pandora's Box", *Journal of Pacific Asia*, Vol. 6 (2000).

while China won praises for not devaluing the yuan (which could have triggered another round of competitive and ruinous devaluation among Asian currencies). It is indeed ironical that Japan was criticized for not doing enough despite its large rescue package while China won praises for not really doing anything at all during the Financial Crisis. The Chinese economy remained resilient during the Asian Financial Crisis and continued its inexorable rise after the crisis was over.

Cognizant that certain politicians and opinion shapers in the US and Japan considered a rising China to be an economic if not a military threat to the region, Beijing shrewdly wooed the ASEAN countries with an FTA. By first proposing an FTA to Southeast Asia in November 2000, then Prime Minister Zhu Rongji sought to demonstrate Beijing's peaceful intent by offering this region a stake in China's expanding domestic market. Zhu's bold proposal may be interpreted as an approach to allay the fears of Southeast Asian countries that China's rise would pose a "threat" to them and obviate any future attempts by the US to mobilize these regional states to contain China in the post-Cold War era.

While the US and Japan share a common concern about the future trajectory of a rising China, Japanese psychology towards their big Asian neighbor is further complicated by the angst that China's phenomenal economic rise is displacing Japan (mired in a decade-long economic stagnation) as the most important Asian power in the region. While many Japanese do not believe that a rising China would pose a direct military threat to their nation, some are anxious that the Mainland's plentiful, skillful and considerably cheaper labor force would hollow out Japan's manufacturing base as factories relocate to China.[12]

Other complicating issues include territorial disputes in the East China Sea and the unresolved issue of history. The Chinese continue to harp on the lack of a sincere apology from Japan over its military invasion and past atrocities in China, the revision of history textbooks, the denial of the Nanjing massacre and top politicians' visit to the Yasukuni Shrine; the Japanese believe that the CCP leadership cynically stirs anti-Japanese sentiments and fans Chinese nationalism to bolster the legitimacy of the communist regime by using the history card. An argument can be made that the fear of China's rise, the angst over Japan's stagnation and the resentment towards persistent anti-Japanese sentiments (despite granting generous ODA to China) have spilled over to Japanese outlook towards China's forays in Southeast Asia, long considered to be Japan's economic backyard.

---

[12] On the so-called China threat to Japan, see Amako Satsohi, ed., *Chugoku wa kyoi ka [Is China a Threat?]* (Tokyo: Keso shobo, 1997). See also Koubun Ryosei and Wang Jisi, eds., *The Rise of China and a Changing East Asian Order* (Tokyo: Japan Center for International Exchange, 2004).

Two additional examples can be cited to illustrate the competitive mentality of Japanese policymakers towards China in Southeast Asia. In December 2003, Prime Minister Koizumi announced that he would sign a key security treaty, the TAC, with ASEAN that had earlier been snubbed. Japan made an about-turn because China signed the treaty in October 2003 at the ASEAN summit in Bali. Observers noted that "Japan eventually decided to sign the treaty partly because it feared its role in Asian affairs would be dwarfed by growing powers India and China".[13] Another episode was concerning the China-ASEAN foreign ministers' meeting in June 2004. The Japanese media noted the foreign ministers' rhetorical declaration of a "strategic partnership" and chose to interpret China's closer ties with ASEAN as an attempt to constrain the US and Japan.[14]

### Japan's Responses to China's Overtures

A caveat should be made about Tokyo's response to Beijing's "smiling" offensive in Southeast Asia. While China's inroads have given impetus to Japan raising its ante in the region, the latter already has established a strong economic and diplomatic presence in Southeast Asia long before Zhu made his bold FTA proposal. Japan's political activism in this region has been driven by at least two factors. First, Tokyo has long desired to play an active political role commensurate with its status as the world's second largest economy. Second, after Japan was roundly criticized for passively offering only money (in actuality a hefty US$13 billion) to the allied efforts in the First Gulf War triggered by Iraq's invasion of Kuwait, Japanese leaders have sought to play a more active role in international affairs.

In the case of Southeast Asia, Japan has participated in UN peacekeeping operations in Cambodia, brokered a peace deal between the then warring parties of co-Prime Ministers Hun Sen and Prince Ranariddh, and offered ODA incentives to the Burmese military junta to release and hold political talks with Aung San Suu Kyi, the Nobel Peace Laureate. In this regard, Japan already has a diplomatic niche in Southeast Asia and, not surprisingly, views China as a challenger to its established position in the region.

In 2001, Zhu Rongji again proposed a China-ASEAN FTA which was well received by the ASEAN countries. It is no coincidence that Prime Minister Koizumi Junichiro proposed a comprehensive partnership between Japan and Southeast Asia shortly after the Chinese proposal. In January 2002, Koizumi in a keynote speech

---

[13] "Koizuni to Ink ASEAN Security Pact", *Asahi Shimbun*, 12 December 2003.
[14] *Sankei Shimbun*, 22 June 2004.

in Singapore marking his trip to a few Southeast Asian countries reaffirmed his country's commitment to Southeast Asia.[15]

The Japanese media commented: "China will be an invisible but prominent guest in Japan's forthcoming talks with the ASEAN countries. It is a matter of prime importance for both Japan and ASEAN to think about what kind of relationship they should build with China, a nation that has emerged as a new power in Asia on the strength of its rapid economic growth. While they are watching its vast market with keen interest, the ASEAN countries remain wary about China. This gives Japan good reason to play a key role in determining what kind of relationship Japan and the ASEAN nations should establish with China".[16] Funabashi Yoichi, a prominent commentator in Japan, quoted a "high-ranking White House official" as saying: "Isn't Japan the one that should be proposing a free trade agreement with ASEAN? The decline of Japanese economic power is causing the expansion of China's economic influence in Asia. Such a trend not only affects Japan but could also undermine US strategic interests in Asia as a whole".[17]

Due to the sensitive nature of agriculture in Japanese domestic politics, Koizumi did not propose any concrete FTAs then and opinion shapers in Japan and Southeast Asia criticized his speech as being rather vague.[18] However, he proposed a partnership which goes beyond trade and extends to cooperation in tackling infectious diseases, the environment, human and narcotics trafficking, intellectual, educational and cultural exchanges, maritime and energy security. Interestingly, Koizumi also noted that Japan is interested in the consolidation of peace or peace building in areas of Southeast Asia which have suffered from domestic ethnic conflict. The Prime Minister specifically mentioned a Japanese diplomatic role in Aceh, Indonesia and Mindanao in the Philippines. Subsequently, Japan would follow up by offering generous ODA as incentives to contestants in Aceh and Mindanao to embark on the path to peace. Japan also organized two international conferences in Tokyo to persuade the Indonesian central government and the Acehnese separatists to embrace peace.[19]

---

[15] See Prime Minister Koizumi Junichiro's speech, "Japan and ASEAN in East Asia: A Sincere and Open Partnership", Singapore, 14 January 2002.
[16] "Koizumi's Asia Trip is Timely", *Daily Yomiuri On-Line*, 9 January 2002, http://www.yomiuri.co.jp/newsse/20020109wo81.htm (accessed 9 January 2002).
[17] Yoichi Funabashi, "New Geopolitics Rages over Various Parts of Asia", *Asahi Shimbun*, 15 January 2002, http://www.asahi.com/ptenglish/ptop-ed/K2002011500444.html (accessed 16 January 2002).
[18] "Koizumi's Japan-ASEAN Vision Too Vague, Say Tokyo Media", *Straits Times*, 21 January 2002.
[19] Lam Peng Er, "Japan's Peace-building Diplomacy in Aceh", *Asian Ethnicity*, Vol. 5, No. 3 (October 2004).

Even though the peace talks on Aceh subsequently collapsed and civil war had erupted again in Aceh, the Japanese attempt to play a mediatory role is significant because it demonstrates that Tokyo is prepared to positively engage in the domestic politics of a Southeast Asian country. While China has caught the eyes of many by offering an FTA to Southeast Asia, it has yet to do what Japan has done in the post-Cold War era — involvement in the domestic politics of Indonesia, Cambodia and Burma.

Moreover, Japan has gone beyond talking about FTAs and has indeed forged its first bilateral FTA with Singapore in 2003. This FTA with Singapore was facilitated by a unique characteristic of the city state — Singapore has no significant agricultural sector and an FTA with Singapore would not provoke a political backlash from Japanese farmers. By late 2004, Tokyo has succeeded in forging FTAs with Mexico and the Philippines, and is in the midst of negotiations with Thailand and Malaysia.

In the FTA between Tokyo and Manila, it was agreed that Filipino nurses and caregivers would be allowed to work in Japan and extend their visas if they can meet Japanese professional standards, while the Philippines agreed to reduce tariffs on steel imports.[20] Moreover, tariffs on bananas and pineapples from the Philippines would be reduced while the sensitive issue of sugar export was side-stepped because of its sensitive nature among the sugar planters of Okinawa. The FTA negotiation between Japan and Thailand is likely to be even trickier because of the issue of rice exports. Ironically, thanks to pressure from China, Japan has done more on FTAs with Southeast Asia than China's track record thus far. While China has only talked about forging an FTA with Southeast Asia only in 2010, Japan has already forged FTAs with Singapore and the Philippines and soon with Thailand and Malaysia. Simply put, Japan has overtaken China on FTAs with Southeast Asia.

Japan has also exercised diplomatic initiatives on the issue of maritime security in the region. The Straits of Malacca is an economic lifeline for a Japan dependent on imported oil and is increasingly important too for China whose economic rise will necessitate an increasing demand for energy imports. In April 2000 and October 2001, Japan organized international conferences in Tokyo on maritime safety in East Asia to address the menace of piracy.[21] Tokyo has also proposed the deployment of its very capable coastguard vessels in collaboration with the littoral states in Southeast Asian waters. The Japanese Ministry of Foreign Affairs also

---

[20] See "Japan, Philippines Clear Steel Hurdle, Reach FTA Accord", *Japan Times*, 20 November 2004.
[21] See The Ministry of Foreign Affairs of Japan, "Japan's Efforts to Combat Piracy and Armed Robbery Against Ships", November 2001, http://www.mofa.go.jp/region/asia-paci/aswan/relation/piracy.html (accessed 21 July 2004).

noted that the Japan Coast Guard has conducted exercises with India, Malaysia and the Philippines.[22] An analyst commented: "In promoting the anti-piracy program, Japan wants to reassert its waning influence in the region as a counterbalance to China".[23]

In November 2004, a new Regional Cooperation Agreement on Anti-Piracy in Asia was forged in Singapore. In actuality, this was a Japanese initiative to combat piracy and deter potential terrorist strikes in the Straits of Malacca and its surrounding seas. The agreement links the ASEAN 10 countries along with Bangladesh, China, Japan, India, South Korea and Sri Lanka through a central network located in Singapore. Apparently, there is the regional concern not only with piracy but also the conceivable possibility that militants linked to Al-Qaeda regional affiliate Jemaah Islamiah could launch an attack on one of the 50,000 ships that pass through the Straits of Malacca every year.[24]

In the 2004 draft of a new National Defense Program Outline drawn up by the Japanese government, it even mentioned that Japan should provide "assistance to Southeast Asian nations to fight terrorism and piracy by selling them old warships".[25] It remains to be seen whether the navies of Southeast Asia do find old Japanese warships to be an attractive buy. Nevertheless, the Program Outline reveals the mentality of Japanese defense planners who take Southeast Asia into their consideration.

One should not exaggerate the geo-political influence of China in Southeast Asia. While China emerges as a great regional power underpinned by its strong economic growth, it has yet to exercise diplomatic initiatives in this region to the extent that Japan has. Other than the presence of the US superpower in Southeast Asia, no countries, including China, can currently rival the manifold influence of Japan in terms of ODA, economic linkages, peace-building in ethnic-torn areas, political bridging role in Burma and Cambodia, and security initiatives such as maritime cooperation in the Straits of Malacca and its vicinity.

While Japan may have the image of a pacifist country framed by the war-renouncing Article 9 in its Constitution, it is for all practical purposes a "normal country" in Southeast Asia. Earlier, Tokyo has already deployed its Self Defense Force (euphemism for a military) for UN Peacekeeping in Cambodia, and it now has a naval presence in the Straits of Malacca and the South China Sea. The *Yomiuri*

---

[22] *Ibid.*
[23] See Mark Valencia, "Joining Up with Japan to Patrol Asian Waters", *International Herald Tribune*, 28 April 2000.
[24] See *Sankei Shimbun*, 23 November 2004.
[25] See "Draft Report Eyes Easing Arms Export Restrictions", *Daily Yomiuri On-Line*, 17 November 2004, http://www.yomiuri.co.jp/newse/20041117wo01.htm (accessed 17 November 2004).

*Shimbun* noted the formidable naval presence of Japan, stretching from Japan to the Persian Gulf: "On March 15 (2004), there was an unusual scene along the Japan-Gulf maritime oil route — ten Maritime Self-Defense Force vessels were on the route at one time. Among them were the transport ship Osumu, carrying about 70 vehicles for Ground Self-Defense Force's use in Iraq, and the destroyer Murasame, which had just arrived in Kuwait. The destroyers Samidare and Myoko and the supply vessel Tokiwa were in the Indian Ocean as part of an international anti-terrorism mission, supplying fuel to the vessels of allied navies. The Hatakaze and two other destroyers were on a training mission in the South China Sea, and two vessels were in the Strait of Malacca on their way to take over the Indian Ocean fuel supply mission".[26]

While Japan can deploy a blue water fleet to the Straits of Malacca and the Persian Gulf, it appears that China presently does not have the desire or capability to project its naval power to the Straits of Malacca. However, in the long term, the Chinese navy will be buoyed by its country's inexorable economic growth and eventually be able to project formidably its power to the Spratlys and the South China Sea.

After the catastrophic tsunami of December 2005 that struck not only the coasts of Indonesia, Malaysia and Thailand in Southeast Asia but also certain Indian Ocean states and the eastern seaboard of Africa, Japan sent around 1,000 troops to Aceh, Indonesia, its largest humanitarian relief deployment ever in the post-war era. Tokyo also committed US$500 million for tsunami relief and reconstruction. Despite the anticipated rise of China in the region, Beijing neither have the financial resources to match Tokyo's financial and material aid to Southeast Asia nor the military capacity and political will to rapidly dispatch troops to the region for disaster relief.

## Between China and Japan: Southeast Asia in the Years Ahead

What are the implications of China's rise on Japan's relations with Southeast Asia? Will China trade places with Japan for influence in Southeast Asia? Conceivably, there are at least five possible scenarios for the distribution of power in Southeast Asia in the next few decades:

- US superpower dominance;
- China as the once and future Middle Kingdom;
- A regional balance of power model;

---

[26] See *Daily Yomiuri On-Line*, 22 September 2004, http://www.yomiuri.co.jp/newse/20040922wo01.htm (accessed 24 September 2004).

- A condominium between China and Japan; and
- Rising multilateralism and a nascent East Asian Community.

*US Superpower Dominance*

Even though China enjoys sustained economic growth, its geo-political influence will remain relatively limited in Southeast Asia because of the strong US military presence and security networks with certain Southeast Asian states. Washington has formal alliances with Bangkok and Manila and a quasi-alliance with Singapore. Given its primary concern on the war against terrorism, the US will support the governments of Thailand and the Philippines if Muslim rebels in southern Thailand and Mindanao are found to have linkages with Jamiah Islamiah and other global Islamic radical networks. Moreover, Singapore offers berthing facilities at Changi Naval Base to American aircraft carriers.

While the US plans to downsize its troops in East Asia, this move will be compensated by revolutionary advances in military technology, including satellite intelligence gathering, command and control systems, increased firepower and better mobility of its forces to regional hotspots. Seen from this perspective, the rise of China will not pose a serious challenge to Japan in Southeast Asia because Japan's ally, the US, will maintain its dominance in this region. While China will undoubtedly become a more important trading partner to the ASEAN countries, these economic linkages will not necessarily translate into geo-political dominance, given the continual US superpower presence in the region.

*China as the Once and Future Middle Kingdom*

In this hypothetical world, the US is no longer the sole superpower. After bleeding from its quagmire in Iraq and subsequently Iran, the US significantly reduced its military presence in East Asia. It quit South Korea and keeps only a small number of planes, ships and marines in Japan. China becomes the main engine of growth for East Asia. Taiwan undergoes de facto reunification with China as its economy becomes intertwined and dependent on the Mainland. Students from Korea, Japan, Taiwan and Southeast Asia flock to study in China rather than the US. Chinese state and society begin to exercise "soft power" in the region as more East Asians become interested in the Chinese language, philosophy, fashion, movies and music. China's burgeoning markets remain open to the US, Japan and the rest of East Asia. Beijing maintains its "peaceful rise" policy towards the region.

It also observes formal equality with the states in East Asia, but the geopolitical reality is that none of the East Asian states including Japan would act against the core national interests of China. Thus China emerges as a regional hegemon and other East Asian states pragmatically accept the new order (*Pax Sinica*) to enjoy the benefits of trade and also because Beijing has no interest in either territorial conquests or interfering in the domestic politics of these countries. Moreover, China does not seek to impose the traditional tributary system of the pre-Western imperial era but accepts the formal equality of states in the Westphalian system. But the reality of power is that China has become the top dog and peacefully displaces Japan as the key player in Southeast Asia.

*Regional Balance of Power*

China becomes the main economic powerhouse in East Asia. Japan remains an important economy though held back by demographic decline and a rise in the number of its retirees. Beijing becomes more assertive over Taiwan and territorial disputes in the East and South China Sea. To ensure the security of its sea lanes and Middle Eastern oil supply, the Chinese navy seeks a naval presence in the Straits of Malacca, the Indian Ocean and the Persian Gulf. Alarmed and suspicious of the intentions of China, the ten ASEAN states band even closer together and seek to strengthen their ties with the US, Japan and India. In this scenario, the rise of China triggers a countervailing balance of other powers. Moreover, the ASEAN 10 encourages Japan to remain engaged in Southeast Asia to balance China's influence.

*A Sino-Japanese Condominium*

A political vacuum emerges in East Asia after the US adopts an isolationist policy in the aftermath of its failed policies in Iraq, Iran and the Middle East. The US-Japan Alliance unravels and Tokyo decides to cast its lot with Beijing. The adage goes: if you can't beat them, join them. Both countries bury their historical hatchets after Japan has built a new war memorial and its top political leaders no longer visit Yasukuni Shrine. China agrees to a future-oriented relationship and no longer whips up anti-Japanese sentiments among its people. Since Japan has played the junior partner to the US before, it is not too difficult to play the junior partner again to the Chinese after a period of psychological adjustment. The Koreans and Southeast Asians fatalistically accept and bandwagon with the new configuration of power in East Asia: a bi-hegemony of China and Japan. To avoid friction, China and Japan have carved their respective spheres of influence in Southeast Asia. Vietnam, Laos, Cambodia, Thailand and Burma belong to the Chinese sphere while Malaysia, Singapore, the Philippines, Brunei and Indonesia fall into the Japanese sphere.

*An East Asian Community*

Contrary to the cynics who believe that an East Asian Community (EAC) is just a piped dream, the East Asian countries doggedly walked the arduous road of institutionalizing a regional community. Even though China and Japan did not forge a historical reconciliation in the early years of community building (like France and Germany did in Europe), they were linked together within the multilateral framework of ASEAN Plus Three. The trade pacts between Japan and various Southeast Asian countries followed by similar ones between China and Southeast Asia enhanced regional cooperation, linked East Asia together and subsequently paved the way for an even closer relationship between Beijing and Tokyo.

Meanwhile, the Democratic Party of Japan (DPJ) ended the one-party dominance of the LDP and succeeded in buying off the shrinking numbers of aging farmers by disbursing direct subsidies to them as compensation for market opening. Moreover, the top leaders of the DPJ did not visit the Yasukuni Shrine and removed a bone of contention between Beijing and Tokyo. Japan also regained its confidence when its economy recovered, in part, due to its rising exports to China, which has overtaken the US as Japan's most important trading partner.

Over the next few decades, China stemmed off rural unrest by improving the lot of the peasants, strengthened its legal system, became politically more inclusive as the level of mass education increased significantly. The Chinese Communist Party even emerged as a dominant party, like the People's Action Party of Singapore, in a Chinese-style, multi-party system. By then, the so-called eight satellite parties have been given a more autonomous role to play. The CCP is communist only in name and has succeeded in representing all important social classes, including the capitalists, the middle class and farmers. Thus the ideological distance between China and other non-communist East Asian countries has narrowed quite considerably. Moreover, the economies of the region become so intertwined that war is no longer a serious option because it will be mutually ruinous. Culturally, the East Asians celebrate their own regional musical potpourri, television serials, fashion, arts and film festivals, karaoke contests and a bi-annual East Asian Games. In this scenario, China's growing ties with Southeast Asia is really not a problem for Japan because it is part of a larger process which underpins the emergence of an East Asian Community.

## Conclusion

The impact of Beijing's overtures to Southeast Asia on Tokyo is contingent on whether the Japanese feel that they are being squeezed in a sub-region long considered to be their economic backyard. The Japanese should feel reassured by the continual presence of its ally, the US, as the dominant military power in the region and the fact that the ASEAN countries generally seek balance in their relationship with the great powers. Given the recent recovery of its economy, Japan remains a key economic player not only in Southeast Asia but also the world.

Insofar as the US remains anchored in the region, international relations are not totally unpredictable. The most likely scenario is a balance of power model in East Asia, with the US-Japan Alliance as its lynchpin against the backdrop of a rising China with the ASEAN countries seeking to be on good terms with all great powers. However, the possibility of an emerging EAC should also not be ruled out because the idea, ideal and identity of an EAC do resonate among many opinion shapers in the region. Most unlikely scenarios will be a Chinese tributary system and a Sino-Japanese condominium because these arrangements are unattractive and unacceptable to Japan and the Southeast Asian countries. One can envisage that if China becomes insensitive to and intrusive towards the Southeast Asian countries, they would instinctively band together and welcome a closer relationship with Japan.

Many Japanese are indeed nervous about the rise of China in general and its expanding role in Southeast Asia in particular. But the reality is that Japan is already so deeply entrenched in this region and will not be easily dislodged by China. While Beijing talked about an FTA, Japan has already forged FTAs with Singapore and the Philippines. Moreover, China still has a long way to go before catching up with the volume of Japanese trade and investments in ASEAN states. Thus far, no Asian powers except Japan have exercised diplomatic initiatives such as peace building, mediation in domestic conflict and maritime security arrangements in Southeast Asia. In the next 50 years, Japan is likely to remain much more affluent and technologically advanced than China. I therefore conclude that Japan will very much remain an important actor in Southeast Asia despite the rise of China.

## Appendix 1: China and Japan's Exports to ASEAN-10, 1996–2002

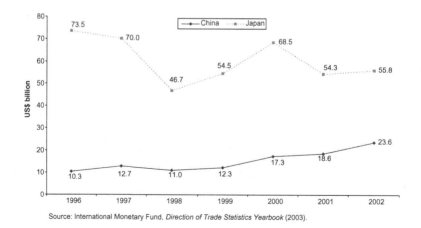

Source: International Monetary Fund, *Direction of Trade Statistics Yearbook* (2003).

## Appendix 2: China and Japan's Imports from ASEAN-10, 1996–2002

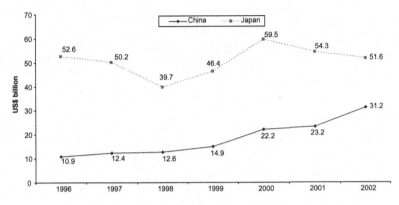

Source: International Monetary Fund, *Direction of Trade Statistics Yearbook* (2003).

# Index

Action Plan, 13, 65, 72, 107, 141, 232, 276, 278
agriculture, 37, 67–69, 95, 100, 107, 117, 119, 168, 188, 299, 300, 303, 304, 310, 317
AIA Framework Agreement, 140, 141, 145, 146, 150, 151, 153
anti-dumping measures, 48, 106
anti-epidemic cooperation, 66
anti-piracy cooperation, 6, 75, 78, 79, 82–84, 86, 88
APEC, 68, 102, 149, 293, 296, 297
arbitral tribunals, 55
arbitration, 53, 55
archipelagic waters, 87, 238
armed robbery at sea, 254, 278
ASEAN Regional Forum (ARF), 83, 289, 311
ASEAN summit, 31, 39, 53, 63, 70, 71, 316
ASEAN Task Force on AIDS, 70
ASEAN+3, 11, 31, 32, 62–67, 69–73
ASEAN-China relations, 11, 27, 33–37, 40–42, 44, 45, 50, 53, 55, 280
Asian Development Bank (ADB), 107, 183, 193
Asian financial crisis, 29, 61, 155, 208, 209, 224, 225, 314, 315
Association of Southeast Asian Nations (ASEAN), 3–14, 17, 19–21, 25–48, 50–55, 59–66, 68–73, 75, 82–84, 86, 89, 93–98, 100, 102–105, 108, 109, 111–119, 123–131, 133–157, 180, 183, 185, 190–195, 199, 200, 203–205, 207–210, 213, 225, 226, 253, 258–261, 263–265, 271–285, 287, 288, 293–299, 303–305 309–317, 319, 321–324
Australia, 30, 112, 113, 124, 153, 154, 184, 205, 211, 253, 257, 293, 296, 300

balance of trade, 172, 175
Bandung Declaration, 10
Bangkok, 35, 63, 65–68, 123, 124, 321
banking, 37, 107, 169, 297
Beibu Bay, 220, 221
Beijing, 12, 17, 31, 35, 59, 62, 65, 67, 68, 70, 71, 139, 245, 309, 313–315, 320–324
best scientific evidence available, 228, 230, 232, 239
bilateral arrangements, 82, 238
bilateral investment treaties (BITs), 141, 144, 146, 150, 151
bio-technology, 37
biodiversity, 180, 188, 228
Boao Forum, 13
border market, 161, 175
border regions, 8, 161–164, 167–171, 174–177
border trade, 8, 161–176
Brunei, 6, 10, 37, 94, 96, 118, 127, 129, 133, 134, 144–146, 150, 156, 207, 208, 221, 223, 246, 248–250, 263, 268, 313, 322
business environment, 163, 298

Cambodia, 5, 6, 9, 29, 37–39, 65, 66, 70, 94, 96, 114, 116, 117, 127, 133, 135, 148, 155, 175, 180–182, 187, 188, 192, 207, 211, 219, 221, 226, 249, 263, 265, 266, 268, 284, 313, 316, 318, 319, 322

Central Party School, 13
Charter of the United Nations, 10, 39, 42, 43, 122, 252
Cheng Cheng-kung, 77
China, 3–14, 17–19, 21, 22, 25–47, 50–55, 59–72, 75–78, 82, 84, 86, 89, 93–98, 100, 102–105, 107–109, 111, 113–119, 123, 125–131, 133, 134, 136, 138–157, 161–170, 172–176, 179, 180, 182–186, 190–195, 199–210, 213, 219–221, 223, 225–227, 233–236, 240–242, 245, 246, 248–250, 253, 255–261, 263–265, 268, 271–274, 276–285, 287, 288, 293–301, 303–305, 309–324
China factor, 310
China National Offshore Oil Corporation (CNOOC), 213, 250
China threat, 13, 29, 32, 40, 115, 134, 147, 155, 298, 315
China-ASEAN Comprehensive Economic Cooperation Agreement, 3, 7
China-ASEAN Free Trade Agreement (CAFTA), 7, 8, 14, 29, 37, 45–50, 52, 96, 111, 113–116, 118–120, 123, 125–131, 133–140, 142, 144, 146, 147, 149–151, 153–157
China-ASEAN relations, 3–5, 8, 10–13, 17, 32, 45, 47, 50–53, 97, 293
China-ASEAN Trade Negotiation Committee, 139
Chinese customs, 13
Chinese delegation, 68, 139, 142, 150
Chinese fishermen, 222, 241–243
Chinese government, 47, 55, 164, 185, 199, 201, 206, 209, 241, 256
Chinese investment, 143, 144, 154, 210
Chinese market, 12, 115, 145, 154, 294
Chinese people, 234, 241, 242
Chinese Taipei, 97, 98
Chinese territory, 180, 192
Chinese-Foreign Equity Joint Venture Law, 200, 201

claimants, 10, 246, 250, 254, 255, 258–260, 314
Closer Economic Partnership Arrangement (CEPA), 31, 98, 104
coastal states, 10, 227, 230, 231, 235, 286–288
Cold War, 12, 71, 79, 133, 224, 311–314
common concern, 35, 315
common interest, 38, 243, 260, 272, 275, 281, 282, 289
communications, 37, 82, 86, 87, 107, 119, 152, 271, 274, 275, 314
compensation, 55, 104, 141, 193, 266, 270, 285, 286, 301, 323
competition, 4, 9, 135, 156, 165, 173–175, 199, 203, 213, 297, 309–311
comprehensive cooperative framework, 4, 61
Comprehensive Economic Partnership Agreement, 153
conciliation, 53, 55
condominium, 252, 321, 322, 324
conflict, 8, 9, 12, 39, 86, 99, 180, 189, 191, 192, 194, 195, 221, 226, 246, 250, 258, 271, 272, 311, 314, 317, 324
consensus, 32, 66, 68, 84, 103, 105, 125, 138, 142, 144, 189, 239, 281, 305
conservation, 10, 183, 184, 219, 220, 222–224, 226–234, 236–243
conservation measures, 233, 234
consultation, 33, 35, 36, 38, 41, 51, 53, 55, 61, 65, 70, 104, 141, 230, 231, 257, 260, 276, 281, 282
consultation partner, 225
continental shelf, 87, 248, 250–253, 259
continental shelves, 220, 221, 235
contracting parties, 41, 44, 45, 96, 99–102, 111, 127, 137, 273
cooperation, 3–6, 8–11, 13, 14, 31–47, 50, 52–54, 59, 61–72, 81–84, 88, 89, 93, 96, 97, 103, 105, 107–109, 113–116, 119, 122, 123, 125, 126, 134–139, 145–147, 168, 176, 177, 180, 183, 192–195, 213, 220, 224–226, 230, 231, 233, 234, 236–242, 248, 251, 253–255, 257, 259,

260, 264, 265, 267, 270–289, 298, 305, 317, 319
countervailing measures, 48, 52, 105, 106
cross-border trade, 162, 175
cross-strait, 98, 220–223, 226, 240–242, 298
customs union, 49, 97, 99, 103, 104

dams, 9, 179–183, 185–188, 190–195
decision-making process, 142, 143
Declaration on the Conduct of the Parties in the South China Sea, 4, 10, 13, 39, 43, 45, 52, 54, 84, 124, 147, 226, 254, 259, 260, 278, 283
democracy, 41, 115
Deng Xiaoping, 202, 255, 312, 313
developing countries, 18, 21, 46, 70, 102, 109, 112, 143, 146, 148, 152, 155, 157, 199, 204, 205, 207, 213, 258, 300
development gaps, 29, 46, 47
development strategies, 17, 20
dialogue partner, 34, 35, 54, 225, 276, 277
diplomacy, 12, 63, 72, 310
diplomatic relations, 134, 199
diplomatic relationship, 34
disease control, 6, 59, 60
dispute settlement, 7, 39, 41, 51–53, 55, 101, 104, 106, 108, 119, 130, 131, 141, 171, 238–240
Disputes Settlement Body (DSB), 49, 104
distressed vessel, 272
domestic laws and regulations, 153, 221, 227, 237
domino effect, 295–297

Early Harvest Program (EHP), 37, 48, 105, 116, 127, 128
East Asia, 7, 12, 17–19, 23–25, 32, 61, 111–114, 155, 246, 250, 255, 260, 261, 295, 297, 298, 305, 306, 309, 311, 314, 317, 318, 321–324
East Asian Community, 310, 311, 322, 323
East Asian economies, 4, 20, 22, 28

East Asian region, 3, 4, 11, 27, 32, 253, 293, 297, 310
East China Sea, 233, 253, 254, 256, 257, 259, 315
East Timor, 253
ecological balance, 236
economic contact, 162, 174, 175
economic development, 3, 4, 8, 13, 14, 20, 46, 47, 86, 87, 109, 134, 147, 151, 154–156, 169–171, 174, 176, 180, 183, 189, 190, 195, 200, 224, 263, 293, 311, 312
economic growth, 3, 4, 14, 17, 19–21, 25–32, 35, 60, 134, 168, 169, 199, 209, 213, 250, 258, 277, 278, 312, 317, 319–321
economic integration, 3, 4, 9, 17, 18, 29, 31, 32, 46, 101, 102, 108, 109, 112, 134, 136, 155, 213, 224, 293, 295, 298, 305
economic interdependence, 21
Economic Partnership Agreement, 294
economic reform, 4, 17, 18, 28
economic relations, 8, 17, 29, 46, 98, 102, 114, 144, 145, 149, 155–157, 175, 177, 199, 201
economic relationship, 154, 199, 241, 242, 309
economic transactions, 52
ecosystem, 193, 219, 223, 228
EEZ, 87, 220, 221, 250, 251
energy, 13, 37, 85, 107, 179, 182, 183, 185, 246, 248, 249, 252, 258, 263, 317, 318
environment, 3, 12, 13, 37, 84, 89, 107, 122, 125, 147, 174, 175, 195, 213, 223, 229, 238, 240, 245, 278, 284, 288, 311, 313, 317
environmental impacts, 9, 195
environmental issues, 9, 179, 183, 185, 189, 194, 195
epidemic containment, 59
equality, 35, 40, 45, 120, 321, 322
EU, 18, 25, 30, 102–104, 125, 143, 205, 269, 270, 286–288, 300
European Parliament, 269, 270

exploitation, 11, 86, 183, 194, 209, 223, 225, 227, 229, 230, 235, 237, 250–252, 255, 258, 259, 286
exploration, 11, 223, 227, 246, 248–251, 255–260, 286
extradition, 6

fact-finding committee, 53
FDI (foreign direct investment), 4, 8, 9, 12, 14, 17, 18, 25–28, 31, 32, 135, 143, 144, 150, 155, 156, 199–211, 213, 294, 297, 309
FDI inflow, 9, 25, 31, 153–155, 200–205, 207–209, 213
FDI performance, 207, 208
finance, 37, 70, 107, 176, 203
financial crisis, 20, 165, 293, 305, 315
Fish Stocks Agreement, 10, 226, 227, 229, 230, 232, 234, 239
fishery, 37, 107, 226, 230–233, 236, 240–243, 256, 269
fishery conservation, 228, 233
fishery management, 220, 223, 231, 233, 236, 256
fishery resources, 10, 219–224, 227–229, 231–234, 236–243, 257
fishing activities, 221, 226, 228, 229, 232, 236–238, 243
fishing entity, 240
fishing grounds, 221, 222, 239
fishing industry, 78, 241, 242
fishing vessels, 220–223, 226, 232, 237, 238, 241, 242, 263, 266, 267
Five Principles of Peaceful Coexistence, 39
flying geese, 21, 311
Food and Agriculture Organization (FAO), 66, 68–70, 231–233, 237, 240
foreign exchange reserves, 18
foreign investment, 136, 145, 153, 156, 169, 170
foreign policy, 40, 305
foreign trade, 25, 67, 78, 95, 139, 161, 165–169, 175, 296
forestry, 37, 107, 183, 203

Framework Agreement on Comprehensive Economic Co-operation (FACEC), 4, 7, 36, 37, 45, 51, 53, 54, 93, 114, 115, 127, 134, 135, 138, 139, 149, 153, 281
free riders, 86
free trade area, 3, 4, 6, 29, 34, 36, 37, 49, 54, 93, 96, 97, 99, 100, 103–105, 108, 112, 114, 134–137, 142, 144, 146, 147, 153, 242
freedom, 36, 40, 124, 272
freedom of navigation, 39, 52

GDP, 18, 20, 22, 29, 60, 87, 108, 114, 169, 207, 209, 210, 300, 301
General Agreement on Tariffs and Trade (GATT), 46, 48, 49, 97, 99–106, 112, 136, 242
General Agreement on Trade in Services (GATS), 46, 101
global economic order, 297
global norms, 305
globalization, 12, 108, 134, 135, 213, 305
good neighborliness, 35, 41, 44, 280
Greater China, 31, 115
Greater Mekong Subregion (GMS) program, 183, 193
Guangxi Autonomous Region, 162, 164, 165
Gulf of Siam, 76, 235
Gulf of Thailand, 78, 253, 261
Gulf of Tonkin, 248, 257, 258

Hainan Island, 82, 248, 250
hard-law, 4, 5, 50, 52
high seas, 79, 80, 220, 226, 228, 230–232, 266
high-tech products, 26, 28
highly migratory fish stocks, 226–228, 230, 231
historic rights, 234, 235, 242
historic waters, 223, 235, 236, 238, 239, 241, 243
HIV/AIDS, 66, 70, 71
Hong Kong, 11, 18, 19, 22, 25–27, 31, 60, 62, 68, 81, 97, 98, 104, 113, 115, 119, 143,

154, 155, 168, 191, 205–207, 211, 242, 250, 263–265, 268, 311
Hu Jintao, 12, 13, 62
human resources, 37, 107, 119, 275, 284, 285
hydro-power development, 179
hydrocarbon resources, 245
hydropower projects, 9, 180, 184, 185, 191, 193

illegal actions, 171
implementation, 7, 8, 13, 36, 38, 42, 48, 54, 55, 63, 65, 72, 93, 96, 116, 119, 126, 131, 140, 164, 201, 226, 227, 233, 238, 251, 254, 258, 265, 267, 269, 270, 275–285, 297, 305
implementation measures, 5, 51–53, 281
independence, 40, 45, 72, 97, 120
India, 6, 30, 43, 54, 76–78, 113, 125, 126, 153, 200, 203, 208, 276, 296, 316, 319, 322
Indian Ocean, 87, 320, 322
Indonesia, 6, 18, 20, 22, 26, 27, 30, 34, 37, 40, 41, 60, 61, 66, 75, 78, 80, 82–88, 94, 96, 123, 127, 129, 133, 144, 149, 199, 207, 208, 210, 211, 219–221, 223, 225–227, 246, 249, 254, 257, 263, 268, 271–273, 296, 311, 312, 314, 317, 318, 320, 322
industrial restructuring, 12, 294
industrialization, 18
infectious epidemic, 62
information, 5, 37, 38, 40, 61–63, 65, 67–69, 72, 82, 83, 101, 107, 112, 119, 174, 202, 226, 228, 229, 237–239, 273, 275, 277, 279, 282, 285, 287, 288
information sharing, 6, 63, 64, 83, 193, 194
Information Sharing Center, 6
intellectual property rights, 37, 48, 105, 106
inter-governmental organization, 123, 231
interference, 40, 45
interim measure, 258
internal affairs, 40, 45, 98
international affairs, 41, 43, 134, 224, 316
international agreements, 34, 42, 120, 122, 123

international community, 69, 98, 224, 231–233, 240
International Convention for the Safety of Life at Sea, 266, 273
International Court of Justice (ICJ), 121, 122, 252
international criteria, 237
international customary law, 44
international economy, 7, 136
international law, 7, 10, 11, 33, 34, 36, 39, 40, 42–45, 47, 94, 98, 103, 111, 120, 122, 123, 125, 127, 129–131, 224, 235, 251, 252, 282
International Law Commission, 120, 123, 129
international legal personality, 7, 111, 120, 125, 129, 131
International Maritime Bureau (IMB), 79
International Maritime Organization, 11, 54, 263, 265, 267
international norms, 10
international organization, 7, 42–44, 54, 67–70, 98, 120–123, 129, 130, 227, 237, 240
international regime, 270
international relations, 42–44, 104, 246, 324
international river, 9
Internet, 19, 79
intra-regional trade, 8, 21, 23–25, 27, 46, 114, 156
investment, 7–9, 14, 17, 20, 22, 25, 37, 46–48, 52, 60, 63, 93, 103, 106, 107, 112, 116, 119, 133–157, 170, 183, 199–202, 207, 209–213, 241, 253, 281, 285, 298, 309, 312, 324
investment agreement, 136, 138, 140–147, 149–157
investment environment, 9, 141, 153, 201, 213
investment negotiation, 133, 136, 138, 142, 146, 150, 152
investment relations, 144
islands, 39, 236, 259, 260

Japan, 6, 11, 12, 14, 17–22, 25–27, 30–32, 61, 62, 65–67, 71, 83, 97, 107, 112, 113, 115, 125, 126, 136, 143, 153, 154, 183, 199, 205, 206, 211, 224, 233, 250, 256, 259, 276, 293–305, 309–324
joint development, 11, 245, 250–261
joint venture, 184, 200–202, 213, 241, 242, 251
jurisdiction, 39, 79, 87, 88, 221, 232, 252, 271

Korea, 17, 25, 27, 30, 31, 54, 113, 125, 183, 211, 233, 293–305, 321
Kuala Lumpur, 34, 36, 42, 62, 64, 68, 79, 112, 124, 278

Lancang Jiang, 179, 182, 184, 191, 192
Lancang River, 9, 190, 193, 284
Laos, 6, 8, 29, 37, 66, 71, 94–96, 133, 148, 161, 162, 166, 167, 172, 174–176, 179, 180, 182, 185, 186, 192, 263, 265, 268, 284, 322
law, 3, 10, 11, 33, 43, 45, 111, 120, 121, 143, 221, 235, 242, 265, 305
law enforcement, 80, 82, 87–89, 237–239, 242, 243
legal basis, 4, 7, 37, 49, 52, 98, 108, 137, 226, 251
legal concept, 11
legal documents, 10, 42, 50, 51, 53, 235
legal effect, 50, 130
legal framework, 3, 5, 34, 45, 51, 136
legal personality, 96, 120–126, 129–131
legal regime, 47, 51, 52, 103, 125, 138, 149, 152
legal rules, 111
legal status, 94, 98, 123, 234, 235, 240
legal subjects, 33
liability, 129, 131, 270, 285, 286
Liberal Democratic Party (LDP), 310, 312, 323
liberalization, 8, 29, 47, 48, 93, 96, 105, 108, 109, 112, 116, 118, 119, 136, 138, 140–143, 145, 146, 148–152, 175, 281, 298, 300, 301, 303

littoral states, 83, 220–227, 234–240, 243, 318
living resources, 223, 224, 227, 231, 233–235, 239, 240, 242, 243, 251
LOS Convention, 251–253
low-tide elevations, 236
Lower Mekong Basin, 180, 187

Macao, 68, 77, 97, 98, 113, 115, 119, 205, 211, 242
Malacca, 76, 82, 84, 85
Malay Peninsula, 76
Malaysia, 6, 10, 18, 20, 22, 26, 27, 30, 34, 36, 37, 40, 42, 60–63, 71–73, 75, 78, 81–83, 85, 88, 94, 96, 97, 117, 123, 127–129, 133, 144, 145, 150, 151, 199, 207, 208, 210, 213, 219–221, 223, 246, 248–250, 255, 263, 268, 271–273, 293, 294, 296, 311–313, 318–320, 322
management, 10, 52, 59–63, 69, 72, 88, 171, 183, 191, 193, 194, 220, 223, 224, 227–234, 236, 239, 240, 251, 256, 257, 260, 281, 282, 285
marine environment, 226–229, 235, 264, 265, 276, 280, 286, 288
marine environmental protection, 39, 226, 254, 255, 278, 281, 282
marine pollution, 267, 269, 270, 275, 276
marine scientific research, 226, 254, 255, 278, 280
maritime accidents, 285, 287
Maritime Assistance Services, 286
maritime boundary delimitation, 226, 250–252, 256, 257
maritime conventions, 275
maritime safety, 11, 263–289, 318
Maritime Safety Agency (MSA), 269, 271, 286, 287
Maritime Safety Committee, 265
maritime security, 3, 6, 9, 11, 81, 83–87, 89, 282, 318, 324
market, 8, 14, 17, 19–21, 25, 28–31, 46, 66, 67, 77, 95, 119, 128, 138, 153, 154, 156, 161–167, 169, 170, 172–177, 183, 199,

200, 203, 209, 210, 241, 246, 294–297, 299, 301, 311, 312, 314, 315, 317, 321, 323
market economy, 55
mass destructive weapons, 224
maximum sustainable yield, 223, 228, 238
mediation, 53, 55, 324
Mekong region, 180, 183, 184, 189, 192
Mekong River, 8, 9, 37, 71, 107, 119, 179–195, 284
member states, 34–39, 42–45, 47, 48, 50, 51, 53, 55, 93–98, 102–104, 108, 116, 122, 123, 125, 127–130, 144, 233, 234, 240, 253, 259, 269, 272, 273, 275–277, 283–285, 287
Mexico, 113, 211, 296, 298, 303, 304, 318
mineral resources, 11, 251, 257, 260
mining, 37, 107, 183, 184, 203
Minister of Public Health, 62
Ministry of Commerce, 118, 139, 143, 202, 209
Ministry of Communications, 82, 282
Ministry of Finance, 143, 300
Ministry of Foreign affairs, 139, 143, 318
Ministry of Public Security, 82, 279
money laundering, 5, 38, 84, 173
multilateral arrangements, 238
multilateral efforts, 71, 72
multilateralism, 12, 72, 102, 111, 128, 146, 293, 297, 302, 305, 306, 321
mutual benefits, 35, 48, 108, 116, 147, 254
mutual legal assistance, 6
mutual respect, 40, 45
mutual trust, 35, 39, 134, 221, 224, 225
Myanmar, 6, 8, 9, 29, 34, 37, 43, 68, 71, 94, 96, 114, 116, 127, 133, 144, 148, 155, 161–164, 166, 167, 169, 171, 172, 174, 175, 179, 180, 185, 208, 211, 263, 268

National Bureau of Statistics, 19, 209
national development, 13, 192
national development strategy, 13
national interests, 125, 246, 313, 314, 321
National People's Congress, 221, 256
National Reference Laboratory, 69
national security, 49, 81, 85, 263
Natuna Island, 223, 235, 246
natural gas, 245, 249, 250
navigation, 11, 80, 84, 179, 185, 186, 188, 226, 263, 266
Navigation Improvement Project, 185, 188, 192
negotiating process, 7, 111, 126, 129
negotiation, 7, 36, 37, 50, 51, 53, 93, 96, 105–108, 111–119, 126–129, 131, 133–144, 146–154, 157, 233, 234, 257, 270, 272, 294, 298, 299, 301–304, 310, 318
neighbors, 11, 13, 14, 29, 38, 72, 115, 147, 200, 210, 302
neutrality, 36, 124
New Zealand, 97, 112, 113, 153, 154, 212, 293
New-Age, 17, 32
newer ASEAN members, 8, 37, 46, 47, 93–96, 116, 118, 161
Newly Industrialized Economies (NIEs), 19–21, 25–27, 199, 311
non-interference, 40
non-living resources, 11, 223, 251, 254, 255, 257, 258
non-tariff barriers, 29, 47–49, 52, 93, 105, 116, 172, 300
non-traditional security issues, 3–6, 11, 37, 38, 42, 43, 45, 51, 59, 75, 84, 126, 279–281, 283
Nuclear Weapon-Free Zone Treaty, 36

OECD countries, 199
offshore oil and gas, 250, 251, 253
oil and gas, 11, 213, 223, 245, 246, 248–251, 256–258, 314
oilfields, 245, 249, 250
One-China principle, 32, 98
OPEC, 246
open-door policy, 4, 17, 18, 26, 312
overlapping claims, 251

Pacific Century, 18
pandemic disease, 5, 6
Paracel (Xisha) Islands, 220, 222, 245, 249
patrol boats, 80, 248
peace and security, 10, 39, 40, 43, 261, 280
peaceful means, 10, 36, 39, 40, 226, 254, 255
peaceful rise, 13, 14, 134, 147, 321
People's Republic of China, 19, 35–38, 41–43, 45, 51, 53, 66, 93, 111, 115, 126, 127, 134, 135, 149, 236, 271, 272, 277, 279–283
petroleum products, 250
phantom ship, 80
piracy, 5, 6, 11, 12, 38, 75–89, 254, 278, 279, 311, 318, 319
Piracy Reporting Center, 79, 82, 84
pirates, 6, 76–82, 88
place of refuge, 286–288
policy makers, 85, 87, 316
policy-making, 143, 237
Politburo Standing Committee, 62
political system, 299, 305
Port of Singapore, 81, 86
port state control, 275, 281, 282, 285, 287
Pratas Islands, 260
precautionary approach, 228, 229, 232, 234, 239
Precautionary Principle, 223, 232, 241
preferential treatments, 128, 169, 175, 206
prescription, 235
principle of most favored nations, 12, 47
principle of special and differential treatment, 47
principles of equality, 35
provisional measure, 257
public health, 13, 59, 66, 69–71, 73
purchasing power parity (PPP), 18

raw materials, 31, 167, 184
Regional Anti-Piracy Agreement, 6, 319
regional cooperation, 6, 10, 11, 17, 59, 69, 73, 97, 135, 161, 209, 219–223, 226, 227, 233, 234, 237–240, 243, 253, 263, 264, 274, 284, 288, 289, 323
regional economic cooperation, 8, 97, 156
regional economic development, 3, 8, 180, 183
regional entity, 296, 297
regional integration, 32, 41, 52, 135, 156, 293, 294, 297–299, 301
regional organization, 67–69, 129, 227, 239, 286
regional peace and security, 10, 38
regionalism, 7, 12, 26, 111–113, 295, 296, 302, 305, 306
regionalization, 112, 258
Renminbi, 18, 29, 168, 172, 173
responsibility, 35, 85, 100, 129, 131, 238, 277, 282
responsible fishery, 232
rise of China, 4, 9, 17–19, 21, 25, 29, 32, 40, 134, 298, 309, 315, 320–322, 324
river bank, 9, 185, 187, 188, 191, 194
rocks, 235, 236
round-tripping, 9, 200, 206, 207
rule of law, 41, 53

safety of navigation, 39, 254, 264, 273, 278, 280, 282, 286
SARS outbreak, 5, 59–66, 68–73
science and technology, 3, 33, 34, 66
scientific research, 39, 203, 227, 228, 235, 286
sea lanes of communication, 85, 314
Sea of Japan, 253
sea-borne commerce, 85
seabed, 251–253
search and rescue, 39, 226, 241, 242, 254, 278, 280
Secretary-General, 35, 124, 126, 213, 264, 289
sediment flux, 9, 185, 186, 194
sedimentary basins, 245
self-restraint, 36, 39, 52, 226
semi-enclosed sea, 9, 10, 219, 220, 227, 235, 236, 239

severe acute respiratory syndrome (SARS), 5, 6, 18, 66, 72, 202
ships, 6, 65, 76, 78–81, 83–86, 185, 221, 224, 248, 265–267, 269, 273–275, 281, 282, 284, 285, 287, 288, 318, 319, 321
Singapore, 3, 6, 18–20, 22, 25–28, 30, 32–34, 36, 37, 40, 52, 59–64, 75, 76, 78, 81–83, 85–88, 94, 96, 97, 107, 111–113, 118, 119, 123–125, 127, 129, 133, 144, 145, 150, 151, 195, 199, 200, 205, 207, 208, 210, 211, 219, 246, 249, 263, 268, 271–274, 294, 296, 298, 302, 303, 311–313, 317–319, 321–324
Singapore Police Coast Guard, 81
single market, 31, 209
slavery, 77
small and medium enterprises, 37
small and medium-sized companies, 298, 303
smuggling, 38, 81, 82, 84, 171, 173, 175
soft power, 13, 321
soft-law, 5, 50, 52
South Korea, 6, 11, 19, 22, 25, 26, 62, 65–67, 200, 256, 259, 311, 314, 319, 321
Southeast Asia, 4, 6, 9, 11, 12, 17, 20, 22, 28, 36, 40, 44, 54, 59–61, 72, 75, 78, 80, 84, 88, 89, 107, 124, 144, 155, 162–166, 169, 174, 192, 199, 200, 203, 209, 210, 263–265, 267, 269, 270, 272, 284, 285, 288, 309–324
sovereignty, 39, 40, 45, 87–89, 143, 192, 220, 225, 235, 236, 248, 271
sovereignty claims, 226, 237
spaghetti bowl effect, 295
spillover effects, 17, 32, 194
State Development and Reform Commission, 139, 143
state practice, 130, 251, 252, 257, 260
straddling fish stocks, 10, 226–231
Strait of Singapore, 86
Straits of Malacca, 12, 75, 76, 80, 83–88, 263, 311, 318–320, 322
sub-regional development, 37, 107

subsidies, 48, 52, 105, 106, 171, 310, 323
sustainable development, 35, 40, 46, 183, 193, 277

Taiping Island, 240
Taiwan, 10, 19, 22, 25–27, 30, 31, 66, 77, 97, 98, 115, 168, 205, 207, 219, 222–224, 227, 232, 235, 236, 240–242, 246, 249, 259, 260, 298, 311, 321, 322
Taiwan issue, 32, 98
Taiwan Strait, 87, 242, 243, 259, 260, 313, 314
Taiwanese government, 207
target reference point, 232
tariffs, 29, 47–49, 52, 93, 97, 100, 105, 112, 116–119, 175, 318
tax evasion, 173
technology transfer, 25, 119
telecommunications, 37, 169, 176, 183
territorial disputes, 10, 39, 52, 221, 246, 254, 258, 261, 315, 322
territorial integrity, 40
territorial seas, 87, 235–237, 271
territorial sovereignty, 234, 235, 259
terrorism, 5, 38, 75, 81, 84, 85, 88, 224, 279, 319, 321
terrorist attack, 5, 83, 224
Thailand, 6, 9, 18, 20, 22, 26, 27, 30, 37, 40, 60–62, 66–71, 78, 94, 96, 107, 113, 117, 118, 123, 127, 129, 133, 144, 179, 180, 182, 183, 185, 186, 193, 199, 200, 203, 207, 208, 210, 211, 219–221, 249, 263, 265, 266, 268, 271–273, 284, 294, 296, 311–314, 318, 320–322
the Philippines, 6, 10, 18, 20, 22, 26, 27, 30, 37, 60, 61, 64, 65, 76, 82, 83, 94, 96, 117, 123, 127, 128, 133, 144, 199, 207, 208, 210, 219–223, 225–227, 246, 249, 253, 256, 259, 260, 263, 266, 268, 273, 294, 296, 311–314, 317–319, 321, 322, 324
the South China Sea, 3, 4, 9–11, 13, 36, 39, 43, 45, 52, 54, 64, 75, 77, 81, 84–87, 109, 147, 180, 219–227, 231, 234–243,

245–250, 252–256, 258–261, 263, 278, 283, 288, 313, 314, 319, 320, 322
the Spratly Islands, 10, 222, 236, 245, 246, 248, 250, 254, 256, 259
the Treaty Establishing the European Community, 121
the United States, 12, 39, 60, 66, 67, 112, 113, 115, 143, 199, 270, 293, 297, 298, 302, 304
third parties, 55
third-party intervention, 55
Tokyo, 12, 282, 295, 309, 313–320, 322, 323
tourism, 13, 30, 37, 60, 61, 64, 86, 107, 169, 176, 183, 184, 192
trade, 3, 4, 7, 8, 12–14, 17, 20, 25–32, 37, 45–50, 52, 55, 63, 64, 76–78, 85, 86, 93, 98–103, 105–107, 111, 112, 114–116, 118, 119, 129, 131, 134–136, 138, 139, 142, 147, 155, 157, 161, 163–165, 167, 170–175, 177, 179, 183, 185, 210, 233, 281, 295, 298, 302–304, 309, 317, 320, 322–324
trade deficits, 30
trade in services, 30, 93, 100, 101, 103, 105, 106, 108
trade partner, 13, 27, 108, 134
trade policies, 8, 171, 175, 302
trade regionalism, 7, 111
trade surplus, 18, 30, 241
trade system, 49, 52, 108
trading system, 7, 97, 100, 111, 112, 305
trans-boundary, 9, 179, 183, 189, 195, 275
trans-boundary environmental issues, 9, 179, 195
trans-national crime, 6, 75, 83, 84, 89
transboundary shared resource, 192
transnational crimes, 39, 225, 254, 278, 280
transparency, 61, 68, 106, 107, 138, 141, 171, 237

transport, 13, 37, 84, 107, 183, 184, 187, 194, 203, 267, 271, 272, 274–277, 281–283, 320
treaties, 11, 42, 45, 50, 51, 53, 95, 120–122, 141, 146, 150, 226, 227, 264, 265, 269
treaty obligations, 7, 130
Treaty of Amity and Cooperation in Southeast Asia, 4, 31, 39, 40, 42–45, 51, 53, 124, 126, 309
treaty-making power, 96, 120–122, 124–126
TRIPS, 48, 95, 105

U-shaped line, 223, 235, 236, 241, 245, 246
UN peacekeeping operations, 316
United Nations Convention on the Law of the Sea (UNCLOS), 10, 11, 36, 39, 43, 45, 54, 79, 221, 222, 224, 226, 227, 231, 234–236, 239, 240, 253, 264, 265, 267, 269, 280, 284, 287
unitization, 251
Uruguay Round, 100, 102, 105, 136, 300, 301
US Geological Survey, 245
use of force, 10, 36, 39, 40, 45, 254

Vienna Convention on the Law of Treaties, 120, 121, 123, 129, 150
Vietnam, 6, 8–10, 29, 37, 60–62, 66, 70, 94–96, 115–117, 127, 133, 144, 148, 154, 161–172, 174, 175, 180, 181, 183, 187, 188, 207, 208, 210, 211, 219–221, 223, 226, 246, 248–250, 256–259, 263, 268, 271, 272, 313, 314, 322

water discharge, 9, 185–187, 194
water quality, 179, 191
Wen Jiabao, 13, 31, 39, 62–64
Westphalian system, 322
win-win situation, 29
World Bank, 18, 20, 22, 70
world community, 5
World Health Organization (WHO), 60, 68

World Trade Organization (WTO), 7, 46, 93, 111, 112, 115, 239
WTO regime, 48, 94, 97–99, 104, 108, 136, 302
WTO rules, 7, 47, 93–95, 98, 101–108, 242

Yellow Sea, 233, 256
Yunnan, 9, 70, 162–170, 173, 175, 176, 182, 192, 193

zero growth, 223, 237
zone of peace, 36, 124